Barbara J. Thompson, Natchimuthuk Gopalswamy,
Joseph M. Davila, Hans J. Haubold (eds.)

Putting the "I" in IHY

The United Nations Report for the International Heliophysical Year 2007

SpringerWienNewYork

Barbara J. Thompson
Heliophysics Science Division
NASA/GSFC
Greenbelt, MD, USA

Natchimuthuk Gopalswamy
Heliophysics Science Division
NASA/GSFC
Greenbelt, MD, USA

Joseph M. Davila
Heliophysics Science Division
NASA/GSFC
Greenbelt, MD, USA

Hans J. Haubold
UN Office for Outer Space Affairs
Vienna, Austria

Softcover re-print of the Hardcover 1st edition 2009

SpringerWienNewYork is a part of
Springer Science + Business Media
springer.at

Typesetting: Thomson Press (India) Ltd., Chennai
Printing: Strauss GmbH, 69509 Mörlenbach, Germany

Cover: "Composite design courtesy of European Science Foundation".

Printed on acid-free and chlorine-free bleached paper

SPIN: 12641345

With 164 Colour and 14 Black & White Figures

ISSN 1868-5307
ISBN 978-3-7091-1726-2 SpringerWienNewYork

SpringerWienNewYork

Studies in Space Policy
Volume 3

Edited by the European Space Policy Institute
Director: Kai-Uwe Schrogl

Preface

Putting the "I" in IHY

This book is about international cooperation. It demonstrates how the power of scientific imagination and investigation can bring together people form all continents in almost all countries around the globe. In presenting this impressive result, we can understand, how much unifying force the quest for understanding our universe and using outer space for that purpose have. Astronomy is far away from being a "political" area of science. But is has enormous political effects – and all of these effects are positive.

This book about the international aspects and achievements of the "International Heliophysical Year (IHY) 2007" can be regarded as a compendium of the fertile impacts of conducting research in this field. The main focus, as the title implicates, is the international cooperation, which has emerged from this grassroots initiative. North and South, industrialized and developing countries have been coordinating their efforts and have been learning from each other in a mutual partnership under a joint understanding of sharing the scientific benefits. Through this, trans-border networks have been created and scientific as well as cultural exchange took place.

Another focus of the book shows, how much astronomy contributes to the basis of knowledge society as today's concept for mastering the future. Astronomy has been and will be attracting large numbers of young people to enter an education and career in science and engineering. Such attractions we desperately need in all countries around the world, and we have to be glad about initiatives like IHY, which are successful in raising awareness, interest and fascination.

This book is therefore particularly well placed in the series "Studies in Space Policy", since it highlights the policy needs for space education as well as international cooperation in a most dedicated and convincing way.

Kai-Uwe Schrogl
Director
European Space Policy Institute (ESPI)

Biographical sketch: Professor Dr. Hans J. Haubold

Education

Diploma, Physics, Technical University Chemnitz, 1974, Germany
Ph.D., Astrophysics, Institute for Astrophysics, 1980, Potsdam, Germany
D.Sc., Astrophysics, Institute for Astrophysics, Potsdam, 1984, Germany
Research Professor, Theoretical Astrophysics, Berlin, 1988, German Academy of Sciences

IHY Roles

Implementation of the United Nations Basic Space Science Observatory Development Programme for IHY 2007

Dr. Hans J. Haubold is pursuing research in astrophysics and co-authored four books, on nuclear astrophysics (with A. M. Mathai, Akademie-Verlag, Berlin, 1988), on the early universe and formation of structure (with S. Gottloeber, J. P. Muecket, V. Mueller, Akademie-Verlag, 1990), on special functions for applied scientists (with A. M. Mathai, Springer, 2008), on the theory and applications of the H-function (with A. M. Mathai, R. K. Saxena, Springer, 2009), and two major review articles on the sun (with A. M. Mathai, Encyclopedia of Planetary Sciences, Chapman and Hall, London, UK, 1997, and Kluwer Academic Publishers, The Netherlands, 2000) and on the structure of the universe (with A. M. Mathai, Encyclopedia of Applied Physics, Vol. 23, Wiley-VCH, New York and Weinheim, Germany, 1998). He is a member of the editorial board of the international journal "Astrophysics and Space Science" and "Earth, Moon, and Planets" (both published by Springer) and STARS (St. Thomas College, Pala, India). He is the (co-)author of more than 250 scientific and technical papers in the fields of astrophysics, mathematics, history of physics, and international cooperation in space science and technology.

Since 1988 he has also been with the United Nations Office for Outer Space Affairs, firstly in New York and latterly in Vienna, Austria, since 1993. At the United Nations he spearheaded the development of education curricula in remote sensing and GIS, satellite meteorology and global climate, satellite communications, and space and atmospheric science for the UN-affiliated Regional Centres for Space Science and Technology Education. Since 1991 he is the principal organizer of the United Nations Basic Space Science Initiative and co-editor of 12 volumes of Proceedings that emanated from this initiative. The decadal report on this initiative titled "Developing Basic Space Science World-Wide" was published by Kluwer Academic Publishers, Dordrecht, The Netherlands, in 2004 (co-edited by W. Wamsteker, R. Albrecht, and H. J. Haubold).

He is a member of the Centre for Mathematical Sciences (Pala and Trivandrum, India). He is also a member of the German Physical Society, German Mathematical Society, American Mathematical Society, Society for Industrial and Applied Mathematics, American Astronomical Society, History of Science Society, Committee on Space Research, and the International Astronomical Union. He is maintaining websites at *neutrino.aquaphoenix.com* and *www.cmsintl.org*.

Biographical Sketch: Dr. Barbara J. Thompson

Education

B.A., Physics and Mathematics, 1991,
University of Pennsylvania
Ph.D., Physics, 1996, University of Minnesota

IHY roles

Director of Operations, IHY Secretariat
IHY Website Editor, IHY Secretariat
Member, IHY-USA Organizing Committee

Dr. Barbara J. Thompson currently an astrophysicist in the Solar Physics Laboratory at the NASA Goddard Space Flight Center in Greenbelt, Maryland. She has devoted the majority of her solar research to the study of coronal mass ejections and their impact on the heliosphere and near-Earth space. She has also done work on the acceleration of particles in the Earth's aurora and the Sun's magnetic structure. She has authored 150 papers on these topics.

Born June 17, 1969 in Austin, Texas, she obtained degrees in both physics and mathematics, with a minor in geology, from the University of Pennsylvania in Philadelphia. She received her Ph.D. from the University of Minnesota, with a thesis topic of the acceleration of particles in the Earth's aurora by waves in the Earth's magnetosphere.

She is a member of the American Astronomical Society, the American Geophysical Union and the International Astronomical Union, and the International Association of Geophysics and Aeronomy (IAGA). She has served as an officer of IAGA and on committees for the AAS and National Science Foundation. Additionally, she was Project Scientist for the Solar Dynamics Observatory and a co-investigator on the Plasma Suprathermal and Ion Composition (PLASTIC) and Sun–Earth Connection Coronal and Heliospheric Investigation (SECCHI) instruments for the Solar–Terrestrial Relations Observatory (STEREO) mission.

She has served as the IHY Director of Operations since its inception as well as the website editor. She has helped to plan many IHY workshops and meetings and was the co-leader of IHY's largest science campaign, the Whole Heliosphere Interval.

Biographical Sketch: Dr. Joseph M. Davila

Education

B.Sc., Mechanical Engineering,
1972, Lamar University
B.Sc., Physics, 1978,
University of California – Irvine
Ph.D., Astronomy, 1982,
University of California – Irvine

IHY Roles

Executive Director, IHY Secretariat
Chair, IHY International Steering Committee
Team Member, IHY-USA Organizing Committee

Dr. Joseph Michael Davila is currently an astrophysicist in the Solar Physics Laboratory at the NASA Goddard Space Flight Center in Greenbelt, Maryland. His research interests have included the linear and non-linear theory of hydromagnetic waves, hydromagnetic instabilities due to energetic particle beams, resonance absorption in inhomogeneous plasmas, the acceleration of high-speed wind streams in solar and stellar coronal holes, and plasma heating in closed magnetic structures. He has authored over 150 papers on these subjects.

Born December 30, 1948 in Port Arthur, Texas, he earned a bachelor of science degree in mechanical engineering from Lamar University in Beaumont, Texas. He then earned a bachelor of science degree in physics from the University of California, Irvine, in 1978 and a Ph.D. in astronomy from the University of Arizona in 1982.

He is a member of the American Astronomical Society, the American Geophysical Union and the International Astronomical Union. He has also published research on the acceleration of cosmic rays, the transport of energetic, particles within the galaxy, the modulation of galactic cosmic rays by the solar wind and the propagation of solar cosmic rays in the interplanetary medium. He was principal investigator for the Solar Extreme-ultraviolet Rocket Telescope and Spectrograph (SERTS), he was the project scientist for the Solar–Terrestrial Relations Observatory (STEREO), and is the lead scientist for Coronagraph-1 (COR1) on the STEREO mission.

He has served as the IHY executive director from its inception, and has organized many of the IHY conferences and committees. He is now focusing on the transition of IHY programs into the International Space Weather Initiative.

Biographical Sketch: Dr. Natchimukonar Gopalswamy

Education

B.Sc., Physics, 1975, University of Madras
M.Sc., Physics, 1978, University of Madras
Ph.D., Plasma Physics, 1982, IIA

IHY roles

International Coordinator, IHY Secretariat
Team Member, IHY-USA Organizing Committee

Dr. Natchimuthukonar Gopalswamy is currently an astrophysicist in the Solar Physics Laboratory of NASA's Goddard Space Flight Center, Greenbelt, Maryland. As a solar physicist and radio astronomer, Dr. Gopalswamy studies solar variability and its impacts on Earth. He is a co-investigator of the Solar Terrestrial Relations Observatory (STEREO) mission. Over his career to date, he has authored or coauthored more than 300 research papers.

Born February 15, 1954 in Chandrapuram, Tamil Nadu, India, he received his B.Sc. and M.Sc. degrees in physics from the University of Madras (1975 and 1977). He received his Ph.D. in plasma physics from the Indian Institute of Astrophysics in Bangalore (1982), followed by a post-doctoral training at the University of Maryland, College Park (1985–1987).

He is a member of the American Astronomical Society, the American Geophysical Union and the International Astronomical Union. He promotes international cooperation in space science, serving as the International Coordinator for the IHY 2007 program. He has organized IHY scientific activities in collaboration with international scientific organizations such as the International Astronomical Union (IAU), the Scientific Committee on Solar Terrestrial Physics (SCOSTEP), and the International Association of Geomagnetism and Aeronomy (IAGA). He serves as a member of the SCOSTEP Bureau, representing IAU.

As international coordinator for IHY since its inception, he has helped to organize many IHY workshops and meetings. He has been a major driver behind the IHY national programs and the U.N. Basic Space Science instrument development program.

Table of contents

1 IHY Overview

"Space is a part of the world's cultural heritage. It has inspired generations of artists, poets, scientists and musicians. Throughout history, societies have admired and searched for meaning in the same night sky."

"Indeed, space exploration can help bring cultures together. Manned space missions today are rarely top-secret national projects. Much more common are international crews, with members from a variety of backgrounds. Crews live together in cramped and challenging conditions for months, sharing experiences, customs and, above all, the enthusiasm for space that brought them together in the first place. Their missions capture the imaginations not only of their native lands, but of people around the world."

"Space is also helping us to address some of today's most urgent problems. Space technology has produced tools that are transforming weather forecasting, environmental protection, humanitarian assistance, education, medicine, agriculture and a wide range of other activities. And, of course, a fascination with space leads many young people to pursue careers in science and technology, helping developing countries in particular to build up their human resources, improve their technological base and enhance their prospects for development."

– *U.N. Secretary – General Kofi Annan,*
on the occasion of World Space Week, 2001

Heliophysical: A broadening of the concept "geophysical," extending the connections from the Earth to the Sun & interplanetary space. On the 50th anniversary of the International Geophysical Year, the IHY activities will build on the success of IGY 1957 by continuing its legacy of system-wide studies of the extended heliophysical domain.

Looking back on IHY: a statement from the IHY Secretariat

The International Heliophysical Year began as a vague concept in late 1999. Over the next few years a series of grassroots community workshops were held, and the IHY began to take shape. In these workshops the concept of Universal Processes was born, and the definition of this new word "heliophysics," coined to imply a parallelism with geophysics, began to take shape. Also four main thrusts of scientific research, observatory development, history, and outreach were articulated.

With this basic organizational structure, the work of planning began. First starting with a few, but then involving more and more space scientists from many countries around the world. Eventually there would be planning committees in nearly 90 nations involving the efforts of thousands of scientists. In March 2007 the IHY was officially opened, and what a success it has been!

A total of 65 Coordinated Investigation Programs were initiated for the IHY. Each program involved a team of scientists working together on problems of mutual interest. The distributed instrument program facilitated the deployment of a wide range of instrument arrays all over the world, and in places that were thought to be unreachable. Eclipses in Tunisia and Libyan Arab Jamahiriya opened the door for scientific collaboration, eventually leading to instrument installations in these countries. More than 150 new instruments were installed in Africa, South America, and Asia. These installations will provide a tangible and lasting legacy for the IHY.

Outreach programs brought heliophysics into the public eye, with public talks, exhibitions, and a worldwide observatory "Open Doors" day. A number of teacher workshops were organized to distribute newly translated teaching materials to educators, and the IHY Schools program developed new curricula in heliophysics. The IHY Gold history program awarded more than 300 certificates to honor the participation of scientists in the IGY fifty years ago, organized workshops in India, Japan and the USA focusing on IGY history, and sponsored retrospective lectures at a number of meetings.

None of these accomplishments would have been possible without the support of funding agencies, scientific organizations and scientists around the world who saw a need, and answered the call. We thank them for their determination, vision, perseverance and friendship.

1.1. IHY's foundation

1.1.1. IHY's heritage

In 1957–1958 a program of international research, inspired by the International Polar Years of 1882 and 1932, was organized as the International Geophysical

Year (IGY) to study global phenomena of the Earth and geospace. The IGY involved about 60,000 scientists from 66 nations, working at thousands of stations from pole to pole to obtain simultaneous, global observations on Earth and in space. There had never been anything like it before.

The International Heliophysical Year was inspired by the 50th Anniversary of the International Geophysical Year (IGY) and 50 years of space exploration. IHY builds on the successes of the International Geophysical Year and the previous International Polar Years. We now extend our global studies out into the heliosphere and incorporate the drivers of geophysical change into the global system. As we approach the "new frontier" of the heliopause and interstellar space, our heritage clearly extends back to previous International Years (see Figure 1 and Appendix V).

The efforts of the past half-century have brought us significant scientific advances, global scientific communication, and an extensive suite of spacecraft and observatories, which some call our "Great Observatory." As we extend the reach of human exploration and prepare for humanity's first encounter with interstellar space, we have expanded our concept of "geophysics" to embrace other planets, interplanetary space, and the Sun itself. The establishment of this new

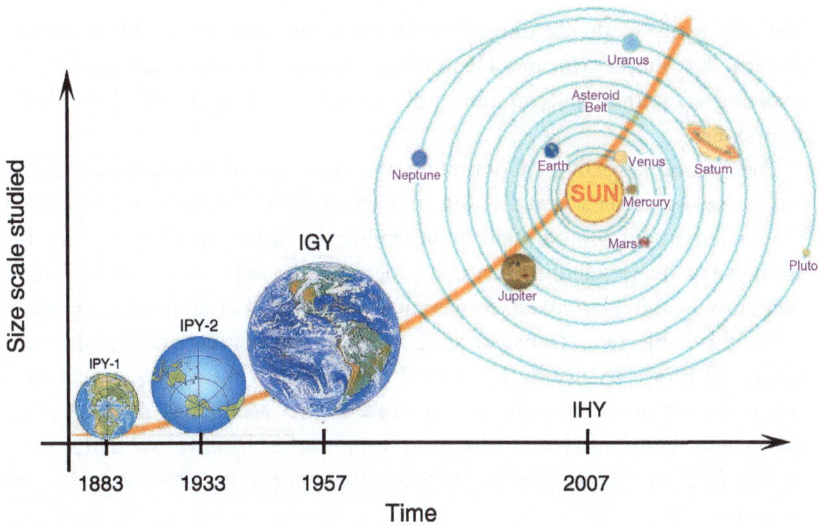

Fig. 1. *The extension of the concept of "geophysical" into "heliophysical". The tradition of international science years began almost 125 years ago with the first international scientific studies of global processes of polar processes in 1882–1883. A second International Polar Year (IPY) was organized in 1932, but a worldwide economic depression curtailed many of the planned activities. IGY in 1957 was an unprecedented success on many levels. IHY continues the legacy of these previous events, extending global synoptic studies and global interconnected processes to the rest of the heliosphere.*

science, heliophysics, is the primary goal of IHY: a system-wide understanding of the entire interconnected heliophysical system.

1.1.2. IHY's goals and objectives

Prior to the launch of IHY, an international planning committee, consisting of scientists from around the globe, launched a series of discussions and meetings to establish the principles of IHY and the means by which we would be able to lay the foundation for the new science of heliophysics. It was determined that the efforts of IHY should encompass *three primary objectives*:

- Advancing our Understanding of the Fundamental Heliophysical Processes that Govern the Sun, Earth and Heliosphere.
- Continuing the tradition of international research and advancing the legacy on the 50th anniversary of the International Geophysical Year.
- Demonstrating the Beauty, Relevance and Significance of Space and Earth Science to the World.

IHY's priorities are set via its three primary objectives. The first objective, "Advancing our Understanding of the Fundamental Heliophysical Processes that Govern the Sun, Earth and Heliosphere" focuses on our scientific activities, involving an international community of scientists spanning all of the heliophysical disciplines.

The second objective, "Continuing the tradition of international research and advancing the legacy on the 50th anniversary of the International Geophysical Year" refers to our commitment to preserving the history and legacy of the IGY through our many scientific research programs, the deployment of new instrumentation worldwide, and our cooperation with other IGY anniversary programs.

The third objective, "Demonstrating the Beauty, Relevance and Significance of Space and Earth Science to the World" reminds us of the many applications of heliophysics that are relevant to our existence on earth, and the tremendous opportunity for outreach and education. From these objectives we derive the *six goals of IHY*, each corresponding to a unique opportunity afforded by the IHY programs:

1. Develop the basic science of heliophysics through cross-disciplinary studies of universal processes.
2. Determine the response of terrestrial and planetary magnetospheres and atmospheres to external drivers.

3. Promote research on the Sun–heliosphere system outward to the local interstellar medium – the new frontier.
4. Foster international scientific cooperation in the study of heliophysical phenomena now and in the future.
5. Preserve the history and legacy of the IGY on its 50th anniversary.
6. Communicate unique IHY results to the scientific community and the general public.

1.2. IHY activities: Science, Observatory Development, Outreach and History

IHY is an integrated program consisting of many diverse activities that are coordinated on an international level to achieve all of the above goals. Because many of the IHY activities correspond to more than one IHY goal, it became clear that their ideal organization would be based on the nature of the effort. As a result, the International Heliophysical Year Program has four main components: *Science, Observatory Development, History and Outreach* (Figure 2).

The four components of IHY are related to the goals of IHY as follows:

Science: IHY's scientific activities are organized via Coordinated Investigation Programmes (CIPs). The CIPs span all the disciplines throughout heliophysics, and coordinate observations and analysis involving scientists and instrumentation around the world. Using the data resulting from the CIPs, we drive towards a more complete understanding of the universal processes that govern the Sun, Earth, planets and heliosphere. Section 3 of this document contains a more thorough description of the science activities and CIP coordination.

Observatory Development: This effort is dedicated to the establishment of observatories and instrument arrays that greatly expands our knowledge of global

Science	Goals 1 – 4
Observatory Development	Goals 3 – 4
History	Goal 5
Outreach	Goal 6

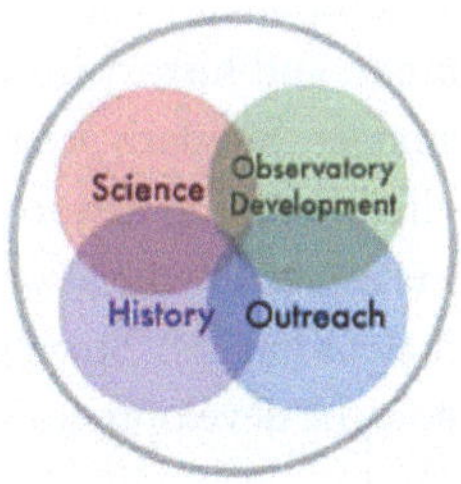

Fig. 2. *The four programmatic components of IHY, and their relationship to the six Goals of IHY. There is a strong overlap between each of the components, indicated in the diagram at right.*

heliophysical processes, with the added benefit of increasing the viability of space science research and education in developing nations and regions that traditionally have not been active in space research. Fifteen instrument programs, led by scientists around the world, have resulted in new and exciting ionospheric, geomagnetic, and solar measurements. Central to this effort are the United Nations Basic Space Science (UNBSS) workshops, which were dedicated from 2005–2009 to IHY instrument development and data analysis. This joint U.N./IHY developing nations program, described in Sect. 4, targets activities which stimulate Space and Earth Science activities at institutions around the globe.

History: IHY is also dedicated to preserving the history and legacy of IGY 1957. Capitalizing on its 50th anniversary, the IGY Gold History Program seeks to identify and recognize planners of and participants in the first IGY, preserve memoirs, articles, photographs, and all items of historical significance for the IGY, make these items available to historians, researchers, etc., spread awareness of the history of geophysics, and plan a series of commemorative special events and "reunions." These activities are described in Sect. 6 of this document.

Outreach: There have been unique opportunities for expanding the education and awareness of Space and Earth Sciences during IHY. As a result, the education programs form a cornerstone of our international initiative. The IHY outreach program focuses on the availability of multilingual materials, so that people from all nations are able to learn about IHY. The IHY Schools Program has given students from around the world an opportunity to attend courses on heliophysics and learn from internationally recognized experts in the field. These efforts are described in Sect. 5 of this document.

Although each of the above items has a unique function, it is clear that the coordination of these activities allows us to successfully implement and maximize the impact of each of the components. For example, the Education and Public Outreach component, when integrated with the science campaigns, observatory development, and history initiatives, is able to reach a greater audience and have a larger impact. Similarly, as we establish a greater presence for Space and Earth Sciences in developing nations and underrepresented areas, we also assist the local researchers establish education and outreach initiatives through our international educational programs. Stronger research programs mean stronger universities, which allow the development of graduate and undergraduate programs, which are key factors in encouraging youth to become interested in the exciting field of heliophysics.

1.3. Schedule and legacy plans

The IHY Opening Ceremony was held on February 19, 2007 in conjunction with the United Nations Science and Technical Subcommittee meeting in Vienna, Austria. A special presentation on IHY plans and activities was made during the Subcommittee session, along with statements of support from the following participating IHY nations: Canada, China, Germany, Greece, Hungary, India, Italy, Japan, Malaysia, Nigeria, Republic of Korea and the United States. Following the subcommittee meeting, U.N. delegates and IHY participants from around the world came together at the Opening Ceremony to formally announce the commencement of the International Heliophysical Year (Figure 3).

The Opening Ceremony was chaired by Dr. Joseph M. Davila, the Executive Director of IHY. It was followed by an international planning workshop hosted by the Vienna Academy of Sciences on February 20, 2007. A special IHY exhibit was on display in the United Nations Rotunda throughout the week, featuring over 50 displays from participating IHY nations and programs around the world. As part of this exhibit, two exciting new items were premiered: the

Fig. 3. *The "SunWorks" art exhibit premiered at the IHY Opening Ceremony in Vienna, Austria. The exhibit featured sculptures, paintings, and photos from artists around the world.*

"SunWorks" collection of original Sun-themed artwork from around the world, and four special artistic posters commissioned by IHY representing the IHY's science activities.

The IHY Closing Ceremony will transpire at the United Nations in Vienna on February 18, 2009. Representatives from all of the IHY regions will be present for the celebration.

IHY would not have been successful had it not been for the efforts of its science and outreach leaders, who initiated planning activities all over the globe. Hundreds of local, regional and international planning conferences and meetings laid the foundation for IHY. While international conferences serve to connect researchers around the world, workshops focusing on the local scale enabled individual organizations and institutions to develop unique IHY programs that suit their own goals and challenges. It is the activities and programs developed by these individual organizations and institutions that form the "building blocks" of the IHY. Therefore, the IHY's international planning activities have focused on the establishment of the four main components of IHY (Science, Observatory Development, History and Outreach) and on the integration of the individual IHY regions and nations to maximize success on all levels.

The true measure of IHY's success is not its activities in 2007–2009, but the lasting impact IHY has on heliophysics science and education. IHY is part of the legacy of the International Geophysical Year; 50 years after the program ended, the IGY is still having a significant impact on the scientific community. Just as the planning stage was the key to the successful launch and implementation of IHY, we require a dedicated legacy effort to ensure that the impact of IHY continues.

The legacy plan involves three steps:

1. determining whether a particular IHY program will continue to have a significant impact on heliophysics science and/or education if supported beyond 2009,
2. identifying a partner organization or creating a legacy organizational structure that will be able to house that activity, and
3. working with the organization to transfer the program's knowledge and management from the IHY Secretariat to this new organization.

For many IHY programs that were partnerships with other organizations, the legacy plan is to transfer the program to that partner organization. The IHY science collaborations with ICESTAR and CAWSES, for example, will continue to grow under the aegis of these organizations. Many IHY outreach activities, such as Sun–Earth Day and Yuri's Night, are ongoing activities that will continue their collaboration with individual IHY institutions. The IGY Gold Program will continue to be supported, on a less formal level, by the IHY Secretariat.

The Observatory Development program, however, requires the most support over the upcoming decades. It will continue as part of the United Nations Basic Space Science program, but UNBSS requires the continued support of the instrumenters and partner institutions. As a result, a core "IHY Africa" planning group will continue beyond 2009, focusing its efforts on the scientific collaborations and educational activities in regions that will benefit from continued attention.

The following is a brief summary of the IHY timeline of activities. A more complete listing of IHY events is included in Appendix IV.

2001–2003: Establishment of IHY Secretariat; establishment of the main elements of the IHY program; initialization of planning activities on all continents.

2004: National and Regional coordination meetings begin to take place; the four essential components of IHY are defined; synergy/coordination discussions with professional organizations; establishment of CIP structure; launch of the IHY UNBSS and IGY Gold History Programs.

2005: Continuation of national and regional coordination meetings; synthesis and coordination from regional to international; precursor activities for each ofthe four main components continue to happen; instrument deployment begins and CIPs proposed by individual community members begin to form the fabric of the IHY science campaigns.

2006: Focus on the implementation of the four main IHY components and on the integration of national and local activities with the international IHY community; prototyping year, particularly for numerous CIPs and outreach activities that serve as trailblazers and/or testbeds.

2007–2008: IHY is launched on March 1, 2007 as an integrated international program. Science, Observatory Development, Outreach and History activities occur around the globe, and the efforts of each individual component and region are multiplied in impact by their coordination with the worldwide effort.

2009–2010: IHY officially ends in February 2009, but many activities continue. Results of the IHY CIPs and science campaigns are analyzed in a wide range of workshops and analysis activities; Observatory Development continues through IHY UNBSS legacy programs; Outreach activities incorporate major scientific results and breakthroughs. Legacy transition efforts begin, wherein active IHY programs are transferred from IHY central organization to IHY partner programs. Central to the legacy efforts is the International Space Weather Initiative, described next.

1.4. The International Space Weather Initiative (ISWI)

To build on the many advances of the IHY, a new United Nations initiative will be launched: the International Space Weather Initiative (ISWI). The purpose of this effort would be to continue to encourage existing and new scientific collaborations, and begin the process of making the data from the new IHY arrays utilized as part of international space weather alert systems.

A special planning meeting will take place during the IHY Closing Ceremony, and a new 3-year work plan will be proposed as an international space weather research program to continue the instrument deployment and operations around the world that was begun during the IHY.

2 Leadership, Coordination & International Programs

2.1. Organizational structure

The IHY program's organizational structure was devised to balance the priorities and influence of the international coordination teams, the working groups, the individual national IHY programs, and the supporting international scientific bodies and consortia.

Figure 1 illustrates how these facets are related; the *Steering Committee, International Advisory Committee* and *Secretariat*, respectively, are charged with the guidance, execution, and coordination of IHY's international activities. Local IHY activities – that is, activities and coordination within a particular region or nation – are represented in the lower portion of the organization chart in Figure 1. Each of the (currently 192) United Nations member states belong to one of the IHY international regions: *Africa, Eastern Asia and the Pacific, Eastern Europe and Central Asia, Latin America, North America, Western Asia and Western Europe.* The internationally coordinated activities, represented on the right side of the organization chart, are grouped by the four IHY components, described in Sects. 3–6 of this document.

2.2. IHY Secretariat

The IHY Secretariat is responsible for the support of IHY committees and international activities. The Secretariat also maintains the IHY website, provides resources for IHY activities, and is responsible for communication within and outside of the participating community. The IHY Secretariat consists of the following:

- Chair of the IHY Steering Committee, who also functions as the Executive Director of the IHY Secretariat
- International Coordinator
- Director of Operations

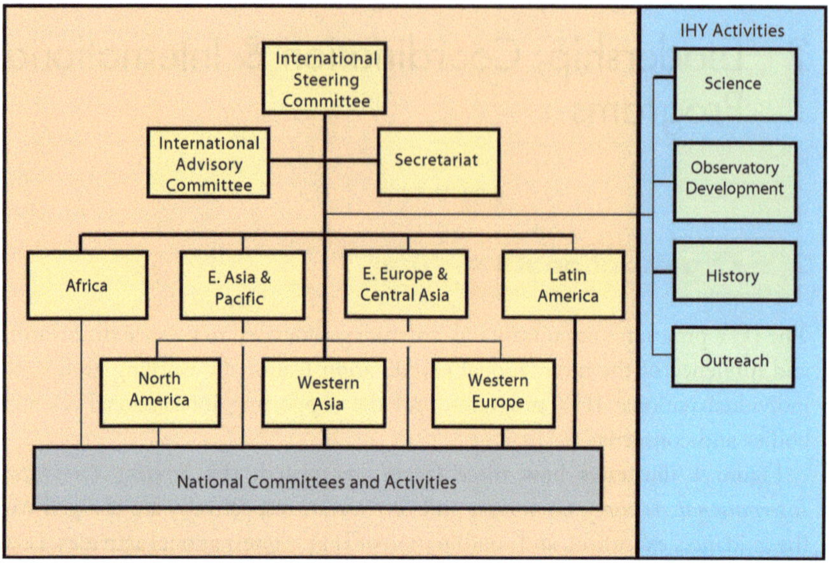

Fig. 1. *IHY organizational structure.*

- Director of the U.N. Basic Space Science Observatory Development Program
- Coordinated Investigation Programme (CIP) Coordinator
- Education and Outreach Coordinator
- International Advisory Committee Chair
- "IGY Gold" History Committee Chair
- Newsletter Editor
- Website Designer and Editor

The individuals who have filled these many roles are located at various research institutions around the world, drawing support from a wide range of international bodies. A list of the people who have served in these positions is included in Appendix I.

2.3. The International Advisory and Steering Committees

As mentioned previously, the guidance, execution, and coordination of IHY international activities are performed by the International Advisory Committee, Steering Committee, and Secretariat, respectively. The membership of the *International Advisory Committee* was chosen to balance geographical distribution, scientific disciplines, and representation from organizations and supporting programs. The International Advisory Committee advises the IHY team on issues of international coordination, planning, and connectivity with international scientific organizations, national bodies, and related activities. This group serves as a consultant body to help optimize the relationship between the goals and efforts of IHY and other international groups and activities.

The IHY *Steering Committee* has the responsibility of ensuring the implementation of the recommendations of the International Advisory Committee, as well as overseeing the operations within IHY. The IHY International Steering Committee is comprised of representatives from each of the seven IHY geographical regions and scientists spanning all of the subdisciplines of heliophysics. The responsibilities of the International Steering Committee include the prioritization of IHY resources and activities, providing guidance and monitoring the progress of the IHY working groups, and ensuring clear communication between the IHY teams, scientific organizations, and constituent bodies. The membership of both of these committees is given in Appendix I.

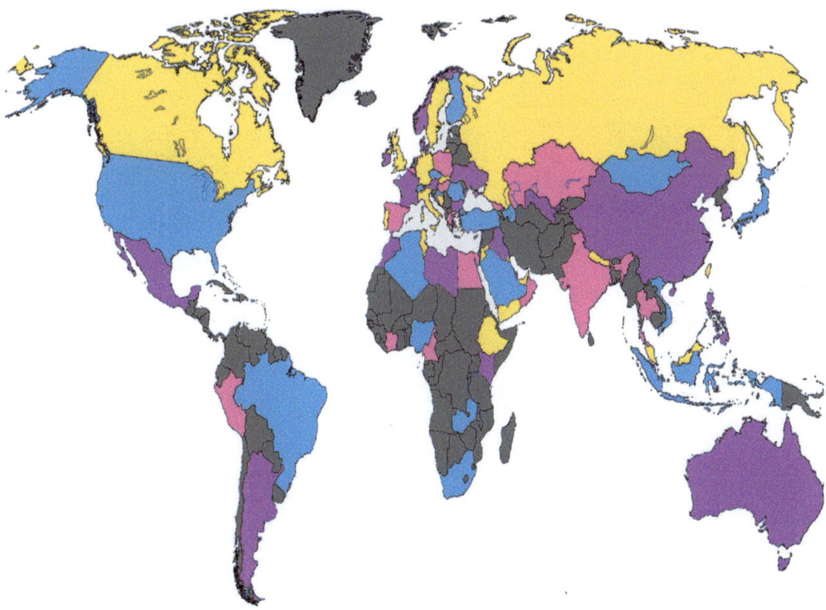

Fig. 2. *There were IHY activities in most of the world's countries, and many chose to form national organizing committees to stimulate local IHY activities. Nations with IHY national programs are shown in color.*

2.4. Regional and national leadership and coordination

There are 78 nations with IHY national committees (see Figure 2), and there are IHY contacts in most of the remaining United Nations member states. The national activities are coordinated within their respective IHY region (*Africa, Eastern Asia and the Pacific, Eastern Europe and Central Asia, Latin America, North America, Western Asia and Western Europe*). An additional network called the "*Balkan, Black Sea and Caspian Sea Regional Network*" was formed, which consisted of nations from both the Eastern and Western European regions. In this document, this additional network is included with the Eastern European region.

Members from each of the regions are each represented on the IHY Steering Committee. Formal coordination between the regions is handled by the Steering Committee. However, the IHY workshops and planning meetings have served as the primary means of coordination and integration between the regional and national programs. It was at the meetings that the leaders could discuss their priorities and resources, and determine the optimal approach and balance between local and international efforts. Brief summaries of the IHY activities for each of the IHY regions, along with reports for individual nations are given in Sect. 7.

3 Science: Universal Heliophysical Processes and Coordinated Investigation Programmes (CIPs)

With the phenomenal improvement of ground- and space-based observations since the International Geophysical Year in 1957, the opportunity to observe heliophysical processes has never been better. Today, ground-based observatories routinely observe most domains of heliophysics, often using local measurements to form a global picture. Observatories in space now provide data from locations and regimes that have only recently become accessible, with drastically improved resolution and range, and large-scale computer modelling has made it possible to assimilate and couple these diverse observational inputs, enabling cross-disciplinary studies which transcend traditional boundaries in space physics.

It is now widely recognized that much of the space physics of the solar system proceeds through a set of *Universal Heliophysical Processes*, i.e. magnetic reconnection, particle acceleration, plasma wave generation and propagation, etc. These processes occur in the atmosphere of the Sun, the Earth, and other planets; in the interplanetary medium; and in the outer reaches of the heliosphere where the solar wind meets the interstellar medium. By facilitating research on these Universal Processes across the many heliophysical regimes that serve as independent sampling sites, we are able to gain insight into these driving forces at a fundamental level.

3.1. Universal Processes: the foundation of heliophysics

The structure of the Universe is determined primarily by the interplay of gravity which is dominant in condensed objects, and the magnetic force which is dominant in the rarefied medium connecting the Sun and planets, among other realms. Each of these forces order the matter into a set of characteristic structures, each with the ability to store and release energy in response to changes in the external environment (Figure 1). For the most part, the storage and release of energy proceeds through a number of Universal Processes. The coordinated study

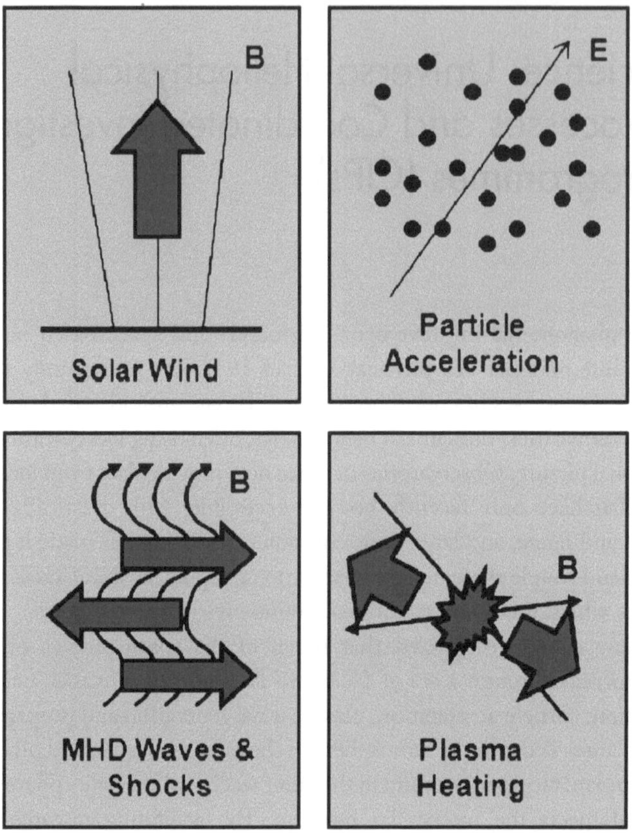

Fig. 1. *The energy stored in magnetic structures can be released as kinetic energy (upper left), energetic particles (upper right), waves (lower left), or heating (lower right). Here "B" denotes magnetic field and "E" denotes electric field.*

of these processes in different settings provides a deeper understanding of the underlying physics governing Universal Processes in geo- and astrophysics.

Magnetic reconnection is one of the best examples of a Universal Process in heliophysics. Reconnection is also observed in the magnetosphere of Earth, particularly in the magnetopause and magnetotail, and the signatures of reconnection are detected throughout the solar wind and it is probable that it will be observed at other planets as well. The energy is released as energetic particles, flows, waves, and plasma heating.

Universal Processes are not confined to the ionized plasma component in the Solar System. The Earth's ionosphere is the boundary between the neutral atmosphere and space. Energy between these two components is exchanged through this boundary. One process by which this energy exchange takes place is through atmospheric gravity waves. Gravity waves are excited in the neutral atmosphere by thunderstorms, mountain ranges, and other processes. These waves propagate upward from their point of origin steepening as they propagate. Eventually these waves become nonlinear and "break" like ocean waves on a beach, depositing their energy in the ionosphere–mesosphere, providing a potentially important coupling between the troposphere where climatic effects are evident and the magnetosphere. Figure 2 shows several images of gravity waves in both Earth's and Mars' atmospheres. The potential importance of these processes for energy transfer has only recently been recognized, and much additional work remains to establish the significance of the Universal Process of gravity wave propagation on the structure of Earth, Mars, and other planetary atmospheres.

The Universal Processes were adopted during the IHY United States Planning Workshop in April 2004, refined in subsequent years, and are described in detail in a new textbook series. Three next textbooks in Heliophysics are created as part of the LWS Heliophysics Summer School 2007–2009, one for each year (see "Publications" Appendix III). A general list of the types of Universal Heliophysical Processes is as follows:

Creation & Annihilation of Magnetic Fields Reconnection Dynamos Diffusion Dissipation	**Generation of Penetrating Radiation** Galactic Cosmic Rays Solar Energetic Particles Anomalous Cosmic Rays Radiation Belts	**Couplings of the Sun, Heliosphere, Galactic Environment & Planetary Climate** Planetary Dynamo Stellar Dynamos Radiative Couplings Electromagnetic Couplings
Magnetic coupling Non-Local (Non-Contact) Flow-Object Cross-Scale Neutral-Plasma Dusty Plasmas	**Explosive Energy Conversions** Solar (Stellar) Flares CMEs Substorms Bursty Bulk Flows	**Spontaneous Generation of Structures & Transients** Flux Ropes-Filaments Current Sheets Cellular Structures Turbulence

The field of comparative aeronomy is leading heliophysics in the examination of Universal Processes, by seeking commonalities in the observations of planets and moons. For example, aurorae (Figure 3) have been observed on Earth, Saturn, and Jupiter, Ganymede, as well as Jovian auroral "footprints" from Io, Ganymede and Europa. The formation of aurorae is observed to be the universal response of a magnetized body in the solar wind. This phenomenon involves several Universal Processes, spanning all of the categories listed above. The cross-disciplinary study

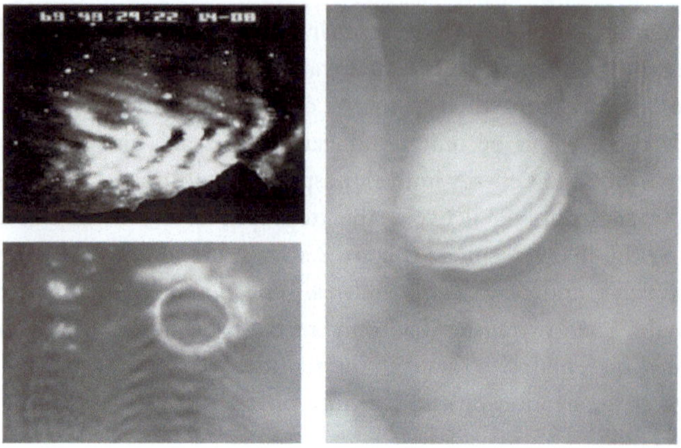

Fig. 2. *Gravity waves in Earth's atmosphere (upper left panel) transfer energy from the troposphere to the ionosphere–mesosphere. Similar waves seen on Mars by the Mars Global Surveyor Mission (lower left and right panels) may couple the lower and upper Martian atmosphere.*

Fig. 3. *Aurorae on Saturn (left), Jupiter (center left), Earth (center right), and on Ganymede (right) are manifestations of universal physical processes at work in the solar system.*

of these processes increases our sample size, treating the whole of heliophysics as our laboratory.

3.2. Scientific disciplines and the Universal Processes

The International Heliophysical Year (IHY) focuses on the cross-disciplinary study of Universal Processes in the solar system, observed in a variety of settings. Studying these Universal Processes together, in diverse environments, and in a comparative way, is leading to new scientific insights.

For the majority of current scientific studies in heliophysics, however, comparison of processes at multiple sampling sites is either secondary or nonexistent. Most of our scientific organizations categorize research according to the regime in which the studies occur (e.g. solar, magnetospheric, atmospheric) instead categorizing by the fundamental physics. To facilitate the transition from regime-dominated science to a focus on Universal Processes, our Coordinated Investigation Programmes (CIPs) are integrated with respect to more "traditional" categories. The IHY disciplines are:

- Solar Physics
- Heliosphere and Cosmic Rays
- Magnetospheres
- Ionized Atmospheres
- Neutral Atmospheres
- Climate Studies
- Meteors/Meteoroids/Dust

Each discipline has one or two scientists who serve as Discipline Coordinators, who review and coordinate the CIPs within their discipline, and work with the other Discipline Coordinators to determine each CIP's relationship to the Universal Processes described above. A detailed description of the CIP process follows in the next section.

3.3. Science campaigns and CIPs

3.3.1. Infrastructure

Three broad imperatives underlay the science coordination approach taken by IHY: first, that it be *decentralized*, driven by grassroots science interests without a central authority to direct the science; second, that it be a *low bureaucracy* approach that provided the maximum benefit with the minimum of formal overhead; third, that it should be *cross-disciplinary* and facilitate investigation of Universal Processes and help researchers to make use of scientific resources with which they were not familiar. The Coordinated Investigation Programme (CIP) mechanism was developed to match these requirements. *CIPs are the fundamental building blocks of IHY science* and provide the primary mechanism for identifying a science activity as contributing to IHY and for linking those activities together.

The model for a CIP was the SOHO Joint Observing Programme (JOP) but with the conscious replacement of "observing" with "investigation" to emphasize that non-observational science activities including theory, modelling and retrospective data analysis were equally welcome. The result was a process that gave open access – anyone could propose a CIP and details of all CIPs are public; and which preserved the autonomy of the activities described in any CIP – the proposer retains control and the IHY infrastructure acts to facilitate and advertise the

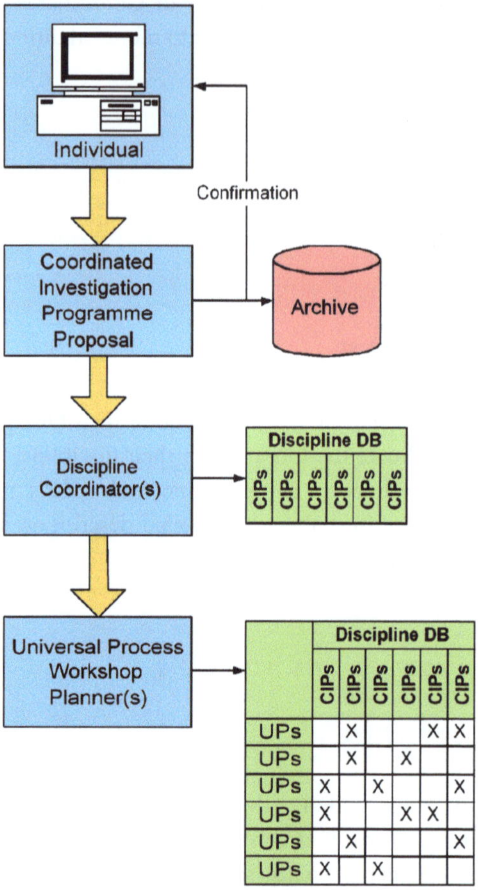

Fig. 4. *"How CIPs work" schematic.*

proposed activities. The CIP process is illustrated in Figure 4. The top level of the figure denotes an individual researcher, accessing the IHY CIP system via the Internet. Any researcher could propose a CIP designed to address a scientific topic according to expertise and available resources.

The second level of Figure 4 illustrates the CIP application and archival process. When a CIP proposal is submitted it is logged and the researcher receives confirmation that the CIP has been placed in the digital archive. CIP descriptions can also be modified, under the control of the original proposer. Notifications of new or updated proposals are also sent to the Discipline Coordinators, who review all CIP proposals and support cooperation and collaboration between related CIPs. An example of how CIPs were clustered in this way is shown in Figure 5. Coordinating the CIPs provides access to a wider range of data that can be assembled and analysed

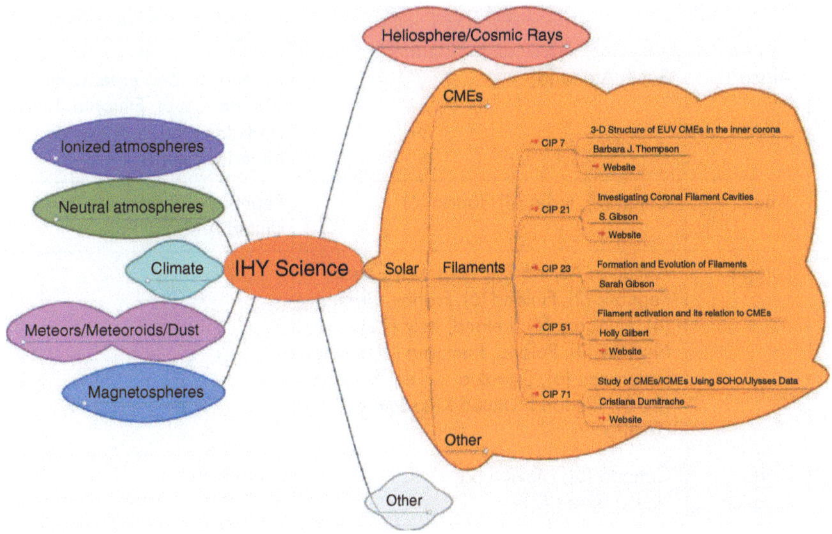

Fig. 5. *Example of CIP browser showing CIPs concerned with Solar Filaments in detail.*

more efficiently. The discipline coordinators will optimize the schedule of these CIPs to maximize scientific benefit and minimize the strain on needed resources.

The final stage of the CIP process occurs months to years later, as the researchers involved in various CIPs participate in one of many workshops to review the outputs of the IHY. This is the process by which the CIPs, organized according to the separate disciplines, are brought into the general paradigm of Universal Processes. A series of cross-disciplinary topical Universal Process workshops, some of are will be through "virtual" Internet communication, will present and discuss the scientific results of the IHY campaigns.

3.3.2. Range and distribution of CIP activities

A total of 65 CIPs were submitted with many spanning multiple disciplines, with the numbers in each discipline being: Heliosphere, 29; Solar, 24; Magnetospheres, 34; Ionized Atmospheres, 24; Neutral Atmospheres, 11; Climate, 4; Meteors/Meteoroids/IPD, 2. They can all be explored online using an online browser tool at *http://ihy2007.org.uk/CIP_browser.shtm*; an illustration of this browser in use is shown in Figure 5, which shows the details of those CIPs relating to solar filaments expanded as an example. Each red arrow is a link to a web page, those labelled "CIP n" to the CIP details and those labelled "Website" to a pertinent external website.

Participation in CIPs has been genuinely multinational with involvement from more than 45 nations in CIPs, either in proposing them or by involvement in their

Participation in the Coordinated Investigation Programme, by state		
Region	Proposing states	Participating states
Africa	South Africa (5)	Algeria, Cape Verde, Côte d'Ivoire, Egypt, Ethiopia, Libyan Arab Jamahiriya, Morocco, Nigeria, Tunisia
Asia	China (2), India (4), Japan (6)	Republic of Korea, Malaysia, United Arab Emirates
Europe	Armenia, Austria (3), Bulgaria, Czech Republic (2), Finland (6), France, Germany (4), Hungary (4), Ireland, Italy (12), Norway (3), Poland, Romania (2), Russian Federation (6), Slovakia, Spain, Sweden (2), Switzerland, United Kingdom (9), Ukraine (2)	
North America	Canada (6), Mexico (2), United States (22)	Costa Rica
South America	Brazil (5)	Argentina, Chile, Peru
Oceania	Australia (2), Indonesia	

implementation, spanning all continents. The table above summarizes this, showing the number of CIPs proposed, or jointly proposed, by each country and also those nations involved in CIP implementation. The geographical spread includes Antarctica, an important location for studying geospace, with 12 CIPs incorporating research and data taken from the Antarctic continent.

Several broad groupings of CIPs can be identified, indicating clusters of interest in research questions or methods of enquiry. Examples of these include:

- Comparative Aeronomy: CIPs such as the "Universality of Auroral Structure" CIP v#3 are spearheading the Universal Processes effort for IHY. CIP #3 takes advantage of our ability to study auroral processes on multiple planets and moons to assess the magnetic environment/energetic-plasma/atmosphere interaction and assess the commonality of auroral manifestations across the heliosphere.
- Cosmic Rays: Fourteen CIPs are concerned with the study of cosmic rays in the heliosphere. They range across studies of the generation of cosmic rays both in the heliosheath and from the Sun, their transport and modulation through the heliosphere, and their impact on the terrestrial ionosphere. The collaboration of this group of CIPs included organizing a workshop in Bad Honnef, Germany in

May 2007 covering a wide range of heliophysics but with a particular focus on the interplanetary magnetic field and modulation of Cosmic Ray fluxes.

- Solar Filaments: Five CIPs were focused on the study of solar filaments including their formation, structure, evolution and their links with Coronal Mass Ejections (CMEs). This cluster of CIPs was instrumental in setting up the most extensive campaign of the IHY science programme – the Whole Heliosphere Interval (WHI).
- CME Initiation and Propagation: A group of seven CIPs are aimed at the study of the origin of CMEs and their propagation outward from the Sun through the heliosphere to impact Earth, the other planets and spacecraft distributed throughout the solar system. These CIPs constituted another major component of the WHI campaign, making use of data from spacecraft including SoHO, Ulysses, Venus-Express and ACE, together with ground-based observations of the impact of CMEs on the terrestrial magnetosphere and ionosphere.
- Incoherent Scatter Radars: Five CIPs made use of the global network of Incoherent Scatter Radars. These instruments are few in number, with fewer than ten worldwide, but provide unique multi-parameter diagnostics of the ionosphere and how its behaviour is driven by coupling to the solar wind. The EISCAT Svalbard Radar ran an unprecedented and essentially continuous year-long operation to support IHY and the IPY from March 2007.

3.4. International science coordination

Many organizations played a fundamental role in coordinating IHY scientific activities. There were IHY special sessions at meetings all around the world, and each IHY workshop and conference would not have been possible without support from research agencies and sponsoring organizations. The United Nations Office of Outer Space Affairs provided the basis for the observatory development programme and many of the scientific collaborations associated with that programme. The programmes that have generously provided support for IHY are too numerous to mention. However, we mention three scientific organizations that have played a significant role in IHY's coordination.

3.4.1. International Astronomical Union

The International Astronomical Union deserves special recognition due to its central role in IHY's scientific activities. David Webb is the IAU's representative to the IHY, and an IAU Working Group on IHY, chaired by Nat Gopalswamy, has pursued research, collaboration and conference opportunities through the

IAU. The IAU IHY activities were first discussed at the 2006 IAU General Assembly in Prague, Czech Republic. "Astronomy for the Developing World", featured a subsession on IHY, and IAU also hosted an international coordination meeting for IHY leaders. The first meeting of the IAU Working Group on IHY also transpired in 2006, and all IHY regions were represented, to make plans for IAU-based activities.

The IAU sent statements of IHY support for each of the United Nations Committee on the Peaceful Uses of Space (COPUOS) meetings, and updates on IAU/IHY collaboration were regularly published in the *IAU Bulletin* and the IHY Newsletter.

IAU sponsored IHY's first conference dedicated to Universal Processes, "IAU Symposium 257: Universal Heliophysical Processes" September 15–19, 2008 in Ioannina, Greece. The conference's unique format consisted primarily of invited presentations that covered all aspects of heliophysics, allowing researchers to discuss the comparative physics underlying the Universal Processes.

At the next IAU General Assembly in Rio de Janiero 2009, there will be a 1.5-day special Joint Discussion titled "IHY Global Campaign – the Whole Heliosphere Interval". The Scientific Organizing Committee consists of IHY scientists from nine different nations, and will highlight the activities of IHY's largest scientific research campaign. There will also be a special session on the "International Year of Astronomy 2009", which will provide an opportunity to discuss the transition of IHY activities into the IYA.

3.4.2. The Whole Heliosphere Interval: a global campaign celebrating IHY and our great observatory network

http://ihy2007.org/WHI

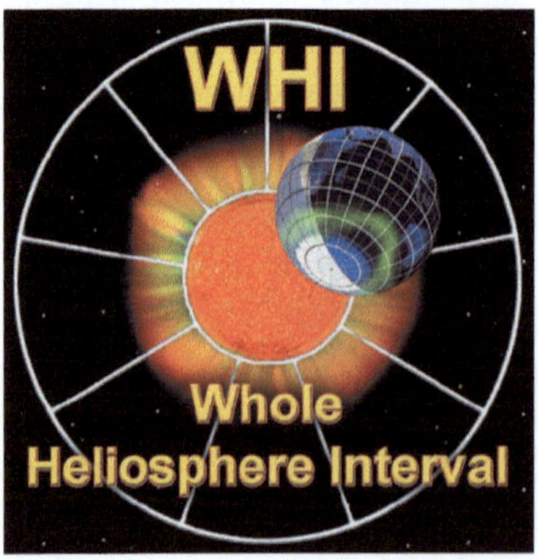

The Whole Heliosphere Interval (WHI) is an internationally coordinated observing and modelling effort to characterize the three-dimensional interconnected solar-heliospheric-planetary system. The heart of the WHI campaign is the study of the interconnected 3-D heliophysical domain, from the interior of the Sun, to the Earth, outer planets, and into interstellar space. Fifty years after the IGY, we have observations of the outer reaches of the heliosphere, and we are poised to make great advances in our understanding of our extended heliophysical domain and our relationship with interstellar space. WHI takes full advantage of the 50 years of scientific progress of IGY by coordinating state-of-the-art models and observations to address the entire interconnected heliophysical system.

The WHI observing campaign focused on the 3-D solar structure from solar Carrington Rotation 2068, which ran from March 20 to April 16, 2008. Observations and models of the outer heliosphere and planetary impacts extend beyond those dates as necessary; for example, the solar wind transit time to outer planets can take months.

WHI corresponds to several CIPs that focus on multi-spacecraft observations, modelling, and Sun–Earth heliospheric structure. WHI occurred during solar minimum, optimizing our ability to characterize the 3-D heliosphere and trace the structure to the outer limits of the heliosphere. Hundreds of scientists have performed observations and models, which were first discussed at the "WHI Data and Modelling Assessment Workshop" in Boulder, Colorado August 25–29, 2008. A special session on WHI happened at the December 15–19, 2008 American Geophysical Union Meeting, and the WHI Science Workshop is being planned for 2009.

3.4.3. ICESTAR: IHY's connection to the International Polar Year

http://www.scar-icestar.org

Interhemispheric Conjugacy Effects in Solar-Terrestrial and Aeronomy Research (ICESTAR) is a Scientific Research Programme endorsed by the Scientific Committee of Antarctic Research (SCAR). According to its name, the ICESTAR community is interested in the Sun–Earth linkage processes with the special emphasis on interhemispheric comparison studies. The programme lasts for 5 years and it started in 2005.

IHY together with ICESTAR coordinated an IPY core project (Cluster 63) under the title "Heliosphere impact on geospace". This initiative includes 29 international consortia, and it links into several IHY CIPs. These groups conduct research under three main themes:

(i) Coupling processes between the different atmospheric layers and their connection with solar activity;
(ii) Energy and mass exchange between the ionosphere and the magnetospher and
(iii) Interhemispheric similarities and asymmetries in geospace phenomena.

The IPY Cluster 63 community runs a large body of instrumentation in both the Arctic and the Antarctic. Many institutes have recently installed new instruments in the polar regions to significantly improve the spatial coverage and resolution and to provide pairs of geomagnetically conjugate observations from both the hemispheres.

ICESTAR and IHY have had complementary roles in the IPY Cluster 63 activities: IHY has arranged overarching synoptic observation campaigns and provided systems and assessment processes in order to facilitate the harvesting of interdisciplinary observations. ICESTAR has led the efforts in establishing Virtual Observatories (V×Os) for various geospace observations. The Gaia V×O for auroral precipitation data and VGMO from magnetic field data are examples of these coordination activities. As scientific output roughly 80 peer-reviewed articles with clear connection to Cluster 63 activities have been published during the period January 2007–June 2008.

The following lists all of the IHY Coordinated Investigation Programmes (CIPs). For a more complete summary of the IHY CIPs, please see Sect. 3.

CIP 1

Programme title: Coronal mass ejection studies: development of algorithms for CMEs

Lead proposer: Professor Richard A. Harrison, Space Science and Technology Department, Rutherford Appleton Laboratory, UK

Other proposers: Chris Davis, Danielle Bewsher, Rutherford Appleton Laboratory, UK; Andrew Breen, University of Wales Aberystwyth, UK; David Webb, Hanscom Air Force Base Research Lab, USA

Scientific disciplines: Solar; Heliosphere/cosmic rays

Scientific synopsis: Using a combination of white light wide-angle imaging of the heliosphere and EUV monitoring of the corona, the aim of this project is to develop specific tools for the prediction of CME onsets and CME arrival at Earth using several spacecraft platforms.

CIP 2

Programme title: Real-time global ionosphere
Lead proposer: Tim Fuller-Rowell, Space Weather Prediction Center, USA
Scientific discipline: Ionized atmospheres
Scientific synopsis: Characterizing the global ionospheric variability.

CIP 3

Programme title: Universality of auroral structure
Lead proposer: Dr. Anil Bhardwaj, Space Physics Laboratory; Vikram Sarabhai Space Centre, India
Other proposer: Marina Galand, Imperial College London, UK
Scientific discipline: Planetary magnetospheres; Ionized atmospheres; Neutral atmospheres
Scientific synopsis: Assessing the magnetic environment, energetic plasma, and atmosphere interaction: how does the magnetic environment interact with the ionosphere? Assess the commonality across auroral manifestations in the heliosphere.

CIP 4

Programme title: Global plasmaspheric dynamics
Lead proposer: Ian Mann, Department of Physics, University of Alberta, Canada
Scientific disciplines: Planetary magnetospheres; Ionized atmospheres; Neutral atmospheres
Scientific synopsis: Characterize plasmaspheric erosion and refilling.

CIP 5

Programme title: State of the LISM at the heliospheric boundary and inside
Lead proposer: Prof. Eberhard Moebius, Space Science Center, University of New Hampshire, USA
Other proposers: Prof. Hans R. Mueller, University of New Hampshire, USA
Scientific discipline: Heliosphere/cosmic rays
Scientific synopsis: How does the interstellar gas in the inner heliosphere change with the solar cycle? What are the defining features of neutral gas–plasma interaction?

CIP 6

Programme title: Environment termination shock and the heliosheath
Lead proposer: Matthew E. Hill, Applied Physics Laboratory, Johns Hopkins University, USA

Other proposers: Mike Gruntman, University of Southern California, USA; William Kurth, University of Iowa, USA
Scientific discipline: Heliosphere/cosmic rays
Scientific synopsis: (A) What is the nature of the heliosphere near the TS? (B) What are the properties of the shock? (C) What is the nature of heliospheric turbulence?

CIP 7

Programme title: 3-D structure of EUV CMEs in the inner corona
Lead proposer: Barbara J. Thompson, NASA GSFC, USA
Scientific disciplines: Solar; Heliosphere/cosmic rays
Scientific synopsis: We will conduct a comprehensive campaign to take advantage of the multiple instruments observing CMEs in the corona and inner heliosphere to determine the relationship between the structure represented by the EUV images (SOHO/EIT, TRACE, PROBA2/SWAP, STEREO/SECCHI) and the coronal white-light structures (Mauna Loa MkIV, SOHO/LASCO, STEREO/SECCHI). We will map the EUV signatures (dimmings, filaments, expanding loops, EIT waves) into the coronal structures as represented by the coronagraphs, with the goal of understanding more clearly the structure and manifestations of CMEs in the low corona.

CIP 8

Programme title: Mid- and high-latitude ionospheric–magnetospheric observations
Lead proposer: Dr. Pierre Cilliers, Hermanus Magnetic Observatory, South Africa
Other proposers: Prof. Harm Moraal, North-West University, South Africa; Dr. Cathryn Mitchell, University of Bath, Bath, UK; Dr. Lucilla Alfonsi, INGV, Rome, Italy
Scientific disciplines: Planetary magnetospheres; Ionized atmospheres; Neutral atmospheres
Scientific synopsis: To better understand magnetospheric and ionospheric variability in the time range from seconds to days and longer and their links to the solar wind and magnetospheric phenomena.

CIP 9

Programme title: Modelling variable cosmic ray fluxes in the dynamic heliosphere, magnetosphere, and atmosphere as well as their imprints on Earth
Lead proposers: Klaus Scherer, Institut fuer Theoretische Physik, Germany; Marius Potgieter, School of Physics, North West University, South Africa

Other proposers: Juerg Beer, EAWAG, Switzerland; Hans-Joerg Fahr, University of Bonn, Germany; Stefan Ferreira, North West University, South Africa; Horst Fichtner, University of Bochum, Germany; Bernd Heber, University of Kiel, Germany; Josef Masarik, Komensky University, Slovakia
Scientific disciplines: Planetary magnetospheres; Heliosphere/cosmic rays; Ionized atmospheres; Neutral atmospheres
Scientific synopsis: The aim of this proposal is to make a synoptic model for the production rates of cosmogenic isotopes, taking into account the cosmic ray transport and its long-term variability, the dynamics of the heliosphere and its variation with interstellar changes, and the transport of cosmic rays through the magnetosphere and atmosphere, to describe the effects on the Earth's environment.

CIP 10
Programme title: Empirical model of ring current distributions
Lead proposer: Anna Milillo, INAF/IFSI, Italy
Other proposer: Stefano Orsini, INAF/IFSI, Italy
Scientific discipline: Planetary magnetospheres
Scientific synopsis: The first aim of this project is to obtain from local ion measurements of the global evolution of the ring current plasma distributions during quite time as well as geomagnetic storm development (real-time global ring current). The second aim is to obtain zero order global plasma distributions by knowing the solar wind parameters at Earth (short-time forecast).

CIP 11
Programme title: Three-dimensional view of inner heliosphere and predicting the arrivals of coronal mass ejections
Lead proposer: P. K. Manoharan, Radio Astronomy Centre, Tata Institute of Fundamental Research, India
Other proposers: M. Kojima, Kyushu University, Japan; B. V. Jackson, University of California, San Diego, USA; J. A. Gonzalez-Esparza, Instituto de Geofisica, UNAM, Mexico; A. Breen, University of Wales Aberystwyth, UK; Xizhen Zhang, University of Colorado, USA
Scientific disciplines: Solar; Heliosphere/cosmic rays; Solar and heliosphere
Scientific synopsis: Coordinated IPS and SMEI observations of the heliosphere to study the three-dimensional view of the solar wind flow and turbulent plasma associated with CMEs; the aim is to predict the arrival of CME at 1 AU.

CIP 12
Programme title: SOHO-Ulysses coordinated studies during the two extended quadratures and the alignment of 2007–2008

Lead proposers: G. Poletto, Arcetri Astrophysical Observatory, Italy; S. Suess, NASA Marshall Space Flight Center, USA
Scientific disciplines: Solar; Heliosphere/cosmic rays
Scientific synopsis: When Ulysses is 90° around the Sun from Earth (in 'quadrature'), the same plasma observed remotely from, e.g., SOHO can later be observed in situ. Ulysses will be in near-quadrature for 5 months in winter 2007 and again in winter 2008. We will use the quadratures to analyze physical parameters (density, temperature, composition, ...) in outflows and transients. In summer 2007, Ulysses passes through a radial alignment with Earth, during which we will analyze composition and transport of energetic particles.

CIP 13
Programme title: High energy processes and dynamics of the low solar atmosphere during explosive events
Lead proposer: Jean-Pierre Raulin, Universidade Presbiteriana Mackenzie, Brazil
Other proposers: V. S. Makhmutov, E. Correia, P. Kaufmann, C. Mandrini, W. Bandeira, M. Luoni
Scientific discipline: Solar
Scientific synopsis: To bring observational clues on the total number of accelerated particles during the impulsive phase of solar flares, to better constraint existing acceleration models. From the gradual development of solar flares we address the dynamics of the low solar atmosphere. Finally we are interested exploring the relationship between the launch of CMEs and the activity close to the solar disk as seen at high frequency radio wavelengths.

CIP 14
Programme title: Ionospheric effects of the solar activity variations, solar flares, lightnings and energetic particle precipitations.
Lead proposers: J.-P. Raulin, E. Correia, V. S. Makhmutov, Universidade Presbiteriana Mackenzie, Brazil
Other proposers: A. A. Pacini, W. M. Carvalho, M. Aparecido
Scientific disciplines: Ionized atmospheres; Neutral atmospheres
Scientific synopsis: This project is concerned with the physical sources of the disturbances in the Earth's ionosphere at different time and space scales and their influence on the characteristics of very low frequency wave propagation.

CIP 15
Programme title: Influence of Sun and other cosmic factors on environment of the Earth

Lead proposer: Dr. Saumitra Mukherjee, School of Environmental Sciences, India
Scientific disciplines: Solar; Heliosphere/cosmic rays; Planetary magnetospheres; Ionized atmospheres; Neutral atmospheres; Climate
Scientific synopsis: It has been observed that earth directed coronal mass ejection from the Sun along with the cosmic influence has the potential to change the environment of the earth temporarily. The changes in the electron flux and planetary indices have been observed before the occurrence of earthquake, tsunami, sudden snowfall, heat wave and cloudburst. Cosmic influence of the sun needed to be observed on continuous basis on the Sun as well as on the earth and other planets of the solar system.

CIP 16

Programme title: Investigation of the high energy particle precipitation in the lower and middle atmosphere in the southern hemisphere magnetic anomaly (SHMA)
Lead proposer: Dr. Luis Eduardo Antunes Vieira, Max Planck Institute for Solar System Research, Germany
Scientific disciplines: Solar; Heliosphere/cosmic rays; Planetary magnetospheres; Neutral atmospheres; Climate
Scientific synopsis: The main goal of the program proposed is the investigation of the cosmic rays modulation of the cloud properties and net radiative flux in the southern hemisphere magnetic anomaly region.

CIP 17

Programme title: Dynamics of the magnetosphere
Lead proposer: Prof. Umberto Villante, Physics Department, University of L'Aquila, Italy
Other proposers: Dr. Marcello De Lauretis, Dr. Patrizia Francia, Dr. Massimo Vellante, University of L'Aquila, Italy; Dr. Roberto Bruno, Dr. Raffaella D'Amicis, INAF-IFSI, Italy
Scientific disciplines: Planetary magnetospheres; Ionized atmospheres
Scientific synopsis: We use measurements of the geomagnetic field variations at both high and low latitudes for a continuous monitoring of the dynamical processes of the Earth's magnetosphere as well as a ground support to space missions.

CIP 18

Programme title: Spatial and temporal TEC characterization of the ionosphere in the South Atlantic magnetic anomaly
Lead proposer: Emilia Correia, Universidade Presbiteriana Mackenzie, Brazil
Other proposers: Edvaldo S. Fonseca Jr., PTR/USP; J. F. G. Monico, Paulo O. Camargo, UNESP/Pres.Pru.; Sonia M. A. Costa, Luiz P. S. Fortes, IBGE;

Eurico R. de Paula, Ivan J. Kantor, INPE; Jean Pierre Raulin, Pierre Kaufmann, CRAAM/UPM; Jose H. Fernandez, UNITAU; Paulo R. Fagundes, UNIVAP; Leonardo C. Oliveira, IME, Brazil
Scientific discipline: Ionized atmospheres
Scientific synopsis: The goal of this work is to characterize the spatial and temporal variations of TEC surrounding the South Atlantic magnetic anomaly (SAMA); during quiet and magnetically disturbed periods. The TEC will be obtained from GPS measurements realized by the Brazilian continuous GPS network.

CIP 19

Programme title: Transport of mass and energy across the magnetopause during periods of northward IMF and their effects on the high latitude ionosphere
Lead proposer: M. F. Marcucci, INAF–IFSI, Italy
Other proposers: E. Amata, D. Ambrosino, M. B. Bavassano Cattaneo, M. Candidi, G. Consolini, I. Coco, S. Massetti
Scientific disciplines: Planetary magnetospheres; Ionized atmospheres
Scientific synopsis: We propose to study the plasma and energy transport processes from the solar wind to the magnetosphere during periods of northward IMF. For that purpose, we will use space borne and ground based data available during IPY. The main emphasis will be put on: CLUSTER, for the high latitude magnetopause, DOUBLE STAR for the equatorial magnetopause, SuperDARN for the polar and auroral ionosphere.

CIP 20

Programme title: Space weather in solar system (SWISS)
Lead proposer: Dr. Larisa Trichtchenko, NRCanada/ISES, Geomagnetic Laboratory, Canada
Other proposers: Dr. Xingqiu Yuan, Geomagnetic Laboratory, Canada; Dr. Konstantin Kabin and Dr. Robert Rankin, University of Alberta, Canada; Dr. Hans de Sterk and Mr. P. Stechlinski, University of Waterloo, Canada; Ms. Zdenka Smith, Dr. Tom Detman, Dr. Murray Dryer, NOAA/Space Weather Prediction Center, USA; Prof. Charles Deehr and Dr. Wei Sun, Geophysical Institute, University of Alaska, USA; Dr. Ghee Fry, Exploration Physics International Inc., USA; Dr. Devrie Intriligator, Carmel Research Center, USA; Prof. Shi Tsan Wu, Dr. Chin-Chun Wu, University of Alabama, USA; Dr. Susan McKenna-Lawlor, Space Technology Ireland Ltd., Ireland; Mr. Brian Choi, Carleton University, Canada
Scientific disciplines: Solar; Heliosphere/cosmic rays; Planetary magnetospheres
Scientific synopsis: To provide information on current, past and future (forecasted) space weather conditions in the Solar system, useful for planned

and current space missions and to provide boundary conditions for more detailed models and case studies. To provide a common frame for numerical models of solar wind-planetary magnetospheres interaction during quiet and disturbed times.

CIP 21

Programme title: Investigating coronal filament cavities
Lead proposer: S. Gibson, NCAR High Altitude Observatory, USA
Other proposers: J. Burkepile, G. de Toma, NCAR High Altitude Observatory, USA
Scientific discipline: Solar
Scientific synopsis: We will gather 3D, multiwavelength observations to constrain forward models of the three-dimensional density, temperature, and magnetic structure of coronal cavities. We will investigate the relationship of cavities to filaments and filament channels. We will also observe the evolution of cavities that erupt as CMEs.

CIP 22

Programme title: CME initiation, evolution, and propagation throughout heliosphere
Lead proposer: John Raymond, Harvard-Smithsonian Center for Astrophysics, USA
Other Proposer: L. Strachan Harvard-Smithsonian Center for Astrophysics, USA
Scientific discipline: Solar
Scientific synopsis: Comprehensive understanding of CMEs. How do CMEs evolve as they propagate from the Sun through the heliosphere? How are CME shocks formed and what is their relationship to SEPs?

CIP 23

Programme title: Formation and evolution of filaments
Lead proposer: Sarah Gibson, NCAR High Altitude Observatory, USA
Scientific discipline: Solar
Scientific synopsis: Understand filaments in the context of the global solar magnetic field. How do filaments form and how do they evolve in time?

CIP 24

Programme title: Effective temperature of the quiet sun
Lead proposer: P. Fox, NCAR High Altitude Observatory, USA
Other proposer: David Rust, JHU/APL, USA
Scientific discipline: Solar

Scientific synopsis: Determine the long term changes in Total Solar Irradiance at Solar Minimum – What are the connections between irradiance, solar radius, and surface magnetic field? What are the effects of coronal holes and chromospheric network on full disk TSI? What is the effective temperature of the Sun (mean and variance)? Are there slowly changing, large-scale variations in the distribution of the sources of TSI on the solar disk? What causes TSI variations that do not correlate with magnetic features?

CIP 25

Programme title: ISPAM
Lead proposer: A. P. van Eyken, EISCAT Scientific Association, Sweden
Other proposers: The Incoherent Scatter Working Group of URSI
Scientific discipline: Ionized atmospheres
Scientific synopsis: ISPAM is a coordinating activity between the high-latitude incoherent scatter radars to provide extensive ionospheric monitoring during the IPY. The resulting data set is widely applicable to studies of solar–terrestrial interactions, atmospheric tides and waves, and modelling (including now- and forecasting) activities.

CIP 26

Programme title: Investigating electromagnetic couplings and plasma environment change in geospace using MAGDAS magnetometer network
Lead proposer: Kiyohumi Yumoto, Space Environment Research Center, Kyushu University, Japan
Scientific discipline: Planetary magnetospheres
Scientific synopsis: (1) By monitoring the global three-dimensional current system, we seek to understand the electromagnetic coupling between high-latitude and Sq current systems. (2) How does the plasma environment change during space storms? (3) How does the polar electric field penetrate into the equatorial ionosphere?

CIP 27

Programme title: Cosmic ray tranport/modulation in the 3-D A-negative heliosphere
Lead proposer: John W. Bieber, Bartol Research Institute, University of Delaware, USA
Other proposers: B. McKibben, Space Science Center, University of New Hampshire, USA
Scientific discipline: Heliosphere/cosmic rays

Scientific synopsis: (A) What is the high energy radiation environment at Earth and in interplanetary space and how does it change over time? (B) What are the causes of GCR modulation? (C) What are the changes of the GCR spectrum? (D) What are the connections between GCRs and climate? (E) How does particle transport work? (F) What is the nature of heliospheric turbulence?

CIP 28

Programme title: Evolution of magnetic fields from the Sun's surface to the heliopause for one solar cycle
Lead proposer: N. Schwadron, Boston University, USA
Other proposers: C. Smith, Bieber, Bartol Research Institute, University of Delaware, USA
Scientific discipline: Heliosphere/cosmic rays
Scientific synopsis: (A) What is the 3D structure of the magnetic field in the quiet heliosphere? (B) What are the heliospheric signatures of the global magnetic field reversal in each 11-year cycle? (C) How do the photospheric, coronal, and interplanetary magnetic fields relate to each other? (D) What are interplanetary signatures and effects of reconnection?

CIP 29

Programme title: Comparative shock study
Lead proposer: M. Lee, University of New Hampshire, USA
Scientific discipline: Heliosphere/cosmic rays
Scientific synopsis: (A) What are the properties of shocks? (B) How do shock properties scale with the shock parameters? (C) How does interplanetary acceleration change with solar cycle? (D) What is the nature of heliospheric turbulence?

CIP 30

Programme title: POLARIS: polarization of the red thermospheric line
Lead proposer: Jean Lilensten, Laboratoire de Planetologie de Grenoble, France
Other proposers: Dr. Joren Moen, F. Sigernes, University of Oslo, Norway; D. Lorentzen, University Centre in Svalbard, Norway; C. Simon, M. Bartholomy, Laboratoire de Planetologie de Grenoble, France; Roland Thissen, Laboratoire de Chimie-Physique, France
Scientific disciplines: Ionized atmospheres; Neutral atmospheres
Scientific synopsis: Half a century ago was performed the last measurement of the oxygen red line by Duncan (1958) during the International Geophysics Year. This measurement has been since then disputed and never confirmed (Chamberlain 1959). We have now the opportunity to perform again this measurement in Svalbard (Norway), after promising theoretical developments on the polarization of the red line made in France. This project will open, as the International Polar

Year is drawing near, a new field of investigation allowing a never-done before diagnostic of the upper atmosphere, which will have a direct impact on orbitography computations. Moreover, almost 50 years after the first measurements, it will build a historical tribute to Duncan's work on the subject.

CIP 31

Programme title: Measurements of the elemental composition and energy spectra of the high-energy cosmic rays and the gradient of the Earth's magnetic field by high-altitude balloon flight around the North Pole
Lead proposer: Prof. Konstantin A. Kotelnikov, Fian, Russia
Other proposers: Prof. Yury P. Tsvetkov, Prof. N. S. Svirzhevsky, Prof. E. V. Vashenyik, Dr. T. I. Sysoeva, Dr. V. S. Puchkov, Dr. S. V. Filippov
Scientific discipline: Heliosphere/cosmic rays
Scientific synopsis: Arrangement of the high-altitude balloon flight around the North Pole. Balloon magnetic field measurements in the Earth's atmosphere with magnetic gradiometer. Measurements of the elemental composition and energy spectra of the high-energy cosmic rays.

CIP 32

Programme title: Decameter investigation of the Sun's activity
Lead proposer: Brazhenko Anatoly Ivanovich, Ukraine
Scientific discipline: Solar
Scientific synopsis: To receive the polarizing characteristics of a sporadic radio emission of the Sun in decameter lengths of waves.

CIP 33

Programme title: IASWR – Antarctic space weather reports
Lead proposer: Phil Wilkinson, IPS Radio and Space Services, UK
Other proposer: Dr. Giorgiano De Franceschi
Scientific discipline: Ionized atmospheres
Scientific synopsis: The principal outcome of this proposal will be increased awareness of the problems inherent in forecasting the high-latitude ionosphere in a region of sparse observations, such as encountered in the Antarctic. The database created during the period will also form a baseline collection against which later forecasting efforts may be compared. To date, no similar data set is readily available, although the task of making space weather forecasts is a well developed, if not a mature enterprise.

CIP 34

Programme title: balloon borne investigation of Earth, atmosphere and space in the polar regions

Lead proposer: Silvia Masi, Physics Department, University La Sapienza, Italy
Other proposers: Paolo de Bernardis, Steven Peterzen, Roberto Ibba, Domenico Spoto, Giovanni Romeo, Kjell Boen, Simona Di Pippo
Scientific discipline: Other (stratospheric balloons and related science)
Scientific synopsis: Stratospheric Long Duration Balloons facility in the polar regions for earth, atmospheric physics and cosmological studies can answer several open questions in science, and can carry instruments to be tested before a satellite mission is launched.

CIP 35

Programme title: Coordinated study on solar wind turbulence during
Lead proposers: Roberto Bruno, INAF-IFSI, Italy; Zoltan Voeroes, Space Research Institute, Austrian Academy of Sciences, Austria; Vincenzo Carbone, Universita della Calabria, Dpt. Fisica, Italy
Other proposers: R. D'Amicis, A. Mura, A. Milillo, S. Orsini, B. Bavassano, M. Cattaneo, P. Veltri, L. Sorriso-Valvo, T. Zhang, H. Biernat, H. Rucker, W. Baumjohann, D. Jankovicova, P. Kovacs
Scientific discipline: Solar; Other (solar wind turbulence)
Scientific synopsis: At the end of August 2007, Venus, Earth and Ulysses will be aligned within a few degrees. This unusual event gives the opportunity to attempt a coordinated study on the radial evolution of solar wind turbulence and coronal transients like CMEs between 0.7 and 1.4 AU.

CIP 36

Programme title: International network for auroral optical studies of the polar ionosphere (short title: auroral optical network)
Lead proposer: Ingrid Sandahl, Swedish Institute of Space Physics, Sweden
Other proposers: Dr. Kirsti Kauristie, Finnish Meteorological Institute, Finland; Prof. Takehiko Aso, National Institute of Polar Research, Japan; Dr. Eric Donovan, University of Calgary, Canada; Dr. Stephen Mende, Space Science Laboratory, UC Berkeley, USA; D. D. Vladimir Safargaleev, Polar Geophysical Institute, Russia; Dr. E. M. Blixt, University of Tromso, Norway; Dr. Dag Lorentzen, The University Centre in Svalbard, Norway; Prof. Joran Moen, University of Oslo, Norway; Dr. Mike Kosch, Lancaster University, UK; Dr. Ian McWirther, University College London, UK; Professor Roger Smith, University of Alaska, USA
Scientific disciplines: Planetary magnetospheres; Ionized atmospheres; Neutral atmospheres

Scientific synopsis: The optical aurora is the only visible manifestation of the physical processes occurring in the Earth's magnetosphere as it responds to the solar wind, the interplanetary magnetic field, and their often dramatic changes. Studies of the optical aurora is an efficient way to learn about both temporal and spatial variations of these processes and is also a powerful tool for monitoring the Sun–Earth interaction. The emissions contain information both about the incident primary particles and about physical and chemical processes occurring locally as the light is produced. The usefulness of ground-based optical measurements is enhanced by the possibility of combining them with information from satellites.

CIP 37

Programme title: Interplanetary causes of intense magnetic storms
Lead proposer: Walter D. Gonzalez, INPE, Brazil
Other proposers: Ezequiel Echer, Alisson Dal Lago, Fernando Guarnieri, Alicia de Gonzalez, Bruce Tsurutani
Scientific disciplines: Planetary magnetospheres; Other (interplanetary-magnetosphere coupling)
Scientific synopsis: During the IHY it is of interest to perform a study about the interplanetary causes of intense storms ($Dst < -100\,nT$) that may occur. The particular position of this year within the solar cycle (start of the ascending phase?) presents an interesting challenge for this research. We expect that the results of this study will present a helpful material for the identification of solar sources of such storms, which could be performed in collaboration with other participant groups, which will be studying the solar activity during the IHY.

CIP 39

Programme title: ULF waves in the magnetosphere
Lead proposer: Balazs Heilig, Tihany Geophysical Observatory, Hungary
Scientific disciplines: Planetary magnetospheres; Ionized atmospheres
Scientific synopsis: (A) Monitoring ionosphere–plasmasphere dynamics at mid-latitudes; (B) Monitoring upstream wave related ULF Pc3 pulsations.

CIP 40

Programme title: Conjugate study on magnetospheric cusp dynamics
Lead proposer: Huigen Yang, Polar Research Institute of China, China
Other proposers: Natsuo Sato, National Institute of Polar Research, Japan; Asgeir Brekke, University of Tromso, Norway; Hu Hongqiao, Zhang Beichen, Liu Ruiyuan, Polar Research Institute of China, China

Scientific disciplines: Planetary magnetospheres; Ionized atmospheres
Scientific synopsis: Investigating dynamics of the magnetospheric cusp. (1) Conjugate study of auroras and ionosphere at cusp latitudes. (2) Ground-based observation of signatures of cusp and solar wind–magnetosphere interaction.

CIP 41

Programme title: Radio tomographic imaging of the Arctic ionized atmosphere by the International Ionospheric Tomography Community (IITC)
Lead proposer: S. Eleri Pryse, University of Wales Aberystwyth, UK
Other proposers: H. R. Middleton, L. Kersley, University of Wales Aberystwyth, UK; J. A. Secan, E. J. Fremouw, Northwest Research Associates, USA; G. S. Bust, Applied Research Laboratories, University of Texas, USA; T. Turunen, Sodankyl Geophysical Observatory, University of Oulu, Finland; E. D. Tereshchenko, Polar Geophysical Institute, Russia; V. E. Kunitsyn, M. Lomonosov Moscow State University, Russia
Scientific discipline: Ionized atmospheres
Scientific synopsis: The project aims to investigate the impact of space weather processes on the terrestrial ionized atmosphere and identify the relative roles of the geophysical processes that influence the large-scale plasma distribution. The basic physics and chemistry of ionization production and loss in the polar regions are understood reasonably well, including effects of large-scale plasma transport and precipitation of charged particles into the atmosphere. However, the actual structure present in any given situation depends on the relative contributions of the various processes and the balance remains an open question that needs to be addressed by experimental observations.

CIP 43

Programme title: SMEI and IPS 3D heliospheric analysis comparison with Ulysses and other NASA spacecraft
Lead proposer: Bernard V. Jackson, University of California, San Diego, USA
Other proposers: David Webb, Hanscom Air Force Base Research Lab, USA; Masayoshi Kojima, Kyushu University, Japan
Scientific disciplines: Solar; Planetary magnetospheres; Heliosphere/cosmic rays
Scientific synopsis: We wish to compare the Solar Mass Ejection Imager (SMEI) 3D data reconstruction analyses (density) and with interplanetary scintillation (IPS) data (to measure velocities) during specific time intervals to compare with the Ulysses spacecraft, and other NASA spacecraft instrumentation (i.e., Mars Global Surveyor magnetospheric measurements at Mars) and perhaps use future Faraday rotation observational data to provide global 3D heliospheric reconstructions of density, velocity and magnetic field in comparison with existing NASA spacecraft resources.

CIP 44

Programme title: Cosmic rays in the heliosphere
Lead proposer: Marius Potgieter, Unit for Space Physics, North-West University, South Africa
Other proposers: Prof. Adri Burger, Dr. Stefan Ferreira, Mr. Jasper Snyman, Mr. Donald Ngobeni
Scientific discipline: Heliosphere/cosmic rays
Scientific synopsis: Numerical and theoretical modelling of the transport, acceleration and modulation of galactic and anomalous cosmic rays in the heliosphere.

CIP 45

Programme title: Imaging the density structure of the inner magnetosphere
Lead proposer: Dr. Endawoke Yizengaw, UCLA, USA
Scientific disciplines: Planetary magnetospheres; Ionized atmospheres
Scientific synopsis: Imaging the structure of electron density profiles of the inner-magnetosphere region to address the following puzzling issues. These include: (1) What is the relative plasmaspheric and ionospheric contribution to the ground based GPS derived TEC changes? Does the plasmaspheric percentage contribution vary as a function of local-time, latitude, season, and geomagnetic and solar activities? (2) How does the density of the plasmasphere behave as a function of altitude, local-time, latitude, season, and geomagnetic and solar activities? (3) How do the dynamics of refilling and erosion of plasmaspheric density behave as viewed from tomographic images of the topside ionosphere and plasmasphere? Is the field aligned damping of plasma to the lower side ionosphere really responsible for the inside the plasmasphere density depletion?

CIP 47

Programme title: VLF and ULF waves and particles
Lead proposer: Arthur R. W. Hughes, School of Physics, University of KwaZulu-Natal, South Africa
Other proposers: Andrew Collier, University of KwaZulu-Natal, South Africa
Scientific discipline: Other (Geospace Physics)
Scientific synopsis: To establish the connection between geomagnetic Pc5, Pi2 pulsations, VLF waves, magnetic sub-storms and particle precipitation.

CIP 48

Programme title: Heavy ion outflow from the dayside ionosphere: multi-point view
Lead proposer: Yulia V. Bogdanova, Mullard Space Science Laboratory, University College London, UK

Other proposers: C. J. Owen, MSSL, UCL, UK; I. McCrea, J. Davies, Rutherford Appleton Laboratory, UK; B. Klecker, MPE Garching, Germany
Scientific disciplines: Planetary magnetospheres; Ionized atmospheres
Scientific synopsis: The aim of this project is a better understanding of the mass-loading mechanism of the heavy ions from the ionosphere into the magnetosphere especially during storm-time. The important questions are: (1) Location of the dayside outflow – is it concentrated near the cusp region or evenly distributed over a broad range of MLT? (2) Question of quasi-neutrality in the region with ion outflow. (3) Latitudinal extent of the region with heating of ions at different altitude (ionosphere level and 3–6 RE level) (4) What is the exact mechanism of energy transfer from the solar wind into the ionospheric ion outflow? (5) What are the locations and properties of ion outflow under northward IMF? (6) Where are the ions going to?

CIP 49
Programme title: COST296/IHY
Lead proposer: Giorgiana De Franceschi, INGV-RM2, Upper Atmosphere Physics Unit, Italy
Other proposers: COST296/IHY Coordination Team Members
Scientific discipline: Ionized atmospheres
Scientific synopsis: Since the 1990s the COST Actions relating to ionospheric physics are particularly useful in creating a critical mass of researchers to increase the knowledge of the ionosphere and its effects on navigation and communication systems. Nowadays COST296 Action (2005–2009) (www.COST296.rl.ac.uk) provides a unique European forum for the progression of such matters. In the IHY frame the idea is to maintain the necessary European cooperation for a multi-instrument monitoring of the middle–upper atmosphere and improving scientific joint efforts to investigate on the interconnection between low, mid- and high-latitude ionosphere. This proposal hopes for an integration of other interested middle–upper atmosphere scientific groups.

CIP 51
Programme title: Filament activation and its relation to CMEs
Lead proposer: Holly Gilbert, NASA GSFC, USA
Other proposers: Terry Kucera, Gary Kilper, NASA GSFC, USA; Rui Liu, Rice University, USA
Scientific discipline: Solar
Scientific synopsis: We will conduct a comprehensive observational study on filament activation, taking advantage of multiple instruments observing filaments

in the chromosphere and CMEs in the corona to better understand how this process works and how it is related to CMEs.

CIP 53

Programme title: Modelling the magnetosheath flow and magnetopause location in the cusp region
Lead proposer: Ermanno Amata, IFSI-INAF, Italy
Other proposers: Monio Kartalev, P. S. Dobreva, Maria Federica Marcucci, D. Ambrosino
Scientific discipline: Planetary magnetospheres
Scientific synopsis: The cusp regions are of great importance for understanding the solar wind influence on the Earth's environment. One outstanding problem in this respect is to explain the detailed features of the flow and the shape and position of the magnetopause in such regions. The model reproduces the cusp indentation and represents a substantial step ahead with respect to previous ones. Moreover it requires only a usual PC, taking reasonable computational time, promising in this way to be a useful tool for data interpretation.

CIP 54

Programme title: Inferring the magnetospheric global potential distribution from SuperDARN data through the IMech numerical model
Lead proposer: Monio Kartalev, Bulgarian Academy of Sciences, Bulgaria
Other proposers: E. Amata, I. Coco, K. Grigorov, M. F. Marcucci
Scientific discipline: Planetary magnetospheres
Scientific synopsis: The ionospheric electrodynamics is a key element of the solar wind–Earth interaction and of the global electric circuit directly affecting the biosphere. To address this issue, a new global model has been developed at the Institute of Mechanics (IMech) of the Bulgarian Academy of Sciences, Sofia. This model, unlike previously published ones, includes field-aligned currents between magnetically conjugate points and treats separately the northern and southern ionospheres, without forcing the potentials at conjugate points to be equal. In order for the model to run, it is necessary to feed it with distributions of the polar field aligned current system calculated for both hemispheres, e.g. by the AMIE model, which is based on a large collection of data from various instruments. In this CIP we plan to test the model by using potential distributions obtained from SuperDARN data through the JHUAPL model.

CIP 55

Programme title: Ionospheric studies through GPS, ionosonde and AWSOME
Lead proposer: Dr. Kamlesh N. Pathak, Department of Physics, S. V. National Institute of Technology, India

Scientific discipline: Ionized atmospheres
Scientific synopsis: The aim of the study is twofold: (1) We are starting postgraduate program in Space Physics where the deployment of various heliospheric instrumentation shall be utilized for hands on experimentation to the students. (2) At Surat which is at 21.16° N and 72.78° E coordinates ionospheric studies are sparse. As the location is just below the anomaly the upper ionospheric observations will throw some light on the characteristics of the ionosphere. Also by the deployment of VLF receiver system like AWSOME a complete ionospheric study capability will be available.

CIP 56
Programme title: Research of highest energy solar cosmic rays
Lead proposer: A. Chilingarian, Cosmic Ray Division, Armenia
Other proposers: G. Karapetyan
Scientific discipline: Heliosphere/cosmic rays
Scientific synopsis: Available experimental data on the GLEs confirm proton acceleration up to $\sim$20 GeV, acceleration to higher energies is not reliably proved. Middle- and high-latitude neutron monitors cannot be used for the reconstruction of the primary energy spectra well above 5 GeV because of very weak fluxes and rather small sizes of the detectors. Along with operation of the AMMM we plan to develop new type world-wide network of particle detectors, measuring neutral and charged fluxes simultaneously and to be installed at middle and low latitudes. New particle detectors will perform monitoring of various species of secondary cosmic rays with different energy thresholds.

CIP 57
Programme title: Solar minimum watching with Italian SINERGIES and more
Lead proposer: Dr. Marisa Storini, INAF/IFSI-Roma, Italy
Other proposers: Researchers from INAF, INFN, INGV, CNR, UNICatania, UNIL'Aquila, UNIPerugia, UNIRm1, UNIRm2, UNIRm3, UNISiena, UNITrieste, UNIUrbino, Italy
Scientific disciplines: Solar; Heliosphere/cosmic rays (terrestrial science)
Scientific synopsis: We use measurements of the different and complementary ground-based instruments managed by research groups of the Italian Astrophysics Community, together with some in situ data, to characterize the solar–terrestrial system during the current solar activity minimum phase (transition from sunspot cycles 23 to 24) in the period in which there is a null sub-Earth point on the Sun.

CIP 58

Programme title: Muon diagnostics of heliosphere and the near-Earth space
Lead proposer: D. A. Timashkov, NEVOD, MEPhI, Russia
Other proposers: N. S. Barbashina, V. V. Borog, A. N. Dmitrieva, R. P. Kokoulin, K. G. Kompaniets, A. A. Petrukhin, D. A. Room, V. V. Shutenko, I. I. Yashin
Scientific disciplines: Heliosphere/cosmic rays; Planetary magnetospheres; Neutral atmospheres
Scientific synopsis: High-energy galactic cosmic rays penetrate the Solar system from all directions and bring to the Earth information about various dynamic processes in the heliosphere. Though primary and solar cosmic rays do not arrive to the Earth's surface, the secondary muons produced in the atmosphere reach the ground level keeping the initial particle direction that allows to "X-raying" the inner structure of heliosphere disturbances and to reveal them from large distances from the Earth.

CIP 59

Programme title: Universal and individual properties of solar eruptions and their interplanetary consequences: dimensionless scaling
Lead proposer: I. S. Veselovsky, Institute of Nuclear Physics, Moscow State University, Russia
Scientific disciplines: Solar; Heliosphere/cosmic rays
Scientific synopsis: The aim of this project is the application of dimensionless scaling to the construction of experimentally based conceptual models of solar flares, eruptive prominences, coronal mass ejections and their interplanetary manifestations.

CIP 60

Programme title: Influence of space weather on micrometeoroid flux
Lead proposer: Dr. Thomas Djamaluddin, National Institute of Aeronautics and Space, Indonesia
Scientific disciplines: Solar; Heliosphere/cosmic rays; Meteors/meteoroids/dust
Scientific synopsis: Preliminary Studies on meteor flux detected by Meteor Wind Radar (MWR) Serpong (near Jakarta, Indonesia) during 1992–1992 and MWR Kototabang (near Bukittinggi, Indonesia) during 2002–2006 indicate that the flux are affected by ecliptic and solar activity variation. Long-term MWR data from all over the world and in situ dust detector on space facility are expected to infer space weather influence on distribution variation of micrometeoroid in Earth orbit. The next question is whether space weather from solar activities affecting the meteor flux be due to atmospheric or interplanetary mechanism.

CIP 61

Programme title: Solar–heliospheric sources of energetic particles
Lead proposer: Peter Kiraly, KFKI Research Institute for Particle and Nuclear Physics, Hungary
Other proposers: Dr. Karoly Kecskemety, KFKI, Hungary
Scientific discipline: Solar; Heliosphere/cosmic rays
Scientific synopsis: (A) Study of spatial and temporal variations of suprathermal and energetic particle fluxes throughout the heliosphere. (B) Study of anisotropy as a function of flux levels. (C) Time-shifted cross-correlation of particle fluxes for various energies. (D) Quiet-time properties of particle populations, including spectral shape, the existence of stable lower bounds, and the dependence on distance from the Sun. (E) Comparison of particle flux characteristics recently measured by Voyager with those measured in the inner heliosphere. (F) Search for nuclear interaction products in heliospheric suprathermal and energetic particle populations.

CIP 62

Programme title: Polar gateways to heliophysical exploration
Lead proposer: John F. Cooper, NASA GSFC, USA
Other proposers: Dieter Bilitza, Nan Lal, Robert E. McGuire, John W. Mitchell, Robert E. Streitmatter, Steven J. Sturner, Matthew E. Hill, Thomas P. Armstrong, Robert B. McKibben, John D. Richardson, Rogelio Caballero Lopez, Harm Moraal, Kirsti Kauristie, Brian Fraser, Phil Wilkinson, Martin Friedrich, Glenn W. Sheehan, Werner Singer, Akira Yamamoto, Koji Yoshimura, Tetsuya Yoshida
Scientific discipline: Heliosphere/cosmic rays
Scientific synopsis: Overall theme: Cosmic ray proton energy flow through the heliosphere to geospace and the polar ionosphere. This investigation addresses the heliospheric flow of cosmic ray energy, mostly in protons, from sources at the Sun, within the heliosphere, and from outside the heliosphere to the geospace environment and into the polar ionospheres. Key questions are (1) the source, loss, and transport processed for cosmic ray protons from tens of keV to GeV and above energies; (2) specification of proton flux spectra averaged over solar cycle time scales for effects on planetary surfaces and atmospheres; (3) specification of the instantaneous and time-averaged spectra for protons in geospace; (4) models for propagation of cosmic ray protons through the geomagnetic field into the polar atmospheric environments; and (5) correlative ground-based measurements for cosmic ray contributions to the polar ionospheres.

CIP 64

Programme title: VLF remote sensing of the daytime and nighttime lower ionosphere
Lead proposer: Umran Inan, Stanford University STAR Laboratory, USA
Other proposers: Phil Scherrer, Deborah Scherrer, Morris Cohen, Sheila Bijoor; Stanford University, USA
Scientific discipline: Solar; Heliosphere/cosmic rays; Planetary magnetospheres; Ionized atmospheres
Scientific synopsis: Using very low frequency (VLF, 3–30 kHz) remote sensing techniques, disturbances in the lower ionosphere can be effectively detected and quantified. Disturbances come from a wide variety of sources, including lightning, magnetospheric electron precipitation, auroral activity, solar flares, solar eclipses and cosmic rays. The goal of this project is to understand the ionospheric effects of these processes on a more global scale, using an array of VLF receivers, known as AWESOME, distributed throughout the planet.

CIP 65

Programme title: Meteors in the earth atmosphere and meteoroids in the solar system
Lead proposer: Kolomiyets Svitlana V., Meteor Radar Centre, Kharkiv National University of Radioelectronics, Ukraine
Scientific disciplines: Heliosphere/cosmic rays; Ionized atmospheres; Climate; Meteors/meteoroids/dust

CIP 66

Programme title: Solar variability impact on the middle atmosphere
Lead proposer: Victor I. Fomichev, Department of Earth and Space Science and Engineering, York University, Canada
Other proposers: Theodore G. Shepherd, Andreas I. Jonsson, Charles McLandress, University of Toronto, Canada; John (Jack) C. McConnell, Stephen R. Beagley, Kirill Semeniuk, York University; Stella M. Melo, Chao Fu, Canadian Space Agency, Canada
Scientific disciplines: Neutral atmospheres; Climate
Scientific synopsis: Solar variability affects the middle atmosphere through both variations in solar spectral irradiance and particle precipitation (Galactic Cosmic Rays, Aurora, Solar Proton evens). The goal of this project is to model the middle atmosphere response to solar variability and build a hierarchy of different forcing mechanisms.

CIP 67

Programme title: Steps of the Bashful Ballerina: global structure of the solar/inner heliospheric magnetic field

Lead proposer: Kalevi Mursula, Physics Department, University of Oulu, Finland
Scientific disciplines: Solar; Heliosphere/cosmic rays
Scientific synopsis: Recently, systematic hemispheric and longitudinal differences have been detected in the properties of the heliospheric magnetic field (HMF) and in the heliospheric current sheet (HCS). HCS has been found to be shifted southwards during a few years in the late declining to minimum phase of the solar cycle. This is related to the simultaneous development of a global quadrupole moment in the solar coronal field. It has also been found that the HMF sector structure depicts an interesting flip-flop cycle of about 3.2–3.5 years, in analogy of a similar cycle of solar active longitudes. This CIP concentrates on these and related issues, aiming to obtain new insight using different solar and heliospheric parameters from several data bases.

CIP 68

Programme title: Modelling of cosmic ray induced ionization of the atmosphere
Lead proposer: Ilya Usoskin, Sodankyla Geophysical Observatory, University of Oulu, Finland
Other proposers: Dr. G. A. Kovaltsov, Ioffe Physical-Technical Institute, Russia
Scientific disciplines: Heliosphere/cosmic rays; Ionized atmospheres
Scientific synopsis: The aim of the proposal is to develop a realistic model of cosmic ray induced ionization of the earth's atmosphere, that should be valid in the whole range of parameters, from the ground to the upper atmosphere, from polar regions to the equator, and from the solar minimum to maximum.

CIP 69

Programme title: Rapid solar wind and its impact on the geomagnetic variability
Lead Proposers: Ovidiu Maris, Institute for Space Sciences (ISS), Romania; Crisan Demetrescu, Institute of Geodynamics of the Romanian Academy, Romania
Other proposers: Georgeta Maris, Venera Dobrica, Institute of Geodynamics, Romania
Scientific disciplines: Solar; Heliosphere/cosmic rays; Planetary magnetospheres
Scientific synopsis: (a) Identifying the high-speed streams (HSSs) in solar wind during the 1996–2006 interval; (b) Spectral analysis of the HSS parameters; (c) Geomagnetic variability analysis (using aa, IHV, IDV, Dst indices); (d) Study of the "cause and effect" relationship between HSS and geomagnetic variability; improvement of forecast methods.

CIP 71

Programme title: Study of CMEs/ICMEs using SOHO/Ulysses data
Lead proposer: Cristiana Dumitrache, Astronomical Institute of Romanian Academy, Romania

Other proposers: Marilena Mierla, A.I.R.A., Diana Constantin, Iulia Chifu, Oana Chiricuta, Adrian Sabin Popescu, Adrian Oncica, Nedelia Antonia Popescu, Vasile Mioc
Scientific disciplines: Solar; Heliosphere/cosmic rays
Scientific synopsis: Study of solar sources of CMEs (polar prominences, active regions). Follow-up of CMEs from SOHO to Ulysses or track back of ICMEs from Ulysses to the Sun

CIP 72

Programme title: Meteors in the Earth atmosphere and meteoroids in the solar system
Lead proposer: Oleg Belkovich, Kasan State University, Zelenodolsk, Russia
Other proposers: Dr. Marina Ishmukhametova, Dr. Vladimir Andreev, Nail Suleymanov, Engelhardt Astronomical Observatory, Russia; Students (5), Kazan State University, Zelenodolsk, Russia
Scientific discipline: Meteors/meteoroids/dust
Scientific synopsis: Application of the modern methods of processing of meteor radar observations made in Kazan, Russia, in the 1960s and 1970s of the last century to find out structures of the main meteor streams (meteoroid flux densities and mass index S as functions of the Solar magnitude). Comparison with results obtained from visual observations.

CIP 73

Programme title: Meteors in the Earth atmosphere and meteoroids in the solar system
Lead proposer: Prof. Nelly Kulikova, Obninsk State Technical University, Russia
Other proposers: Dr. Valentina Tishchenko, Dmitriy Kalinin, Nicolay Polyakov, Students (3); Obninsk State Technical University, RUSSIA; Dr. Valentina Chepurova, Moscow State University, Russia
Scientific discipline: Meteors/meteoroids/dust
Scientific synopsis: Studies of meteoroid danger on Earth and in space. Investigation of the formation and evolution of meteoroid complexes in comet disintegration by probability methods and stochastic approaches. The criteria of comet stream and flux generality. Development of a computer technology.

4 IHY/UNBSS Observatory Development Program

4.1. Introduction

The purpose of the Observatory Development component of the IHY is to develop activities and facilitate partnerships that stimulate Space and Earth Science activities throughout the world, particularly in developing or undersampled regions, such as the establishment of ground-based instrument arrays and research programs. This includes the deployment of small, inexpensive instruments such as magnetometers, radio antennas, GPS receivers, all-sky cameras, etc. around the world to provide global measurements of ionospheric and heliospheric phenomena. This joint program, a collaboration between the IHY and the United

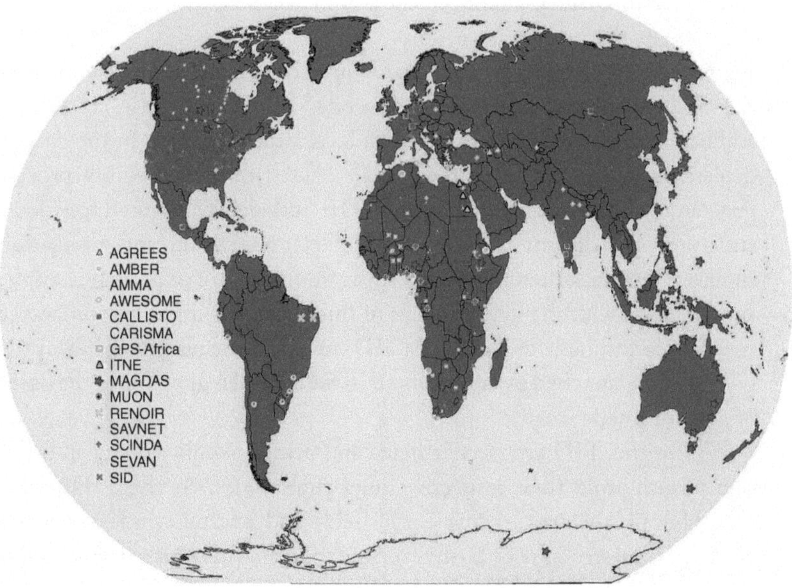

Fig. 1. *Map indicating instrument deployments supported as part of the IHY/UNBSS program. Please note that the map does not include the majority of the observatories participating in IHY, only those participating in the instrument development program. Many of the national reports in Sect. 7 contain summaries of observatories that participated in IHY campaigns but are not shown on this map.*

Nations Basic Space Science (UNBSS) Initiative, centers around a series of annual workshops hosted in varying international locations.

Fifteen instruments have participated in this activity, and as a result of this program more than 150 new instruments and nearly 300 educational SID instruments (see Sect. 5) have been deployed worldwide, with a significant fraction from Africa. The map in Figure 1 indicates the location of the instruments deployed as part of IHY.

The Observatory Development program has focused on facilitating partnerships between instrument providers and instrument host institutions. The lead scientist, or principle investigator (PI) provides instrumentation (or fabrication plans for instruments) for the array; the host country provides the workforce, facilities, and operational support to obtain data with the instrument, typically at a local university. The instrument hosts then participate in research data analysis activities culminating in the UNBSS workshops.

Exemplary Science on a Global Scale: Connecting Local Ionospheric Disturbances to Global Processes

Figure 2 shows the effects on the naturally occurring ionospheric emissions caused by an instability process generated at the magnetic equator in addition to a geomagnetic storm. The local structures, seen in the data as depletions in the airglow intensity, are caused by an instability process generated at the magnetic equator. The turbulence within these local structures can disrupt transionospheric communication and navigation signals. The local structure typically drifts from west to east. In this example, simultaneous with the development of this equatorial instability process, a traveling ionospheric disturbance (TID) propagates equatorward from the polar region, launched by energy input in the auroral region due to the onset of a geomagnetic storm.

Within this TID are electric fields and neutral winds which can differ significantly from their respective quiet-time values. As the TID passes over Hawaii, the perturbed electric fields and neutral winds affect the observed local structure by both reversing the drift direction to the west and initiating the development of secondary instabilities on the eastern edge of the primary local structure.

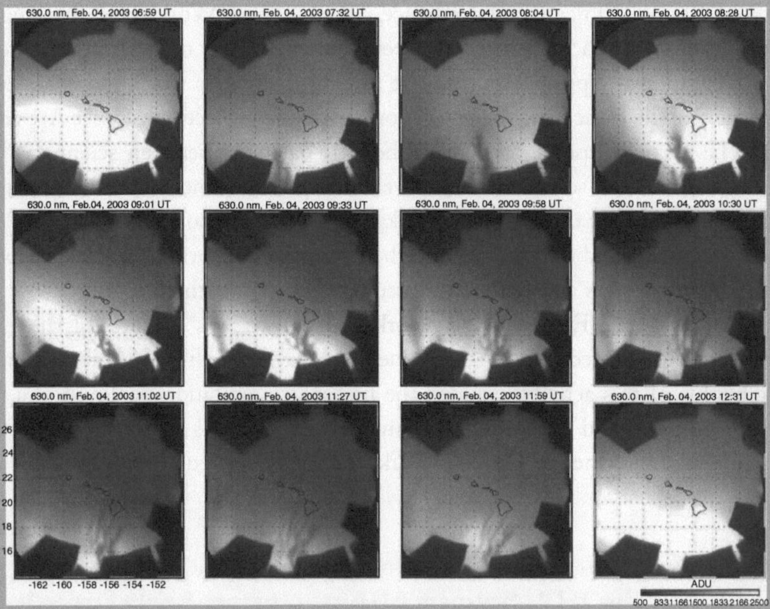

Fig. 2. *A series of images capturing the development of structures, seen as dark regions, associated with an equatorial instability process. These structures are further modified by the passage of a traveling ionospheric disturbance, seen as an enhancement traveling from northeast to southwest, associated with energy input in the auroral region. The images are of the 630.0-nm emission that occurs naturally in the Earth's ionosphere/thermosphere system and were recorded from the site of the Haleakala Volcano on Maui, Hawaii.*

The instabilities introduced by TIDs and other ionospheric phenomena impact our ability to communicate through the ionosphere (e.g. GPS, SATCOM). The lack of ability to predict such phenomena leads to unanticipated transionospheric communications outages that negatively impact everyday aspects of life in the 21st Century. In order to mitigate those outages, we need a *global* predictive capability. A *global* capability requires comprehensive and extended observations that can resolve both the fine scale structures as well as the global coupling effects that influence the development, structure, and impact of ionospheric disturbances on transionospheric radio signals.

4.2. The U.N. Workshops on Basic Space Science

The United Nations, in cooperation with the European Space Agency, initiated in 1990 the organization of annual Workshops on Basic Space Science through the

United Nations Office for Outer Space Affairs. These Workshops, focusing on astrophysics and space science, have been held in a variety of locations, balancing the geographical representation of the program activities (for a more complete summary of the Basic Space Science Workshops and projects, please see Appendix II). The participation in these workshops is balanced between scientists from developed nations, scientists from developing nations, and scientists from geographical areas near the site of the workshop. The workshops from 2005 to 2009 have been dedicated to the IHY, and their format has provided an ideal venue to bring instrumenters and potential instrument hosts together.

The 2005 IHY/ESA/UNBSS Workshop, which was the first of the U.N. workshops dedicated to IHY, was held in Al-Ain, United Arab Emirates. Highlights of this activity included the first instrumentation deployed as a part of this program and the initiation of many collaborations that resulted in other instrument deployments. The 2009 workshop will be hosted by Republic of Korea.

4.2.1. Brief Summary of the UNBSS/ESA Workshop on the IHY, held in Al-Ain, United Arab Emirates, 20–23 November 2005

The 4-day workshop was hosted by the United Arab Emirates University, under the patronage of H. H. Sheikh Nahayan Mabarak Al-Nahayan, Minister of Education and the Chancellor of the United Arab Emirates University, and on behalf of the United Arab Emirates (UAE) Government. The meeting was co-organized by the Committee on Space Research (COSPAR), the International Astronomical Union (IAU) and the National Astronomical Observatory of Japan (NAOJ) and the Department of Physics at the UAE University. The local organization was also provided by the Department of Physics at UAE University.

Workshop participants represented 44 United Nations member states, including a significant portion of North Africa, the IHY-West Asia region, as well as leadership from the remaining six IHY international regions. There were special sessions on IHY instruments and host institutions, as well as IHY science, global scientific initiatives, education programs, astrophysical research in Arab nations, and the 2005 World Year of Physics. Special discussions also included planning for the IHY-Africa initiative, education and outreach activities. The primary achievement of this meeting was the initiation of many IHY UNBSS instrumentation activities, whereby an instrument provider is able to find host sites in other nations to increase the coverage of vital scientific measurements.

4.2.2. Brief summary of the UNBSS/NASA on the IHY, held in Bangalore, India, 27 November–1 December 2006

The Indian Institute of Astrophysics hosted this 5-day meeting. There were 95 participants from 30 different nations. Presentations were given on IHY science and instrumentation, and there was a special working group session on outreach in India and internationally. This workshop built on the success of the previous year. The instrumentation programs initiated during the first workshop in the UAE were developed and a survey was performed of potential host institutions. Scientists representing the host institutions were invited to participate in the meeting, and much of the discussion centered on the instrumentation programs.

4.2.3. Brief summary of the UNBSS/ESA/NASA/JAXA workshop on the IHY, held in Tokyo, Japan, 18–22 June 2007

The next UNBSS/IHY workshop was held at the National Astronomical Observatory of Japan in Tokyo. Nearly 80 participants from 28 nations participated in this workshop, which focused on the continued support of the instrumentation/host programs, as well as the development of scientific collaborations and analysis efforts for the new IHY instrument arrays. Special discussions focused on the participation of developing nations in IHY, the operation of new facilities in developing nations (see the BSS "Tripod" below), access to data of the Sun–Earth system from ground-based and space-borne facilities through data archives and virtual observatories.

4.2.4. Brief summary of the UNBSS/ESA/NASA/JAXA workshop, "First Results of the International Heliophysical Year," held in Sozopol, Bulgaria, 2–6 June 2008

This workshop was held in Sozopol, Bulgaria, hosted by the Solar-Terrestrial Influences Laboratory of the Bulgarian Academy of Sciences on behalf of the Government of Bulgaria. There were three major topics around which the meeting was structured. Two of the topics continued the efforts of previous workshops: "Participation of nations in project development for international heliospheric

Brief Summary of the IHY-Africa Space Weather Science and Education Workshop held in Addis Ababa, Ethiopia 12–16 November 2007

The annual IHY UNBSS workshops have been extremely successful in stimulating space science instrumentation in developing areas, particularly Africa. A special IHY-Africa focus group identified a need for a workshop to support African research activities and to further develop research partnerships in conjunction with the instrumentation program. In response to this need, more than one hundred international scientists met in Addis Ababa, Ethiopia, in November 2007, for a special IHY workshop focused on Space Weather Science and Education. The meeting was the culmination of more than 2 years of planning, spurred on by the IHY-United Nations Basic Space Science (UNBSS) Initiative to promote African space weather science and further deployment of instrument arrays in developing countries.

The IHY-Africa workshop had two overall objectives:

- To study space weather science at mid and low latitudes in the African longitude sector.
- To support African space science and education.

The success of the workshop was greatly assisted by the generous support by several national and international organizations including the National Aeronautics and Space Administration (NASA), National Science Foundation (NSF), European Office of Aerospace Research and Development (EOARD), International Center for Theoretical Physics (ICTP), Air Force Office of Scientific Research (AFOSR), Office of Naval Research (ONR), Air Force Research Laboratory (AFRL), Climate and Weather of the Sun–Earth System (CAWSES), and Committee for Space Research (COSPAR). We were also fortunate that representatives of many of the sponsoring agencies were able to attend the meeting.

The science focus was built around three major themes:

1. Ionospheric Irregularities.
2. Ionospheric Total Electron Content.
3. Equatorial Electrodynamics and Plasmasphere/Ionosphere Coupling.

The ionospheric irregularities session was an extension to the 1-day SCINDA workshop held on the Sunday preceding the IHY workshop.

There was also a special GIFT Teacher workshop (see Sect. 5) sponsored by NASA and the Houston Museum of Natural History, as well as an AWESOME educational space weather monitor training session.

Another workshop, building on the results of this workshop, is being planned in Zambia in June 2009.

space missions and supporting low-cost ground-based instrument array initiatives for world-wide studies in space science" and "Access to data of the Sun–Earth system from ground-based and space-borne facilities through data archives and virtual observatories." Due to the success of these two efforts, a third topic could now be raised: "First results of the International Heliophysical Year." A total of 150 scientists representing 36 different nations participated in this meeting.

4.3. Instrument programs and status

Drawing on 17 years of experience, U.N. Basic Space Science Initiative developed the "Tripod" concept, an implementation model for the accelerated establishment of basic space science associated activities in developing nations. The Tripod concept, identified in the very first UNBSS workshop in India in 1991, is described as follows:

1. The availability of research tools of a level where meaningful science can be made, but at a level where the national socio-economical infrastructure can maintain functionality in the university/research laboratory environment, e.g. a small telescope facility or instrument array.
2. Teaching materials allowing basic space science to be introduced at the teaching level of fundamental mathematics, physics, chemistry and astronomy courses in middle and higher education.
3. Application materials for original research in basic space science, such as observing programs or software development.

The Tripod concept has been central to the selection of the instrument programs for the IHY. Several initiatives that had reached a mature stage of development were selected for the first phase (2005–2006) of the joint IHY/UNBSS Observatory Development activity. Following the early success of this phase, more instrument programs were added for a total of 15, including GPS and ionospheric networks, magnetometers, radio observatories, and particle and muon detectors. The instrument programs were prioritized based on suitability to accomplish the joint goals of the IHY and UNBSS programs:

- Production of scientifically significant and publishable results pertaining to the objectives of the IHY activities;
- Identification of activities which can be performed in developing nations (many of which are near the equator); and
- Costs and technical requirements must be compatible with the resources available in the participating nations.

Additionally, the following factors were viewed as desirable:

- Legacy Potential: leading to a beneficial relationship for the participants in developing nations.
- "Tiered" Technology: instruments with several designs corresponding to increasing design complexity.
- Educational components, including university level (such as the AWESOME SID monitors described in Sect. 5).

The efforts of the IHY activity have focused on finding host sites for the observatories and instruments and assisting in the identification of technical and financial resources. The following sections give summaries of the IHY observatory deployment programs.

4.3.1. IHY UNBSS instrument programs

4.3.1.1. GPS & ionospheric networks

4.3.1.1.1. African GPS Receivers for Equatorial Electrodynamics Studies (AGREES)

PIs: Endawoke Yizengaw and Mark Moldwin, University of California at Los Angeles

In looking at a map of the world showing the location of ground-based space physics instrumentation (radars, magnetometers, ionosondes, GPS dual frequency receivers, and lidars), one quickly recognizes Africa's relative lack of space physics research infrastructure. This has created a gap of global understanding of the physics behind the evolution and formation of plasma irregularities in the equatorial region, which imposes severe limitations on ionospheric and plasmaspheric density modeling efforts. Knowledge of ionospheric and plasmaspheric (inner magnetospheric) density structure is required to create weather dependent density model and mitigate the navigation and communication degradation. Understanding the source and loss processes of plasma during magnetically active and quiet periods also requires knowledge of inner magnetospheric density structure. Knowing the topside ionosphere and plasmasphere density structure is also essential for obtaining the density contributions of these regions to the ground based total electron content (TEC), which is a long-standing problem. The question is how do we obtain reliable density structure of the inner-magnetosphere as a function of local time and magnetic activity? The scientific community has been developing and using different means to gather density information on the ionosphere and plasmasphere. This includes: incoherent scatter radar probing, coherent scatter radar observations of under dense electron density irregularities, observations using topside sounders onboard satellites, just mention few. However, they are sparsely located around the globe and unable to provide the global understanding of the inner-magnetospheric density structure and dynamics.

The ground- and space-based GPS receivers continuously provide the integrated total electron content (TEC) between the satellite and receiver on the ground, which is one of the most important quantitative characteristics of the ionosphere and plasmasphere. However, the uneven distribution of ground-based GPS receivers hinders our ability to obtain a global understanding of the dynamics and structures of the inner magnetosphere. Although ground-based GPS receivers are located in dense regional arrays, they are primarily only in North America and Europe. In regions like Africa, observations of the inner-magnetosphere are not possible due to lack of ground-based instruments. The density structure over this region has been traditionally estimated by using vast areas of model interpolation. Therefore, to have a complete and global understanding of the electrodynamics of the inner-magnetosphere in a cost effective way, deployment of small instruments, like ground-based GPS receivers, is essential. Therefore, the AGREES (African GPS Receivers for Equatorial Electrodynamics Studies) GPS receivers network in Africa will perform vital studies of the fundamental governing electrodynamics of equatorial ionospheric motion. As shown in Figure 3 (blue dots), the AGREES network will contain five GPS receivers which will be deployed at mid- and low-latitudes in the African continent. The AGREES network will be deployed based on two main priorities that are required for this study. These include:

(1) to routinely monitor the strength and evolution of EA region by filling the largest land-based gap of GPS receiver coverage in Africa,
(2) to understand the electrodynamics that governs the equatorial ionospheric motion by combining data from GPS receivers network with the already available magnetometer (MAGDAS and AMBER) data.

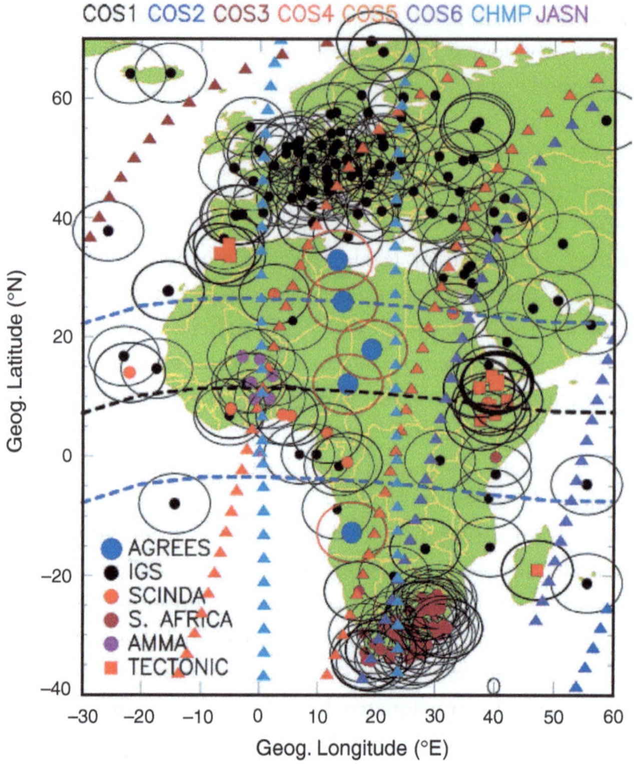

Fig. 3. *European and African ground-based GPS receivers' network along with proposed AGREES network (blue dots). The circles in each station depict the elevation cut off (15° above the horizontal). The 2-h coverage of eight LEO satellites ground tracks in the region is depicted by different colored triangles.*

The ground-based GPS receivers, including AGREES, in the African region will be augmented by data from GPS receivers on board Low-Earth-Orbit (LEO) satellites that are equipped with dual-band GPS receivers. The 2 h coverage of eight LEO satellites (COSMIC, CHAMP, and Jason) in the region is shown in Figure 3. Therefore, the simultaneous application of tomographic reconstruction technique on both ground- and space-based GPS TEC data will allow us to monitor the structure and dynamics of inner-magnetospheric density. This will provide a great opportunity to the scientific community to clearly quantify the dynamics of magnetosphere–ionosphere (M–I) coupling phenomenon.

The geographic locations of AGREES stations will be 12.73° S, 15.78° E; 12.11° N, 15.05° E; 17.92° N, 19.12° E; 26.00° N, 14.15° E; and 32.90° N, 13.18° E. Three African countries (Angola, Chad, and Libyan Arab Jamahiriya) will participate in the AGREES project by hosting the GPS receivers. Therefore,

AGREES project will have direct impact on space science research and education both in African and in the United States by fostering international scientific cooperation. In addition to providing world-class science data to the scientific community, AGREES infrastructure will also provide a self-sustained training and research opportunities in the African universities where space science research activities are rare. In the United States AGREES project will create opportunities for graduate and undergraduate students to do research that will provide them high quality research experiences. The experience will include expertise in data analysis, remote sensing, computer modeling, data assimilation, and grid-based computing and data retrieval.

4.3.1.1.2. AWESOME (Atmospheric Weather Educational System for Observation and Modeling of Effects)

PIs: Umran Inan, Morris Cohen, Deborah Scherrer and Sheila Bijoor, Stanford University

The AWESOME instrument is an ionospheric monitor that can be operated by students around the world. The monitors detect solar flares and other ionospheric disturbances. The key elements of the AWESOME are the computer, the Stanford Monitor, and the antenna.

The science of AWESOME is central to IHY. The ionosphere is the region of the upper atmosphere where ionization occurs due to solar and cosmic radiation. In the D-region of the ionosphere (50–90 km altitude), the ionization levels are dominated by X-ray flux from the Sun in the daytime, and ambient cosmic-ray flux at night. The D-region is sufficiently ionized as to strongly reflect radio waves in the ELF/VLF range (300–50 kHz), and since the Earth is also highly conducting at these frequencies, these signals can propagate efficiently around the planet in the so-called “Earth-ionosphere waveguide”. Natural ELF/VLF signals are dominated by the impulsive radiation from lightning strikes, known as radio atmospherics, and can be picked up around the Earth from the source.

Since the propagation of these waves in the Earth-ionosphere waveguide is strongly dependent on the ionospheric conditions (being reflected off of the ionosphere), the propagation properties of these waves intrinsically respond to changes in the ionosphere, and are therefore a good diagnostic of ionospheric disturbances. In particular, the nighttime ionosphere is highly variable and poorly understood, but being at a height too high for balloons, too low for satellites, and too lightly ionized for radar measurements, radio measurements such as ELF/VLF constitute one of the few effective means of remote sensing. The ionosphere is known to react to a broad array of geophysical phenomenon, including solar flares, lightning, sprites, lightning-induced electron precipitation, cosmic gamma-ray bursts, earthquakes, and more. Studying the ionosphere therefore enables analysis of a broad array of geophysical phenomenon, especially when combined with data from other instruments.

AWESOME sensitively detects these ELF/VLF radio waves enabling detailed quantitative analysis of ionospheric conditions. The figure here shows sample ELF/VLF data taken with the AWESOME network. The key elements of the AWESOME are the computer, the Stanford Monitor, and the antenna. An internet link is important; otherwise a good quality DVD burner can be used.

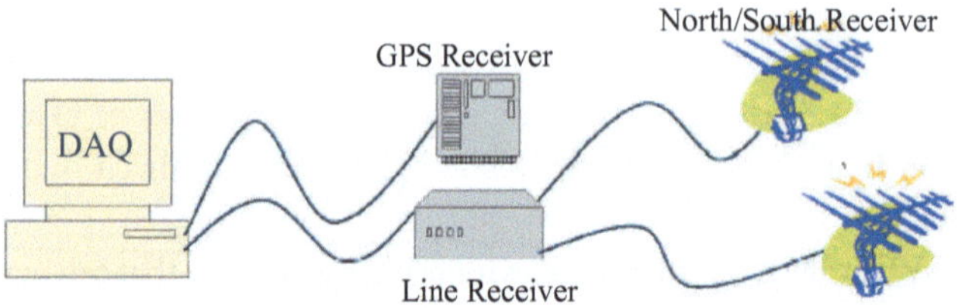

Fig. 4. *The AWESOME data acquisition system, which works in conjunction with a TrueTime GPS receiver, the line receiver and, the North/South and East/West receivers.*

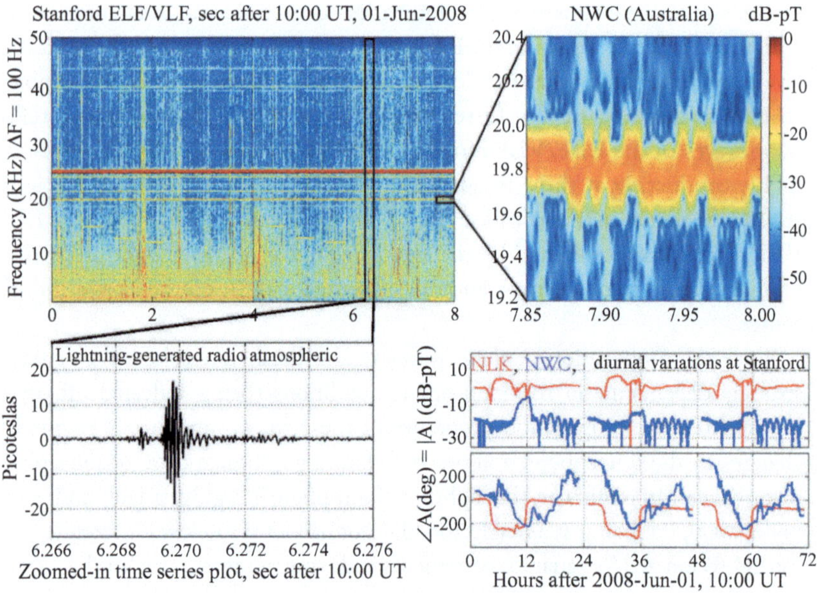

Fig. 5. *Data from the AWESOME instruments allow studies of ionospheric phenomena that would be impossible without ground-based arrays.*

The setup is illustrated in Figure 4. The line receiver gets VLF signals from two antennas. There is usually one antenna in the North/South orientation and another in the East/West orientation. These signals are sent to a 200 kHz ADC card attached to the PCI slot of the computer. The ADC will capture data from each of the two antennas at 100 kHz each. The timing signal from the GPS is also fed into the ADC card, allowing for very precise acquisition of data. Now in development is a USB interface to replace the ADC card, which will enhance the ease of use, and substantially reduce the cost.

In Figure 5, the top left plot shows the full 0–50 kHz spectrum, with the color indicating the strength of the magnetic field. The bottom left plot shows an example of a naturally-generated radio atmospheric, originating from a lightning strike which could be anywhere on Earth. The top right plot shows a zoom-in of a VLF transmitter signal from Australia, received at Stanford, California. The digital communication modulation signal is visible in the frequency variations. These steady transmitter signals are excellent diagnostic tools for the ionosphere, since changes in the ionosphere are reflected in the amplitude or phase of this signal. The bottom right plot shows an example of how the transmitter signal, monitored over 3 days, clearly indicates a diurnal pattern dictated by the day/night ionospheric variation along the transmitter-to-receiver path.

Because of its broad application for studying a wide variety of geophysical phenomenon via the method of VLF remote sensing, relative ease of data analysis,

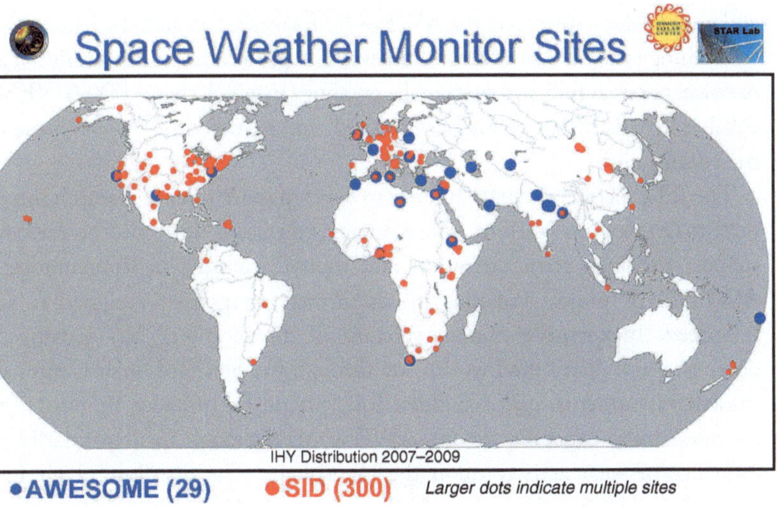

Fig. 6. *Map of AWESOME and SID monitors. Ten additional monitors are expected to be complete by the middle of 2009.*

and its comparatively low-cost within the realm of scientific instruments, the AWESOME is well suited for extension to a global network under the IHY program. Under the umbrella of the IHY program, 29 AWESOMEs have been distributed worldwide, with another 10 planned by the middle of 2009. Figure 6 shows these sites (blue dots represent AWESOME placements).

We build on this global AWESOME network to establish radio science expertise at underdeveloped sites by (1) creating a collection of online tools to enable advanced data analysis, and (2) developing initiatives to promote mentorship and collaboration amongst sites. The first of these tools is the AWESOME Online Data Viewer, a web-accessible interface that allows users to map ELF/VLF wave propagation along global paths, graph data, and easily share and download images. Another tool, known as the AWESOME Collaborative Forum, is an online forum for international researchers to collect scientific information, share discoveries, and foster new collaborations. We are currently developing a web-accessible AWESOME ELF/VLF Science Tutorial Series to educate new scientists about the breadth of VLF research topics.

In Fall 2008, Stanford conducted a series of regional Workshops dedicated to promoting science cooperation amongst sites. The Workshops consisted of invited research lectures, hands-on data tutorials, collaborative research activities, and interactive networking sessions to fully enable new ELF/VLF researchers to utilize their VLF data for advanced and meaningful scientific research. The goal of these workshops is to establish collaborative projects resulting in published papers in refereed scientific journals. With the first papers already in development, and the current and next round of site deployments now developing their own expertise and project ideas, we expect the AWESOMEs impact on global science to grow in the coming years. We intend for these workshops to culminate in a series of annual or twice-per-year fully international workshops in which these IHY/AWESOME collaborative links can be further nurtured, and new/exciting results presented to the community.

The IHY recognizes that, in order to develop space science research infrastructure in Developing Nations, space science education must also be improved to support the long-term operation and use of scientific instrumentation. The AWESOME project includes an educational component designed to address this need. Inexpensive student versions of the monitors, less sensitive, were developed and distributed worldwide through the Stanford Solar Center. These student instruments, called Sudden Ionospheric Disturbance monitors (SIDs), provide similar capability to the AWESOMEs though focus primarily on the influence of solar activity upon the ionosphere. The distribution kit includes extensive curricula and backgound information which were designed to inspire and engage the next generation of space scientists. Over 300 SID instruments have been distributed for the IHY (red dots on map). The SIDs are described in more detail in the education section (Sect. 5) of this report.

4.3.1.1.3. GPS in Africa

PIs: Christine Amory-Mazaudier, Monique Petitdidier, and Paul Vila (CNES, France), Tim Fuller-Rowell (University of Colorado and NOAA Space Environment Center), Sunanda Basu (Boston University).

Much attention in space weather has been devoted recently to the large changes in total electron content (TEC) over the American sector during geomagnetic storms. The attention has been fueled by the ability to map the electron content using networks of ground-based, dual-frequency GPS receivers. Figure 7 shows a large gradient in TEC stretching across the Eastern United States during an ionospheric storm, driven by electrodynamics. Steep gradients in TEC can disrupt operational systems such as the Wide Area Augmentation System (WAAS), which is utilized by the Federal Aviation Administration (FAA) for aircraft positioning. This gradient severely disrupted the FAA use of WAAS over its service volume for periods totaling 26 h navigation errors sometimes exceeded 50 m more than 10 times typical uncertainties of 2–3 m.

The Geomagnetic Field: There has been speculation that the large change in plasma density and electrodynamics in the American sector is related to the unusual configuration of the geomagnetic field, including the South Atlantic Anomaly (SAA) and large values of declination. The speculation has spurred the need to target other geographic regions with significantly different geomagnetic field configurations. In contrast, over Africa the magnetic equator is parallel to the geographic equator and more uniform in magnitude. This region, however, does

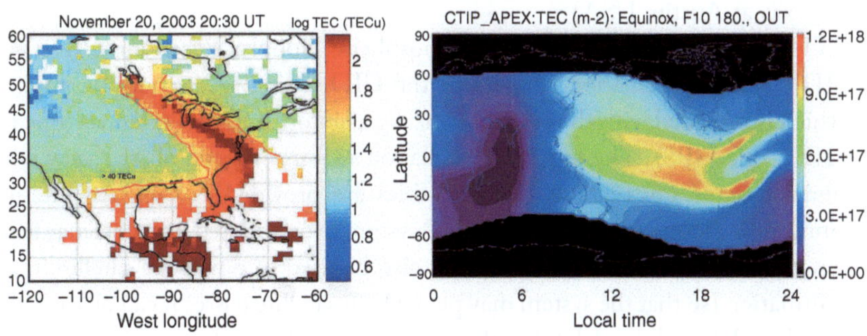

Fig. 7. *Left: Example of a plume of storm enhanced density derived from GPS "imaging" (J. Foster, private communication). Right: Illustration of the distortion of the geomagnetic field between the American and African sectors.*

not have a dense network of ground-based GPS receivers available to perform a complementary study, nor does it have sufficient magnetometer chains to monitor the dayside plasma drift.

Objective: The observations over Africa will be used to determine the source of the plasma bulge at midlatitudes that provide the seed for the plume seen in Figure 7. The plasma bulge can either be created in-situ as a natural consequence of the expansion of the high latitude magnetospheric convection pattern to low latitudes, or is formed by transporting plasma from low latitudes. The latter mechanism relies on strong meridional and zonal plasma transport from the equatorial ionization anomaly (EIA) in the presence of the distortion of the geomagnetic field. Observation over Africa, with the much more uniform magnetic field, provide an excellent test for the two theories. The space science community is therefore exploring ways to increase the observational infrastructure at mid and low latitudes over the African continent. The GPS in Africa project is therefore collaborating with other activities in the region, such as the Multidisciplinary Analysis of the Monsoon in Africa (AMMA), the CARISMA Project (Ian Mann) and the SCINDA project (Keith Groves), in order to provide the necessary spatial coverage. In addition, we would like to encourage scientists in the African Universities to become interested in the science objectives and be willing to host a dual-frequency GPS system at their Universities. Unraveling the physical processes controlling the redistribution of plasma requires coordinating the ionospheric observations over the African continent and assembling the information in a common database. The observations from the common database can be used to address a range of science questions under the IHY-Africa initiative.

4.3.1.1.4. ITNE (Ionospheric Tomography Network of Egypt)

PIs: Ayman Mahrous (Helwan University, Egypt) and Trevor Garner (University of Texas at Austin, USA)

ITNE is a network of Coherent Ionospheric Doppler Receivers (CIDR). A CIDR system installation consists of the CIDR receiver, a control computer (which is provided with the system, typically a laptop) and two antennas (one for CIDR, one GPS). The antenna installation requires a good all-sky view with minimal or no obstructions. 100-m cables are provided as well. An internet connection allows each individual CIDR system to be accessed by the science team remotely. The internet connection will also be used to download satellite track information (so that the system may plan which satellite passes it can observe and record) and the collected data is uploaded to archives for use by the science team. If there is a broadband connection, individual satellite passes can be remotely monitored via a web interface.

As part of the IHY effort to increase the ionospheric instrumentation in Africa, Helwan University in Helwan, Egypt and the Applied Research Laboratories: The University of Texas at Austin (ARL:UT) have teamed up with other Egyptian universities to develop the Ionospheric Tomography Network of Egypt (ITNE). ITNE is a chain of three UHF/VHF Doppler receivers distributed at different universities throughout Egypt (see Table 1). It is a roughly north–south chain (like much of Egypt, it is aligned with the Nile river). ITNE stretches from $\sim 24^{\circ}$ to $\sim 15^{\circ}$ magnetic latitude, and sits under the northern peak of the equatorial anomaly. The first receiver system is operational with the two additional systems (already in Egypt) coming on line in the near future. INTE provides a new and valuable data set to study the equatorial anomaly in the African sector.

Coherent Ionospheric Doppler Receiver systems will be the primary observing system for this study. CIDRs, developed at ARL:UT, are radio receivers designed to measure the line-of-sight, relative TEC using the 150 and 400 MHz radio beacons on board LEO satellites. Figure 8 shows the standard CIDR system. The fundamental CIDR measurement is the Doppler shift in each channel. By down-converting each signal to a common frequency and subtracting the Doppler shifts, the rate of TEC (ROT) change, dTEC/dt, can be derived and integrated to produce a relative TEC. These are line-of-sight measurements with measurements from other receivers can be combined through different tomographic and data assimilative techniques to reconstruct two- or three-dimensional electron density distributions.

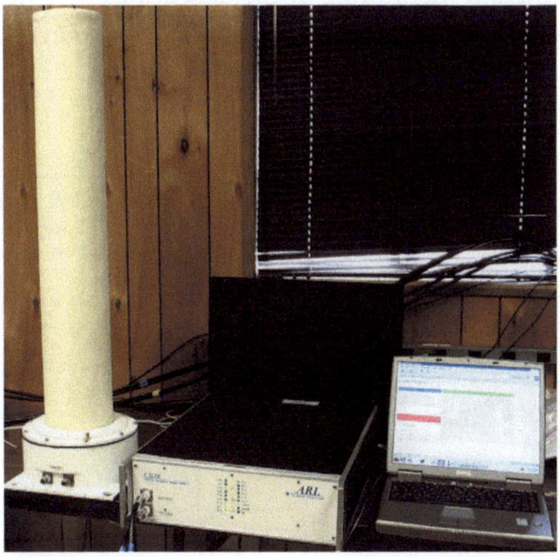

Fig. 8. *A photo of a CIDR system with its individual components. From left to right, the components are the UHF/VHF antenna, the receiver, and the laptop.*

Since the low-Earth orbiting (LEO) spacecraft (800–1000 km apogee) cross the receiver's field-of-view in only a few minutes (typically 20 min), these measurements fundamentally show spatial variations in the electron content. In contrast, GPS satellites take ~8 h to cross a ground receiver's field-of-view, making GPS TEC measurements a better measurement of temporal changes in the ionosphere than of spatial changes. In addition, LEO spacecraft fly completely within the ionosphere so the observed TEC consists solely ionospheric plasma (GPS TEC measurements contain plasmaspheric plasma as well). The data rate in CIDR equipment is higher than in typical GPS receivers. CIDRs typically take measurements at 1 Hz and can take data as high at 1 kHz, a higher data rate than a typical GPS receiver (every 30 s). With fourteen available beacon satellites, CIDR systems operate nearly continuously with typical data gaps of less than 60 min. As such, CIDR systems will provide a high quality data set for observing the equatorial ionosphere.

A CIDR is capable of tracking up to three different beacon satellites with different frequency offsets at one time, including the Navy Ionospheric Monitoring System (NIMS) operational and maintenance offsets, and any LEO satellite operating on the "geodetic" channel. Currently, there are four NIMS satellites and ten geodetics spacecraft that CIDR can track. In addition to tracking a large number of satellites and the ability to track multiple satellites simultaneously, CIDR has a number of other capabilities including:

- User selectable sampling rates as high as 1 kHz;
- Instrument error of <0.1 radians of phase (5.0×10^{13} el/m^2 for 150 MHz);
- Remote control and monitoring of the data acquisition system; and
- Data acquisition software written entirely in open-source Linux.

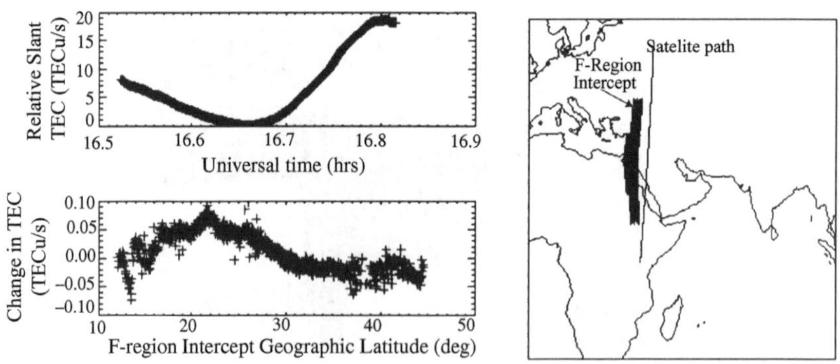

Fig. 9. *A sample of CIDR observations from Helwan.*

Figure 9 shows a sample of CIDR observations taken by the CIDR at Helwan, Egypt. This plot demonstrates several of features of the CIDR systems. These observations are taken at 1 Hz and show the equatorial fountain peak near 22° geographic latitude. The upper left plot shows the relative slant TEC as a function of UT for a NIMS pass. The lower left plot shows the rate of change in the TEC as a function of the F-region intercept, while the right plot is a map showing the relative position of the satellite orbit and the 350 and 100 km pierce points of the radio rays. Because this pass has a high elevation angle, the 350 pierce points nearly overlap the 100 km intercepts (the most westward line). This satellite pass is from north to south so that the relative TEC enhancement at 16.8 h corresponds to the rate of TEC change peak at 22°. Both sets of observations indicate that the northern equatorial anomaly peak (EAP) is located at 22° geographic latitude. The increased spread in the ROT measurements equatorward of the EAP corresponds to an increase in. the ionospheric irregularities. The data has been manually quality checked. The signals come from a NIMS spacecraft traveling in a near polar orbit from north-to-south.

4.3.1.1.5. Low cost ionosonde for IHY/UNBSS

PIs: John Bradford, Chris Davis, Richard Stamper, Rutherford Appleton Laboratory

During the IGY in 1957, a number of ionosonde stations were opened, but in subsequent years, most of these have ceased operation and the remaining stations

Fig. 10. *Previous ionosonde design; design studies have shown that a lower-cost version can perform with little or no loss of data quality.*

are largely at mid and high latitudes. There are scientific and operational reasons for studying the equatorial ionosphere and the IHY provides an opportunity to do this; the Rutherford Appleton Laboratory (RAL) is proposing to develop a new ionosonde for this purpose as part of UNBSS (Figure 10).

An important design driver is cost and it is anticipated that this can be an order of magnitude lower than that of commercially available ionosondes. This will be achieved by using commercial components wherever possible; for example by using standard PCs rather than dedicated DSP devices and RF components originally designed for ham radio applications.

The instrument design will be based on meteorological radars recently designed by RAL; these are now operational instruments and a good deal of the signal processing system design and software will be directly applicable to the proposed ionosonde.

The design will be upgradeable by, for example, adding receiver channels to provide directional information or by using GPS time and frequency references to operate a pair of ionosondes in a bistatic mode. It is proposed that when the design and prototyping have been completed, the construction and installation of the instruments would to a large extent be carried out by the host organizations themselves.

4.3.1.1.6. RENOIR (Remote Equatorial Nighttime Observatory for Ionospheric Regions)

PI: Jonathan J. Makela, University of Illinois at Urbana-Champaign

RENOIR is a suite of instruments dedicated to studying the equatorial/low-latitude ionosphere/thermosphere system, its response to storms, and the irregularities that can be present on a daily basis. The occurrence of equatorial plasma instabilities, commonly referred to as equatorial spread-F, equatorial plasma bubbles, or depletions, can cause radio signals propagating through the disturbed region to scintillate. This results in a fade in received signal power translating to a loss of the signal. Scintillations on frequencies from several GHz and below are known to occur and are a concern to many sectors. Through the construction and deployment of a RENOIR station, it is possible to achieve a better understanding of the variability in the nighttime ionosphere and the effects this variability has on critical satellite navigation and communication systems.

A typical RENOIR station involves the following:

(1) An array of single frequency GPS scintillation monitors. These provide measurements of the irregularities present, their size, orientation, and speed.

(2) A dual-frequency GPS receiver. This provides measurements of the total electron content of the ionosphere. If a site could be located that already fields a dual-frequency GPS receiver, this would not be needed.
(3) An all-sky imaging system (PICASSO). This measures two different thermosphere/ionosphere emissions from which the two-dimensional structure/motion of irregularities can be observed. The data can also be used to calculate the density and height of the ionosphere.
(4) Two miniaturized Fabry-Perot interferometers (MiniME). These provide measurements of the thermospheric neutral winds and temperatures. The two FPIs are separated by approximately 300 km or so, allowing bistatic, common-volume measurements.

Deployment of RENOIR stations is being planned in collaboration with the IHY/UNBSS Initiative. The first two RENOIR stations will be deployed in Brazil in early 2009. The instrumentation that make up a RENOIR station have all been used in the field in previous experiments and are at a moderately mature level of development. The optical systems can be housed in individual, self-contained housing units, requiring very little infrastructure. If an optical facility is available at a host institution, the optical equipment could easily be modified to interface with available optical domes. The facility should be located in a region with relatively dark skies (away from any major cities) and away from any tall structures (buildings and trees). If two Fabry-Perot interferometers are to be fielded, the second system should be located approximately 300 km away from the main site.

The dual-frequency GPS receiver is quite rugged and simply requires a location to mount the antenna and minimal space to locate the control computer. The array of single-frequency GPS scintillation monitors requires a space of approximately 100 m × 100 m over which to space the five antennas in a cross formation. Again, minimal space is needed to locate the control computers for each receiver. The facility should be located away from any tall structures (buildings and trees). For an example of RENOIR's science capabilities, please see "Science on a Global Scale" at the beginning of this section.

4.3.1.1.7. SCINDA (Scintillation Network Decision Aid)

PI: Keith Groves, Hanscom Air Force Research Laboratory

SCINDA is a real-time, data driven, communication outage forecast and alert system. Its purpose is to aid in the specification and prediction of communications degradation due to ionospheric scintillation in the equatorial region of Earth. UHF and L-band scintillation parameters are measured, modeled, and propagated in time to provide a regional specification of the scintillation environment in an effort to mitigate the impacts on the satellite communications (SATCOM) community.

Equipment at the remote sites record scintillation parameters from available UHF Fleet Satellite Communication System and L-band (Geostationary Operational Environmental Satellite, GPS) satellite links and measure ionospheric drift velocities. The data drives a semi-empirical model that produces simple three-color graphical representations of large-scale equatorial scintillation structures and associated communication impact regions.

Ionospheric disturbances can cause rapid phase and amplitude fluctuations of satellite signals observed at or near the earth's surface; these fluctuations are known as scintillation. Scintillation affects radio signals up to a few GHz frequency and seriously degrades and disrupts satellite-based navigation and communication systems. SCINDA consists of a set of ground-based sensors and quasi-empirical models, developed to provide real-time alerts and short-term (<1 h) forecasts of scintillation impacts on ultra high-frequency satellite communication and L-Band global positioning system signals in the earth's equatorial regions.

The SCINDA system (see Figure 11) concept is presently being demonstrated using eight equatorial stations in South America, Southwest Asia and Southeast Asia. Scintillation parameters from available UHF (FLTSAT) and L-band (GOES, GPS) satellite links and ionospheric drift velocities are measured and recorded at the remote sites. The scintillation maps are available to users for prototype operational support via a secure network. Analysis of data collected during the recent solar maximum period (2000–2002) indicates that both single and dual-frequency GPS receivers are subject to significant errors during severe scintillation events. All SCINDA sites are now equipped with GPS scintillation monitors and model development is in progress. Following the solar cycle, L-band scintillation activity will decline over the next few years and should remain relatively benign until around 2008. The goal is to have accurate GPS navigation error products available to support the operations before the next solar maximum.

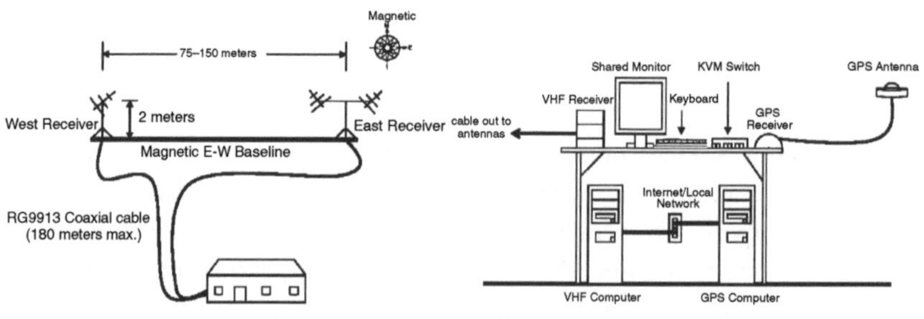

Fig. 11. *VHF antenna set-up (left) and VHF receiver chain and data acquisition system (right).*

The SCINDA Workshops

The SCINDA program, headed by Keith Groves, has augmented their instrumentation program through a series of special workshops that allow scientists to learn about the operation of the SCINDA array and to introduce new team members to SCINDA's scientific research topics.

The first of these workshops was held in Sal, Cape Verde from July 10–14, 2006. The overall goals of the workshop were to establish space science expertise and to install SCINDAs across Africa following the geomagnetic equator. The workshop provided instructions on the deployment, operation and interpretation of data. These instructions were accompanied by talks by the participating scientists, and meal-time discussions on how to advance space science research and education in Africa in the future. At the end of the workshop, the participants learned to set up a GPS TEC/scintillation system online at their home institution. The workshop squarely addressed space weather impacts by providing the requisite training to install and operate sensors that provide real-time warning from SCINDA sensors.

The second SCINDA workshop was held November 11–16, 2007 in Addis Ababa, Ethiopia, in conjunction with the IHY-Africa Space Weather and Education Workshop. Scientists from many nations including Ethiopia, Nigeria, Ivory Coast, Cameroon, Cape Verde, USA and Malaysia participated in the workshop. The workshop had similar objective as the previous workshop, and with an abbreviated schedule.

The next SCINDA workshop will take place in Livingstone, Zambia on June 7–8, 2009. The purpose of the workshop is to continue facilitating scientific interaction and promoting space science and education in Africa, particularly relating to understanding the global ionospheric response to magnetic storms and forecasting space weather impacts including the occurrence of ionospheric disturbances that affect radio wave propagation.

4.3.1.2. Magnetometer networks

4.3.1.2.1. AMBER (African Meridian B-Field Education and Research)

PIs: Endawoke Yizengaw and Mark Moldwin, University of California at Los Angeles

It was quickly recognized that the observation of ionospheric processes over Africa represented the best opportunity for new IHY science. In order to have a

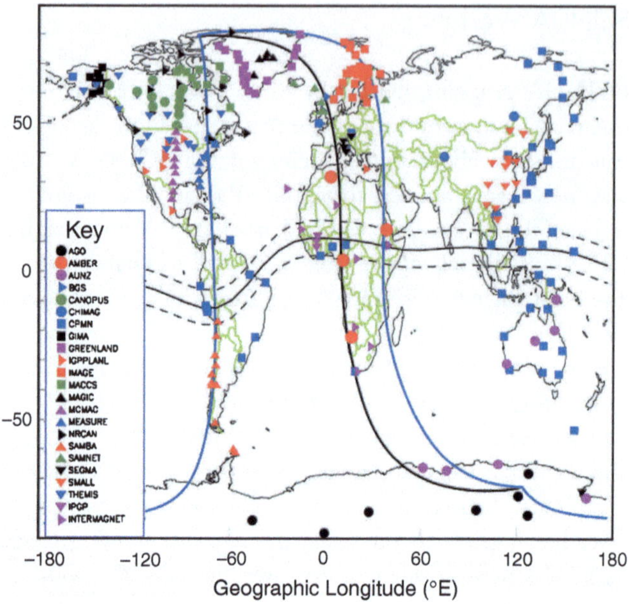

Fig. 12. *Global magnetometer coverage, after the installation of the AMBER array (red dots).*

complete *global* understanding of equatorial ionosphere motions, deployment of ground-based magnetometers in Africa is essential. The NASA-funded AMBER (African Meridian B-Field Education and Research) magnetometer array is one of those that are deployed in Africa. The AMBER magnetometer array is comprised of four magnetometers. Three of them are already stationed in Adigrat in Ethiopia, Algiers in Algeria, and Yaounde in Cameroon. The remaining one is in process to be deployed in Windhoek in Namibia. Figure 12 shows the global magnetometer network, including the AMBER array (red dots). Polar cap to equatorial coverage will exist in the American meridian with the deployment of McMAC array linking the expanded Canadian CARISMA (described below) and the SAMBA arrays. However, the Europe/Africa meridian coverage essentially ends with SEGMA in Italy. The AMBER array is now connecting the European magnetometer array to low latitudes in Africa by filling the largest land-based gap in global magnetometer coverage, as shown in Figure 12. AMBER stations in Algiers, Yaounde, and in Windhoek will be used for connecting IMAGE-SAMNET-SEGMA array to low and dip-equator latitudes, and link up with South African Intermagnet and Antarctic magnetometers in the southern hemisphere so that we can have complete meridian observation in the region.

In addition to filling the largest land-based gap in global magnetometer coverage, the AMBER array will address two fundamental areas of space physics: (1) the processes governing electrodynamics of the equatorial ionosphere as a function of latitude (or *L-shell*), local time, longitude, magnetic activity, and season, and (2) Ultra Low Frequency (ULF) pulsation strength and its connection with equatorial electrojet strength at low/mid-latitude regions. In coordination with GPS receivers in Africa, the AMBER magnetometer array will provide a great opportunity to understand the electrodynamics that governs equatorial ionosphere motions. Combined observations with the other magnetometer arrays provide an enormous opportunity to understand the unique equatorial ionospheric structures in the African sector that have been often detected by satellite observations.

Data from the AMBER magnetometer array will be directly accessible to space weather forecasters and the space science community at large. In addition to new scientific discoveries and advancing the space science research into Africa by establishing scientific collaborations between scientists in the de-

Fig. 13. *Typical fluxgate magnetometer sensor head mounted in the concrete base of its protective silo (site: GULL).*

veloping and developed nations, the AMBER project also contributes to developing the basic science of heliophysics through cross disciplinary studies of universal process. This includes the creation of sustainable research/training infrastructure in partnership with African universities. This will then create opportunities for undergraduate students to enhance their inspiration to space science and perform research activities in the future within their own countries.

4.3.1.2.2. CARISMA (Canadian Array for Realtime Investigations of Magnetic Activity)

PI: Dr. Ian Mann Program Manager: Dr. David Milling, Space Physics Group, University of Alberta

The original CARISMA component of the Canadian GeoSpace Monitoring (CGSM) array prior to April 2005 (the CANOPUS era) consisted of 13 NAROD S100 fluxgate magnetometers, returning data with a 5 s resolution (providing there was no interruption to the satellite connection data delivery). As a program of upgrades since then, the array has been updated to include local data logging capabilities to prevent data loss, improved connectivity using a high-speed satellite broadband internet link and return of data at the fluxgate magnetometers native rate (8 Hz) in real-time. As part of the upgrade, the array is also being expanded with the addition of 15 new sites, each containing a fluxgate magnetometer (shown in Figure 13 with locations shown in Figure 14). Of the

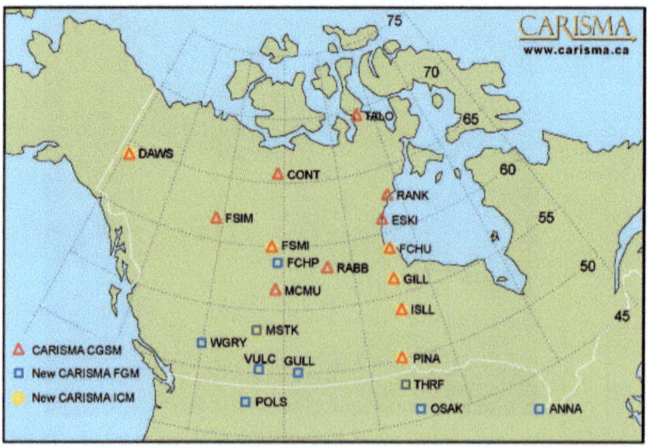

Fig. 14. *Map showing the sites in the CARISMA array.*

Fig. 15. *CARISMA CGSM site at Rankin Inlet.*

new sites, 10 were installed during IHY, with nine currently recording data (Figure 15).

In addition to the installation of the new sites, induction coil magnetometers – with a higher frequency response – have been deployed at eight sites to complement the measurements of the fluxgate magnetometers. At each site – Dawson City, Fort Smith, Fort Churchill, Gillam, Island Lake, Ministik Lake,

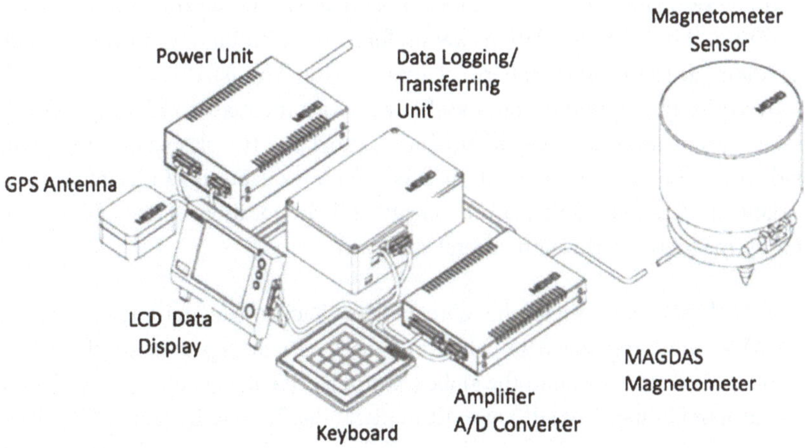

Fig. 16. *Details of the MAGDAS/CPMN magnetometer system for real-time data acquisition.*

Pinawa and Thief River Falls – two coils were installed aligned with magnetic North and East.

4.3.1.2.3. MAGDAS (Magnetic Data Acquisition System) Project

PI: Kiyohumi Yumoto, Space Environment Research Center, Kyushu University

The MAGDAS is being deployed for space weather studies during 2005–2008, overlapping heavily with the IHY/UNBSS program. The project aids the study of dynamics of geospace plasma changes during magnetic storms and auroral substorms, the electro-magnetic response of iono-magnetosphere to various solar wind changes, and the penetration and propagation mechanisms of DP2-ULF range disturbances from the solar wind region into the equatorial ionosphere. With the help of MAGDAS data, one can conduct real-time monitoring and modeling of (1) the global three-dimensional current system and (2) the ambient plasma density for understanding the electromagnetic and plasma environment changes in geospace (Figure 16):

Global 3-D current system: The MAGDAS data will be used to map the ionospheric equivalent current pattern every day. The current and electric fields at all latitudes are coupled, although those at high, and middle and low latitudes are often considered separately. By using the MAGDAS ionospheric current pattern, the global electromagnetic coupling processes at all latitudes will be clarified.

Ambient plasma density: New MAGDAS magnetometers will be deployed at several pairs of stations along the 210° magnetic meridian to observe the magnetic field line resonance (FLR) pulsations. Each pair will be separated in latitude by ~100 km. The FLR oscillations are useful for monitoring temporal and spatial variations in the magnetospheric plasma density. The MAGDAS data will be analyzed by the amplitude-ratio and cross-phase methods to identify the FLR events and measure their eigenfrequencies, providing the plasma density varying with time. These measurements will be highly valuable in understanding the variations of the ambient plasma density and the location of the plasmapause during magnetic storms and auroral substorms.

MAGDAS will utilize the Circum-Pan Pacific Magnetometer Network (CPMN) involving the 210° MM and the magnetic dip equator chains in a number of countries around the globe (Australia, Brazil, Canada, China (Taiwan Province of China), Côte d'Ivoire, Ethiopia, India, Indonesia, Japan, Micronesia (Federated States of), Nigeria, Peru, Philippines, Russian Federation, South Africa and United States) (Figure 17).

MAGDAS/CPMN
(MAGnetic Data Acqusition System/Circum-pan Pacific Magnetometer Network)

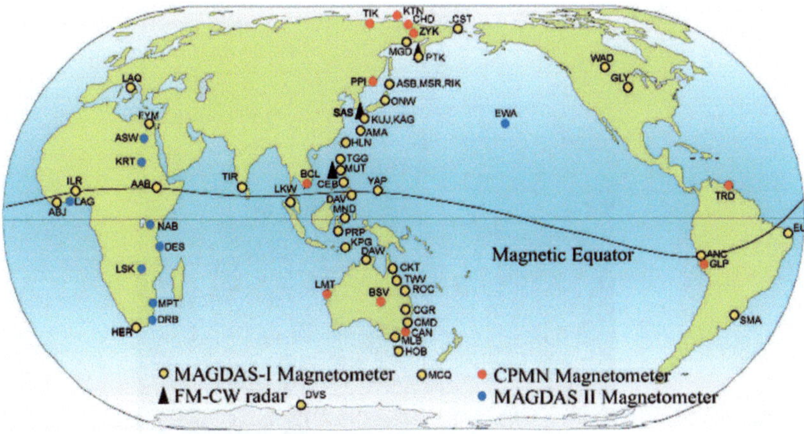

Fig. 17. *Stations of the Circum-Pan Pacific magnetometer network.*

4.3.1.3. Radio observatories

4.3.1.3.1. CALLISTO (Compound Astronomical Low-cost Low-frequency Instrument for Spectroscopy and Transportable Observatory) Frequency Agile Solar Spectrometers

PI: Arnold Benz and Christian Monstein, Institute of Astronomy, ETH-Zentrum in Zurich

The e-CALLISTO network aims at 24-h coverage of the radio emission of the Sun. The goal is to spread identical spectrographs in the meter wave and low decimeter radio band around the globe and connect them through the internet. The e-CALLISTO project is based on a spectrometer unit, provided by ETH Zurich, as well as an antenna and a PC interface to the internet, provided by the local partners. Here we report on the international set-up of a distributed net of CALLISTO spectrometers, data acquisition by the international net and first results of the system (Figure 18).

The spectrometers have been shipped or carried by an ETH engineer to the host institutes. An important first step was a site evaluation, revealing sometimes unexpected interference by nearby radio transmitters. A spectral overview was made using a special function of CALLISTO for every host site. It allows the measurement of the whole frequency range from 45 to 870 MHz in steps of 62.5 KHz leading to potentially 13,120 channels. This high-resolution spectrum is then used to create a dedicated frequency program avoiding channels with terrestrial interference.

Fig. 18. *Logarithmic periodic antenna of e-CALLISTO at UNAM, Mexico.*

The CALLISTO spectrometer is composed of several standard electronic components available from the consumer market and few others from eBay assembled on a single PCB (printed circuit board). This PCB fits into a standard aluminum box and has connectors for the antenna, computer, power supply, focal plane unit and – as an option – to an external 1 MHz clock source. The main specifications are listed in the table below. Up to now, almost all CALLISTO spectrometers have been assembled by young apprentices (mechanics) of the ETH Physics Department. Each CALLISTO was completely tested using an auto-

Main specifications of the CALLISTO instrument

Parameter	Specification
Frequency range	45.0–870.0 MHz (in 3 sub-bands)
Frequency resolution	62.5 KHz
Radiometric bandwidth	300 KHz at –3dB
Dynamic range	~ 50 dB at –70 to –30 dBm maximum rf level
Sensitivity	25 ± 1 mV/dB
Noise figure	<10 dB (measured at the rf input connector)
Max sampling rate	Internal clock 800 s/sec, external clock 1000 s/sec
Number of channels	Selectable 1–500, nominal 200 frequencies per sweep
Power supply	DC 12 ± 2V/225 mA
Weight	~800 g
Dimensions	110 mm × 80 mm × 205 mm
Material cost	<200 US$
Input data	3 files (configuration, frequency, scheduler)
Output data	2 files (1 FITS-file per 15 min and 1 log file/day)

Three CALLISTO printed circuit boards assembled by mechanics apprentices at ETH Zurich

Fig. 19. *C. Kathiravan of IIA at the CALLISTO console in the Gauribidanur radio observatory near Bangalore in India. The CALLISTO radio spectrometer is positioned to the left of the screen.*

mated test setup, controlled by a PC connected to a programmable radio signal generator via IEEE488 interface bus (Figure 19).

The e-CALLISTO system had its first success in December 2006, when the last large flares of Solar Cycle 23 occurred. These X-class flares where observed by the then newly launched Hinode satellite and may remain the only large flares observed by this satellite in the near future. Thus they have been studied by several groups around the world. None of the flares were observed during the entire length from the ground due to the short December daylight in the Northern hemisphere. However, e-CALLISTO, then operating in Switzerland, India and Siberia, covered more than 60% of the time.

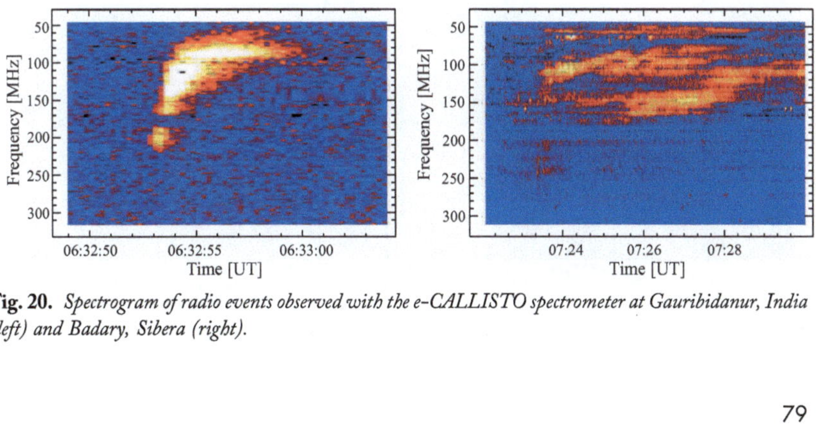

Fig. 20. *Spectrogram of radio events observed with the e-CALLISTO spectrometer at Gauribidanur, India (left) and Badary, Sibera (right).*

Figure 20 (left) shows the solar flare radio emission of May 20, 2007 observed by the e-CALLISTO spectrometer element in Gauribidanur (India). The data is displayed as a spectrogram, intense emission presented bright, no emission black. There are 200 channels displayed horizontally, time progressing to the right. The pseudo three-dimensional representations of time, frequency, and intensity produce an image that can be interpreted more easily. The frequency is given in MHz, increasing downwards. As the emission is proportional to the density (plasma emission), the vertical axis also represents altitude in the solar atmosphere, comprising about one to two solar radii above the photosphere. As density decreases with altitude, height increases upward in the picture. The radio emission is tilted towards the right, indicating that an exciter is moving upwards. Assuming a density model of the corona, the speed can be estimated. It amounts to about a third of the speed of light. The emission is therefore interpreted as the signature of an electron beam escaping from the Sun. Such events are known as Type III radio bursts. Note the low level of interference. Only few terrestrial emissions are seen as horizontal lines.

The spectrogram at right in Figure 20 displays the radio emission of a flare on May 23, 2007. The emission drifts again to the right, indicating an exciter moving upwards in the Sun's atmosphere. However, the speed is much slower in this case and of the order of 1000 km/s. This is a typical velocity of a shock wave in the corona. Such radio emissions, called Type II radio bursts, are often seen with coronal mass ejections, which disturb the whole heliosphere and are of great interest for near-Earth space weather.

4.3.1.3.2. Low-frequency radio antenna arrays

PIs: Justin Kasper (Massachusetts Institute of Technology), Robert MacDowall (NASA Goddard Space Flight Center)

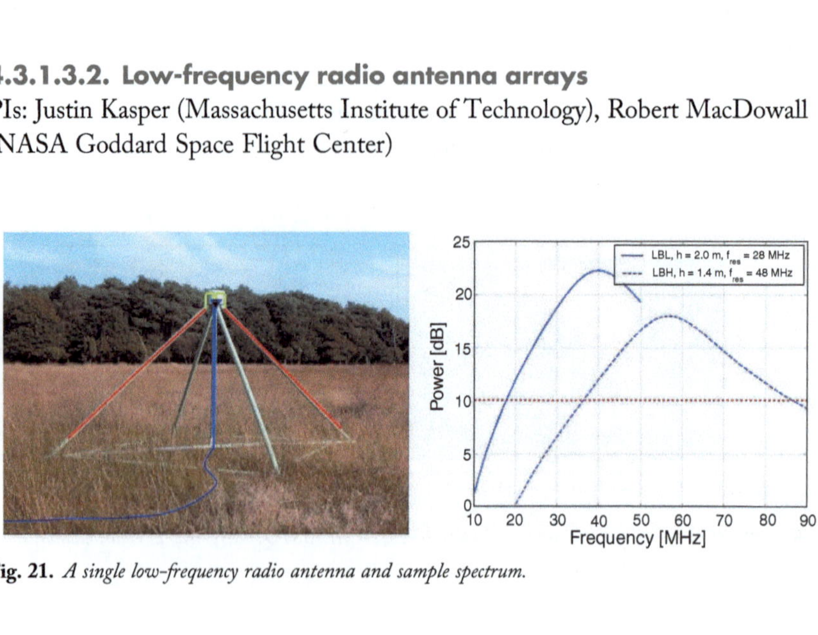

Fig. 21. *A single low-frequency radio antenna and sample spectrum.*

This network employs a new, inexpensive radio receiver technology to conduct low-frequency radio observations. A single unit consists of an antenna, receiver, computer and processing board. The receiver unit could either produce 10–100 MHz power spectra of the total power incident on the receiver, or whole imaging arrays distributed at locations around the globe can be assembled to provide spatial resolution to act as all-sky monitors and provide 24-h low-frequency coverage of the Sun. This investigation could also be used to track Jovian radio emissions and could provide prompt afterglow observations from gamma ray bursts (Figure 21).

The low-frequency radio arrays which be deployed at two levels: Option (1) low-frequency monitoring of solar radio bursts with single dipoles; Option (2) 8–16 element arrays for all sky monitoring Opportunities are being explored to install a low frequency Radio telescope in Gauribidanur radio telescope site in India to work in conjunction with CALLISTO in Ooty, India.

4.3.1.3.3. SAVNET (South Atlantic VLF Network)

PI: Jean-Pierre Raulin, Universidade Presbiteriana Mackenzie

SAVNET is an international project led and funded by Brazil, which includes collaboration with Argentina and Peru. SAVNET is part of a large international effort for heliophysical research studies for IHY. It is formally supported by the United Nations Basic Space Science (UNBSS) program. SAVNET involves about 10 international research groups, about 20 researchers and many students from all these institutions also participate. SAVNET is a set of seven Very Low Frequency (VLF) receiver bases located in the South American continent.

SAVNET development began in September of 2006 and the instrument was totally installed and fully operational in December of 2007 (Figure 22).

Fig. 22. *Example of a SAVNET receiver base, at Punta Lobos, near Lima, Peru.*

Scientific objectives of SAVNET

The main goals of SAVNET are (i) the monitoring of solar activity on short and long time scales related to solar flares and to variations induced by the solar activity cycle, respectively; and (ii) the study of the South Atlantic Magnetic Anomaly (SAMA). Other objectives include the study of atmospheric (mesospheric) dynamic phenomena and to provide low-ionospheric diagnostics of seismic-related electromagnetic effects.

SAVNET: outreach and education

Two schools were organized and related to scientific issues involving the SAVNET instrument. They were hosted by CONIDA, the Peruvian Space Agency, in Lima, Peru, July 2007 and July 2008.

Three students are involved in master degree researches directly focused on SAVNET scientific questions. Eight students are performing scientific initiation programs, also related to ionospheric studies.

One of the main goals of the IHY/UNBSS project is also to achieve good quality science. Since December 2006, two SAVNET papers were published in international referred journals. Four more papers have been submitted. Thirty-one SAVNET related studies were presented in 11 international conferences, mainly student papers.

4.3.1.4. Particle detectors and muon networks

4.3.1.4.1. SEVAN (Space Environment Viewing and Analysis Network)

PI: Ashot Chilingarian, Aragats Space Environmental Center of the Cosmic Ray Division of Alikhanian Physics Institute

A network of middle- to low-latitude particle detectors has been enacted in the framework of the IHY, to improve space weather forecasting research and alerts. The network is designed to be flexible to detect more than one species of particles and provide directional data.

Particle beams, accelerated at the Sun, are superimposed onto the uniform and isotropic Cosmic Ray (CR) background from galactic and extragalactic sources. Space-borne spectrometers measure the time series of the changing fluxes with excellent energy and charge resolution. Surface detectors measure time series of secondary particles born in cascades originated in the Earth's atmosphere by primary ions. Studies of these particles shed light on the high-energy particle acceleration mechanisms by flares and Coronal Mass Ejection (CME) driven shocks.

Time series of intensities of high-energy particles can provide highly cost effective information on the key characteristics of the interplanetary disturbances. Because cosmic rays are fast and have large scattering mean free paths in the solar wind, this information travels rapidly and may prove useful for space weather forecasting. The size and occurrence of the southward magnetic field component in Interplanetary CMEs (ICMEs) is correlated with the modulation effects the ICME poses on the ambient population of the galactic cosmic rays during its propagation up to 1 AU (at Earth's orbit). On the way to Earth (15–50 h, depending on the velocity of the ICME) the magnetic cloud and shock modulate the Galactic Cosmic Ray (GCR) flux, making it anisotropic. Surface monitors located at the Aragats Space Environmental Center (ASEC) on Mt. Aragats at 2000 and 3200 m altitudes (40°30 N, 44°10 E. Cut-off rigidity: 7.6 GV) detect charged and neutral components of the secondary cosmic rays with different energy thresholds and various angles of incidence (see Figure 23 for a schematic view of the new detector at ASEC). This richness of information, coupled with the simulation of the physical phenomena, can be used to estimate the shock size and the magnetic field "frozen" in the ICME. Consequently, one can predict the upcoming geomagnetic storms hours before the ICME arrival at the magnetometers on ACE and SOHO. The half-hour lead time provided by the L1 monitors is a bit short to take effective mitigation actions and protect satellites and surface industries from harm of major geomagnetic storms. To identify the major sources of error in the predictions, we need to measure, simulate and compare: (1) time series of neutrons, the low energy charged component (mostly electrons and muons) and the high-energy muons); (2) the correlation between changing fluxes of various secondary particles; and (3) directional information.

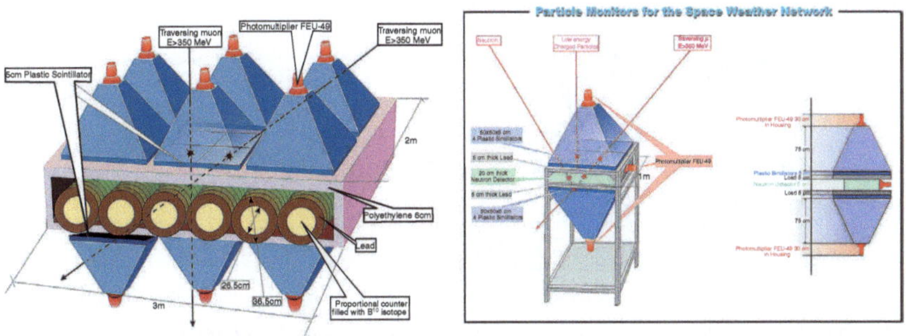

Fig. 23. *Schematic diagram of the basic SEVAN detector for muons and neutrons installed at the Aragats Space Environmental Center, Armenia and two modules of prototype detectors for the space weather network (from A. Chilingarian). Left: One section of the Nor Amberd Multidirection Muon Monitor (NAMMM). Right: Compact prototype detector for the space weather network.*

Based on our experience with correlation analysis of multivariate time-series from ASEC monitors, we designed and fabricated several new-types of particle detectors to meet the above goals. In order to keep the instrument inexpensive the options are kept flexible by using modular designs. The price of a fully autonomous single unit is not expensive, so that the network of nations involved in space research can be significantly expanded after IHY. At any time one can cascade these single units to achieve additional functionality: for example, add several new observational directions. Like the world network of neutron monitors, the new monitors measure the neutron fluxes and in addition they will measure charged particle fluxes with different energy thresholds, thus allowing the investigation of additional populations of primary ions.

The network is installed at middle and low latitudes and will be compatible with the currently operating high latitude particle detector network *Spaceship Earth*, coordinated by the group from Bartol Research Institute of University of Delaware in USA and with the Muon Detection Network (described below) coordinated by the group from Shinshu University in Japan. There are four SEVANs operational in Armenia, two in Croatia and Bulgaria, and in 2009 SEVANs will be in stalled in Slovakia and India.

4.3.1.4.2. Muon Detection Network

PI: Kazuoki Munakata, Shinshu University

The Muon Detection Network collaboration consists of nine institutes from seven countries (Japan, USA, Brazil, Australia, Kuwait, Armenia, and Germany). Many of the countries are already operating muon detectors, and some have recently installed them.

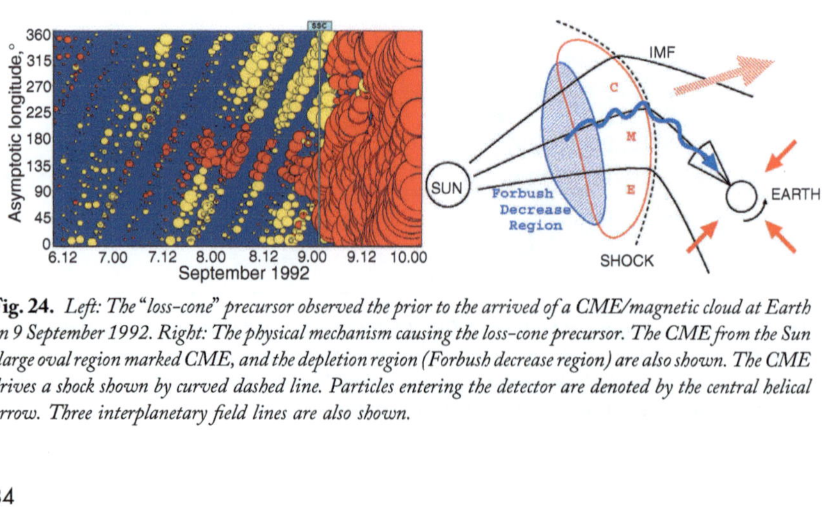

Fig. 24. *Left: The "loss-cone" precursor observed the prior to the arrived of a CME/magnetic cloud at Earth on 9 September 1992. Right: The physical mechanism causing the loss-cone precursor. The CME from the Sun (large oval region marked CME, and the depletion region (Forbush decrease region) are also shown. The CME drives a shock shown by curved dashed line. Particles entering the detector are denoted by the central helical arrow. Three interplanetary field lines are also shown.*

The utility of the muon detector for detecting interplanetary CMEs is shown in Figure 24. Each circle represents an hourly measurement by a single telescope as a function of time (day of year on the abscissa) and the asymptotic longitude of the viewing direction (in degree on the ordinate). Lighter and darker circles represent, respectively, an excess and deficit of cosmic ray intensity relative to the average, and the size of each circle is proportional to the magnitude of excess or deficit. The precursory decrease (darker circles) of cosmic ray intensity from $\sim 135°$ longitude (sunward direction along the nominal IMF) is clearly seen more than 1 day prior to the storm sudden commencement (arrival of CME-driven shock at Earth). The physical mechanism for the precursory decrease is illustrated in Figure 24, in the right panel). A CME propagating away from the Sun with a shock ahead of it affects the pre-existing population of galactic cosmic rays in a number of ways. Most well known is the Forbush decrease, a region of suppressed cosmic ray

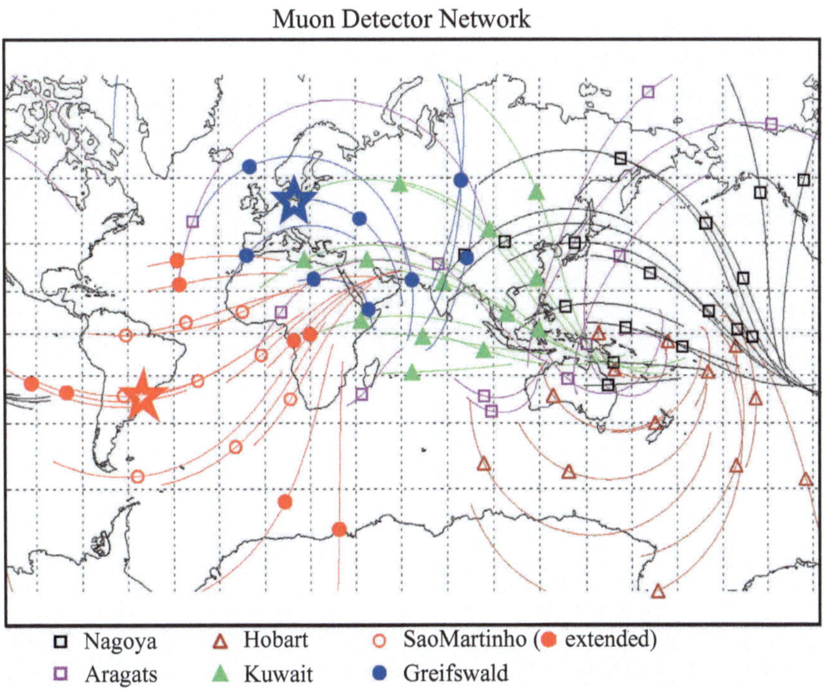

Fig. 25. *The geographic location of each detector is indicated by a big star and distinguished by shape coding. Each of the symbols (squares, triangles and circles) shows the asymptotic viewing of a particle incident on each telescope with the median primary rigidity. Open symbols display the existing viewing directions, while full symbols represent the directions to be added by the planned installation and extension of detectors. The track through each symbol represents the spread of viewing directions corresponding to the central 80% of each telescope's energy response.*

density located downstream of a CME shock. Some particles from this region of suppressed density leak into the upstream region and, traveling nearly at the speed of light, they race ahead of the approaching shock and are observed as precursory loss-cone anisotropy far into the upstream region (Figure 25).

In March 2006, this worldwide network of muon detectors was upgraded with both an enlargement of a detector in Brazil and the installation of a new detector in Kuwait City, Kuwait.

5 Education and Public Outreach

The IHY presents unique opportunities for expanding the education and awareness of Space and Earth Sciences. As a result, the education program forms a cornerstone of our international initiative. Although each of the four IHY components (Science, Observatory Development, Education and History) has an unique function, it is clear that the coordination of these activities allows us to successfully implement and maximize the impact of each of the components. For example, by integrating the Education and Public Outreach (EPO) component with the science campaigns, observatory development, and history initiatives, will be able to reach a greater audience and have a larger impact.

As we establish a greater presence for Space and Earth Sciences in developing nations and underrepresented areas, IHY works to assist the local researchers in establishing education and outreach initiatives through our international educational programs. Stronger research programs mean stronger universities, which foster the development of graduate and undergraduate programs that are key factors in encouraging youth to become interested in the exciting field of heliophysics.

The focus of the Education and Public Outreach (EPO) Program is the IHY basic objective: "Demonstrate the beauty, relevance and significance of Space and Earth Science to the World," although in the long run it support the other two basic objectives as well. However, the requirements of an effective outreach and education program vary from region to region. The IHY activities must therefore be adapted for different cultures, and correspond directly to the educational goals and resources of a given region. After reviewing the many possibilities for outreach activities, it was determined that the IHY Outreach and Education Program should focus on:

- Developing new and exciting outreach programs that provide unique opportunities for the global community;
- Increasing the visibility and accessibility of existing outreach programs;
- Developing partnerships between existing programs and activities to broaden their scope and impact;
- Determining the need for multi-lingual adaptations of educational resources and facilitating their translation.

The program aims to inspire the next generation of Space and Earth scientists and explorers; and spread the knowledge of our solar system and the exciting process of scientific exploration to the people of the World. Encouraging young people to pursue careers in science and technology helps secure and, in the case of developing nations, build up human resources, which will improve the technological base and enhance the prospect for further development. IHY has presented unique opportunities for expanding the education and awareness of Space and Earth science, as the outreach program was developed in conjunction with the United Nations and hundreds of observatories and institutions participatedin its activities.

The IHY's EPO program has many facets:

- Summer schools on Heliophysical topics for college undergraduate and graduate students;
- Educational activities with the Observatory Development program that provide students with a deeper challenge and connection to global space science;
- "Moving exhibitions" or kiosks that can be deployed to reach a wider audience;
- Multi-lingual adaptations of educational resources;
- Public events, such as eclipse viewings, lectures, and the "World Space Party";
- Development of tools and resources for educators;
- Development of educational websites and online activities;
- Distribution of scientific instruments for student research;
- Activities with amateur astronomy clubs, planetariums, science centers, museums and other established programs;
- Historical commemorations and anniversary events.

IHY established a greater presence for research in Space and Earth Science in developing countries through the IHY/UNBSS Observatory Development program, and several accompanying educational activities were deployed to provide a base for future growth. This provided opportunities for undergraduate and graduate students to participate actively in international cutting-edge research projects. To reinforce education at college level, IHY created the "IHY Schools Program", presenting university-level topics in heliophysics at different locations throughout the world with the purpose of educating students in heliophysics. The variety and diversity of activities in this report demonstrate the success of the IHY EPO grassroots program worldwide. Its success was due to the effort and dedication of the IHY regional, national and EPO coordinators and their teams. The following sections give a brief summary of IHY EPO activities; while this does not represent a comprehensive survey of the IHY educational activities, it does capture the breadth and diversity of the efforts. Descriptions of national outreach events are included with the national reports, in Sect. 7.

5.1. Organizational structure

The organizational structure of the IHY EPO program closely mirrored the overall organization of the IHY program: a central coordinator and educators serving as national coordinators for EPO (see table in Appendix I). The central coordinator, Maria Cristina Rabello-Soares, was part of the IHY secretariat and served as a contact point and facilitator for the IHY team members and their EPO activities. The EPO national coordinator for a given country was appointed by the nation's IHY national coordinator and acted as the liaison between those working on education and public outreach in his/her country and the IHY secretariat, as well as to the coordinators of the other nations. Each national coordinator for EPO was responsible for promoting local activities and inviting exemplary EPO programs in their country to participate in the IHY.

Most of the IHY EPO activities were developed as partnerships with other programs or institutions. Chief among these collaborative opportunities are participating IHY institutions and observatories, scientific organizations, educational programs of government agencies, and of course planetariums, museums and science centers. In some cases, IHY worked to expand participation and awareness of an activity, such as "Sun–Earth Day" and the Yuri's Night World Space Parties. In other cases, the IHY assisted a program or institution with the development of a new component or activity, such as new exhibits for museums and science centers, or events hosted by amateur astronomy clubs. There are also, of course, many activities unique to IHY, like many of the educational components of the IHY UNBSS Observatory Development Program.

This section explains events and activities occurring as part of the international IHY outreach program. Many nations hosted exciting events, such as IHY-Yemen's monthly workshops, IHY-Thailand's Science and Technology Fair, and IHY-Tunisia's documentary on the eclipse research expedition.

5.2. Forums: meetings, special sessions, and planning discussions

To engage the international community in the IHY EPO Program, a number of splinter meetings and official sessions were organized at relevant international meetings. These forums provided a great opportunity to discuss plans and pathways to foster new, mutually beneficial international collaborations in heliophysical science education and public outreach in the context of IHY. Some of them are described here.

At the 37th Committee on Space Research (COSPAR) conference in July 2006, a very productive and active splinter meeting was held entitled "Globalizing Space Science Education and Outreach" where scientists and educators laid the foundation for IHY education and outreach activities. The theme of this splinter meeting was to discuss pathways to foster new, mutually beneficial international collaborations in space science education and public outreach. It was organized by M. Cristina Rabello-Soares (IHY EPO Coordinator), Cherilynn Morrow (IHY EPO Coordinator for the United States) and Barbara Thompson (IHY Director of Operations) and had over 20 participants from around the world, including Argentina, Australia, Germany, Italy, Israel, Japan, Mexico, Norway, Russian Federation, South Africa, Slovakia, and United States.

An IHY EPO Subcommittee meeting was convened at the United Nations in Vienna by Dr. M. C. Rabello-Soares on February 19, 2007 simultaneous with the meeting of the Science and Technical Subcommittee of the United Nations. There were 30 participants from around the world representing IHY: Austria, Bulgaria, Czech Republic, France, Germany, Italy, Netherlands, Poland, Romania, Russian Federation, Slovakia, South Africa, Thailand, Turkey, United Kingdom, and United States (Figure 1). It also had the participation of Dr. Emily Cobabe-Amman, the Electronic Geophysical Year EPO Coordinator, and representatives of the International Year of Planet Earth (IYPE). After the meeting, the participants proceeded to the IHY Opening Ceremony and Reception.

Morrow also organized a session at the AGU Fall meeting 2006 (11–15 December 2006) in San Francisco, CA, USA entitled "Perspectives on Cross-Cultural Collaboration and Communication in Science Education and Public Outreach". The dissemination worldwide of EPO resources during IHY has raised questions about how to identify products that best communicate science to people of different cultures. This session featured invitees and called for contributors who have experiences, lessons learned, and best practices to share in the development of exemplary cross-cultural educational products and events.

Fig. 1. *IHY EPO Subcommittee participants at the United Nations in Vienna.*

At the AGU Joint Assembly 2007 (22–25 May 2007) in Acapulco, Mexico, there was a session entitled "Engaging Scientists and Educators in Education and Public Outreach Associated With IHY", wherein papers from scientists and educators that describe ongoing, innovative EPO activities that are related to IHY themes were presented.

A working group on education met during the IHY-Africa Space Weather Science and Education Workshop held in Addis Ababa, Ethiopia (12–16 November 2007), to discuss IHY EPO in Africa. It was organized by the IHY EPO coordinator for South Africa, Dr. Lee-Anne McKinnell, and Dr. Abebe Kebede. There were participants from many African nations, as well as other continents.

5.3. Public events and exhibitions

5.3.1. IHY "Open Doors Day": June 10, 2007

IHY-Europe launched a special event called "Open Doors Day", where IHY institutes and observatories around the world invited the public to learn about IHY and heliophysics. Eighty institutes in Europe participated, including Czech Republic, Finland, France (5 institutes), Germany (5), Hungary (6), Italy (9), Ireland, Slovakia (4), Spain (11), Sweden, Switzerland (4), and the UK (3). "Open Doors Day" drew over 20,000 participants across Europe (Figure 2). Results from each nation's Open Doors Day are given in the national reports (Sect. 7).

5.3.2. Passport to Knowledge

The "Passport to Knowledge" program has partnered with IHY to document many activities and produce video programs to spread the knowledge of heliophysics.

Fig. 2. *The SWEETS mobile exhibition helped attract thousands of visitors to the Belgian Open Doors Day at the Royal Observatory of Belgium, the Belgian Meteorological Institute, and the Belgian Institute for Space Aeronomy.*

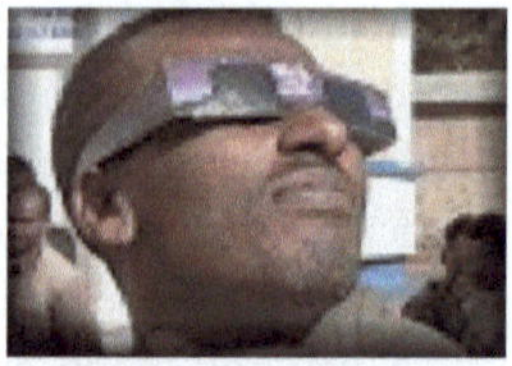

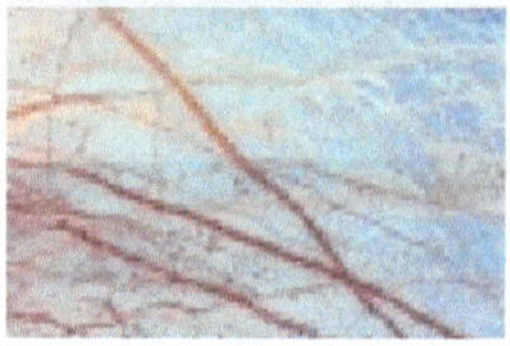

The documentary "From the Sun to the Stars" will be released in early 2009. Additionally, Geoffrey Haines-Stiles and the Passport to Knowledge (P2K) team sent videographer Art Howard to IHY meetings and conferences, to film the workshops and talk with educators and scientists, including the IHY-Ethiopia Teacher Training workshop in November 2007, and the project will continue to follow the progress or the SID space weather monitors in action in Ethiopia and elsewhere in Africa. According to Haines-Stiles, "The teachers' faces clearly show their appreciation for the materials and information you brought to them". The podcast "Opening Eyes to the Sun–Earth Connection" is available on the P2K website, *http://passporttoknowledge.com*.

P2K also traveled to experience the Arctic Sunrise in Barrow Alaska, and film the interactions between heliophysicists and local residents. The podcast "Science in Barrow – from IPY1 to the first IHY" is also available on the website, as well as "Heliophysics Below Zero" and "Earth's Poles as Analogs to Other Icy Worlds". The "Polar Palooza" project celebrates the anniversaries of IGY and IPY and the international science years through a series of activities, live broadcasts and informational shows.

5.3.3. Sun–Earth Day

Sun–Earth Day is an annual event comprised of a series of programs and activities that occur throughout the year, culminating with a celebration on or near the spring equinox. The specific date of Sun–Earth Day is chosen to coincide with either the equinox or another significant geophysical event, such as an eclipse, conjunction, or the launch of a spacecraft. Sun–Earth Day's focus is to engage K-12 schools and the general public in space science activities, scientfic demonstrations, and interactions with space researchers.

In collaboration with partners that include science centers and museums around the world, spacecraft, and online science forums, Sun–Earth Day (*http://sunearthday.nasa.gov*) produces webcasts, multi-media, and print resources for use by school and informal educators nationwide and internationally. A number of

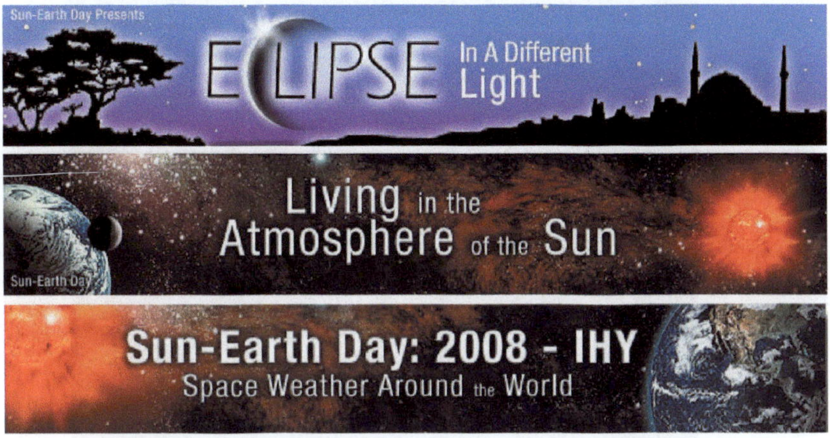

Fig. 3. *Sun–Earth Day banners for 2006 (top), 2007 (middle) and 2008 (bottom).*

events occurring at museums and science centers have maximized worldwide participation in IHY through the use of printed materials, webcasts and online activities.

For the years 2006–2008, IHY partnered in the execution of this exciting event. The 2006 Sun–Earth Day, held March 29, centered around a total solar eclipse (see next section). It attracted participants in 175 countries, more than 135,000 teachers and students, and an audience of more than 1.5 million for the webcast stream. 2007 Sun–Earth Day, "Living in the Atmosphere of the Sun", held on March 20, was subtitled "Celebrating International Heliophysical Year" and heralded the official opening of IHY. In addition to events held around the world, it attracted a web stream audience of nearly 2 million. 2008 Sun–Earth Day focused on the effects of the interconnected heliophysical system by launching events about "Space Weather Around the World". It attracted similar audiences, and featured a supplemental event in August 2008 called "Eclipse – Live From China". The connections forged between the Sun–Earth Day program and the educators throughout the international IHY community ensure that future Sun–Earth Days will have an even greater audience, and the worldwide impact will continue to grow (Figure 3).

5.3.4. IHY viewing stations along the path of the total solar eclipse of 2006

Solar eclipses provide excellent educational opportunities, and the annual "Sun–Earth Day" event for 2006 was held on the day of a total solar eclipse on

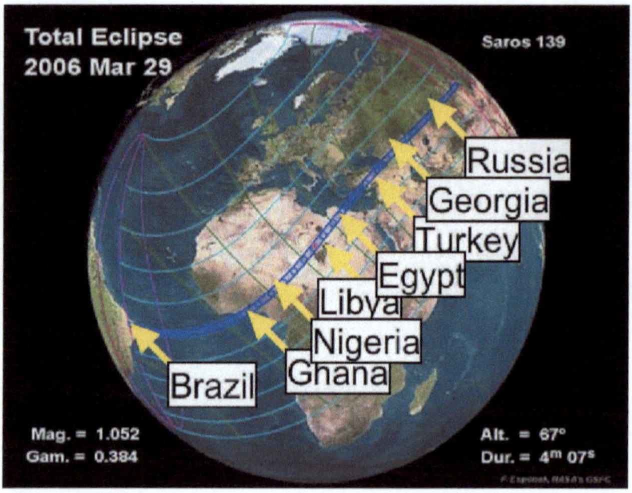

Fig. 4. *IHY Eclipse Viewing Stations along the path of totality allowed people the opportunity to safely observe an eclipse, receive educational materials, and discuss the event with researchers.*

March 29. The eclipse path began in Brazil, traversed the Atlantic, Northern Africa, Central Asia, ending at sunset in western Mongolia. Both educators and researchers participated in eclipse events, which was also targeted by the African Scientific Network. IHY outreach activities included the establishment of 12 eclipse "viewing stations" along the path of totality (see Figure 4) where "eclipse kits" containing eclipse viewing glasses, eclipse posters, eclipse information, IHY materials were distributed. The items were translated into four languages commonly spoken along the eclipse path: English, Arabic, French and Spanish. Eclipse activities also included the deployment of the AWESOME

Fig. 5. *IHY eclipse viewing station in Kumasi, Ghana.*

Space Weather Monitor in Tunisia, a special release of "Solar Facts", and space weather educational products.

Scientists were available at many of these viewing stations to conduct the observations and answer questions about solar eclipses, the Sun, and space research. A large number of children and adults attended the event in all participating countries (see Figure 5). Live webcasts of the total eclipse were also available from many sites in different languages.

Due to the effort and dedication of the IHY viewing center coordinators and their team, the 2006 total solar eclipse became one of the first IHY success stories. Taking advantage of this spectacular natural phenomenon, we were able to spread the knowledge of our solar system and the exciting process of scientific exploration to the general public, especially to young people. Several of the national reports in Sect. 7 highlight eclipse outreach activities.

5.3.5. Yuri's Night world space parties

Yuri's Night is an international celebration held on April 12 every year to commemorate the flight of the first human in space, Yuri Gagarin on April 12, 1961, and the first Space Shuttle launch on April 12, 1981. Locations have included all seven continents and the International Space Station.

The goal of Yuri's Night is to increase public interest in space exploration and to inspire a new generation of explorers. Driven by space-inspired artistic expression and culminating in a worldwide network of annual celebrations and educational events, Yuri's Night creates a global community of young people committed to shaping the future of space exploration while developing responsible leaders and innovators with a global perspective. These global events are a showcase for elements of culture that embrace space, including music, dance, fashion, and art. All are welcome to organize Yuri's Night events. Some have hundreds of participants while others are more small and informal (Figure 6).

Through the IHY Outreach and History programs, IHY team members around the globe partnered with Yuri's Night to bring greater awareness to past and current events in space. This annual event is particularly significant to IHY in that it is one of our activities that reaches beyond space science and focuses on the development and exploration of space in general. Through an IHY partner program, we were able to sponsor more than 50 IHY Yuri's Night parties and

Fig. 6. *Yuri's Night parties in Azerbaijan included the performance of a national Azerbaijani dance (left) and activities for children, such as artwork (right).*

expand the reach of this unique activity. There were IHY parties in many nations, including Cameroon, Nigeria, Azerbaijan, Bosnia, Kuwait, Germany and Yemen, and 17 Bulgarian parties in 2007 alone. The success of this partnership with this unique activity will continue into the future.

5.3.6. World Space Week

At the February 2007 IHY EPO Subcommittee meeting at the United Nations in Vienna, IHY team members met with the leaders of the World Space Week program, and it was agreed that a special IHY activity would be launched during World Space Week. The Center for Science Education at the University of California, Berkeley and the Stanford Solar Center at Stanford organized, in collaboration with IHY, a series of web-based activities and discussions on space weather during the World Space Week October 4–10, 2007 (*http://cse.ssl.berkeley.edu/segway/WSW.html*). It was organized by Dr. Laura Peticolas (University of California, Berkeley) and Dr. Deborah K. Scherrer (Stanford University) in collaboration with Dr. M. C. Rabello-Soares and Dr. B. Thompson.

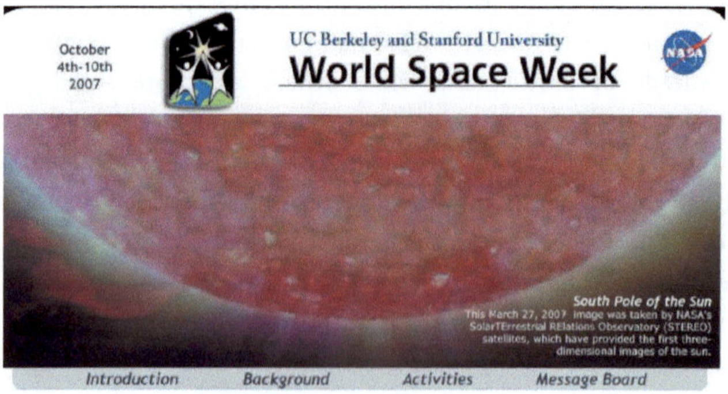

The website provided two main science activities for students and the general public with the objective to demonstrate the importance of our relationship with the Sun: (1) "Exploring the effect of Coronal Mass Ejections (CMEs) on Earth's magnetic field in four steps" and (2) "Tracking solar flares to Earth's ionosphere". The main goal of the first activity was for students to understand the connection between the magnetic Sun and the magnetic Earth. By going through various steps, students had the opportunity to practice with actual images of the sun from the SOHO and STEREO missions, as well as real-time magnetosphere data from the THEMIS mission to predict auroras. The second activity had students searching for solar flares in SID Space Weather Monitor data (described later in this section). Students were encouraged to participate in a message board that allowed students and teachers to discuss their findings after doing their activities with the help of moderators in the space research community. It was also open to ask questions about space science where scientists were available to post answers to the questions in the message board.

5.3.7. SWEETS (Space Weather and Europe – an Educational Tool with the Sun)

Traveling exhibitions are an excellent way to reach a wide, diverse audience. The SWEETS (Space Weather and Europe – an Educational Tool with the Sun) program was a major effort within IHY spanning 11 nations, and it had an enormous impact on the public awareness of space weather in the areas it visited. The project, sponsored by the European Commission and COST Actions 724 & 296, concluded at the end of 2007.

SWEETS activities included a space weather-on tour mobile exhibition that traveled through Austria, Belgium, France, Germany, Latvia, The Netherlands, Norway, Poland, Slovakia, and Portugal since it started its journey in June 2007. It contained a 13-panel, bi-lingual space weather poster exhibition available in several languages, an interactive exhibition including video presentations on space weather, personal computers with near-real time access to space weather observing satellites, an optical telescope and a radio telescope for solar observations by the visitors. The tour through the bus was guided by different high-level and outreach-educated European

Fig. 7. *Bilingual poster panels inside the bus explain the nature and impact of space weather. Translations for eight languages were available.*

Fig. 8. *The SWEETS "Space Weather Bus" was designed to be attractive and versatile. It traveled more than 23,000 km on multiple sites in 10 different nations.*

space weather scientists. The bus quickly generated interest, including an estimated 10,000 people attending the IHY "Open Doors Day" in Brussels (Figures 7 and 8).

Other elements of SWEETS include a Space Weather DVD in seven languages (Dutch, English, French, German, Polish, Portuguese and Slovak), a European web quiz on space weather, a film "The Breath of a Star", a "Solar Storm" dance show, and organization of science festivals in Europe including a Space Weather Forum in Schwerin Palace (19 November 2007) and a Space Weather Fair in Greifswald (20–21 November 2007). More details may be found on the SWEETS web page: *http://www.sweets2007.de* (Figure 9).

Fig. 9. *SWEETS used many innovative approaches to reach a wide audience. Shown is a momentous scene from the "Sonnensturm" (Solar Storm) dance show.*

SWEETS Deliverables:
Space weather DVD in 7 languages (Dutch, English, French, German, Polish, Portuguese and Slovak)
Space Weather-on-Tour mobile bus
Space Weather Forum, 19 November 2007, Schwerin Palace and Space Weather Fair, 20–21 November 2007, Greifswald
Video streaming including Videolink to Australia
Dance show "Sonnensturm" (Solar Storm)
SWEETS organized and participated in Science Festivals in several countries
Space Weather Web Quiz
Rocket Lauch
Movie "Breath of a Star"
Space Weather Planetarium shows in Portugal and Poland
Forecast display maps
Spark Chamber
TV Mirror System

5.3.8. Hartmann event: "Du Soleil à la Terre"

The exposition "Du Soleil à la Terre" ("In the Fire of the Sun" is the English translation) was produced by a commercial company (Hartmann Events, GmbH, Au, Switzerland), explicitly in view of the IHY. The scientific collaborators were mainly from ETH Zurich (led by Arnold Benz) and the Observatory of Paris, Meudon (led by Brigitte Schmieder). The exposition appears in three languages (German, French, and Italian), covering an area of up to 200 m^2 depending on the exhibit space. It consists of a movie theatre, posters, hands-on experiments, and a children's corner. The exhibit includes educational panels on a wide range of heliophysical topics, a 3D mini-film, hands-on activities, and scientific items such as fully functional CALLISTO radio spectrometer (see description in Sect. 4), a spectrometer component demonstration, spacecraft models and a meteorite. There were four interactive stations featuring magnetic fields, colour mixing, light prisms, and fold-out panels (Figure 10).

The exhibit was designed for large audiences, and the display is optimized for busy public areas such as shopping malls, as well as museums (such as Turin in Italy and Orléans in France), schools and fairs. The exhibition was displayed for 6 weeks at the Observatoire de Paris in November/December 2007 and the visitors could benefit of the presence of å professional astronomers. The tour will continue at least into 2009, the International Year of Astronomy. On the opening day at a given location, it is often inaugurated by a press conference.

The artistic design of the exhibit and the interactive displays have attracted many viewers since its premiere in Zurich in January 2007. As of November 2008,

Fig. 10. *Layout of the Hartmann Event "Du Soleilàla Terre" exhibit. It is available in French, German and Italian.*

the exhibit has travelled approximately 20,000 km, has been displayed in 35 European cities, and has been seen by 2,500,000 visitors and students.

5.4. Scientific research programs

5.4.1. The Space Weather monitor project

To bring hands-on science to high school and undergraduate university students, Stanford's Solar Center, Electrical Engineering Department, and local educators developed inexpensive Space Weather Monitors that students around the world can install and use at their local high schools and universities. By tracking the signal strength of VLF transmissions, the instruments detect changes to the Earth's ionosphere caused by solar flares and lightning and related phenomena. Two versions of the monitors exist – the SID monitors for high schools and the AWESOME monitors, which are more sensitive research devices, for universities. Through the IHY EPO program, SID student monitors were placed in high schools throughout the world. Through the IHY/UNBSS Program, the research AWESOME monitors were deployed to universities in developing nations. Both SID and AWESOME programs bring real scientific instruments and data in a cost-effective way to students and researchers. Instruments meet the objectives of being sensitive enough to produce research quality data yet inexpensive enough for widespread placement (Figure 11).

To participate in the SID project, students must "buy in" by constructing their own antenna, usually simple and inexpensive, and by providing access to a computer for data collection. In partnership with the Chabot Space & Science Center, Oakland, CA, USA, classroom and educator support materials were developed to accompany the distribution. Materials are culturally sensitive and portions were translated into the six official languages of the United Nations (Arabic, Chinese, English, French, Russian, and Spanish). The SID monitors were provided free of

Fig. 11. *Students at Deer Valley High School in Antioch, CA, USA experiment with their SID monitors.*

charge to developing nations and could be set up anywhere there is access to power. Stanford is providing a centralized data repository and a blog site where students and researchers can exchange information and discuss the data.

5.4.1.1. SID student network focuses on Sun monitoring in developing nations

The student-oriented SID monitors respond primarily to solar-induced changes to the Earth's ionosphere. Students were provided with research guides and

Fig. 12. *Students from the Nigeria IGS School are instructed on use of their SID monitor.*

Fig. 13. *SID Monitor Workshop in Addis Ababa, Ethiopia, given by Deborah Scherrer and Cristina Rabello-Soares.*

teachers were provided with training material to bring them up to speed on the necessary solar and ionospheric science. The SID program set up access to scientist mentors, arranged joint observing sessions such as for the 1 August 2008 solar eclipse, and provided a centralized database and communication channels so that teachers and students worldwide could share their data (Figure 12).

To support the IHY goal of "Inspiring the next generation of space and Earth scientists as well as spreading the knowledge, beauty, and relevance of our solar system to the people of the world," SID student monitors were deployed to high schools and universities around the world, with a focus on developing nations, particularly Africa.

The further enhance the ability to place monitors in Africa, in conjunction with the IHY Africa Space Weather Science and Education Workshop, held in Addis Ababa, Ethiopia in November 2007, Deborah Scherrer and IHY International EPO Coordinator Cristina Rabello-Soares gave a 2-h SID workshop to 50 conference attendees, all African researchers interested in hosting SID monitors at their universities and/or local high schools. Half of the attendees received their SID monitors during the workshop, the rest were provided by mail. Enthusiasm for receiving the instruments was very high! Currently, there is a network of over 70 monitors on the African continent (Figure 13).

5.4.1.2. AWESOME distribution focuses on research in Africa and Asia

The chief goal for placing AWESOME monitors is to perform a quantitative comparison of local ionospheric disturbances, magnetospheric activity, and thunderstorm intensity across much of the globe through the method of Extremely Low and Very Low Frequency (ELF/VLF) monitoring. Although such VLF

Fig. 14. *Newly installed AWESOME antenna in Uzbekistan.*

monitoring methods are widely employed, a number of scientific questions and techniques can only be answered with the existence of a worldwide network of instruments. The most intense region of lightning activity on our planet is central Africa. This lightning region has been comparatively poorly studied due to a lack of on-the-ground instrumentation. Asia is also home to a region of intense thunderstorm activity stretching from India to northern Australia. Much of this lightning activity occurs over water, which has been much less studied than the effects of land-based lighting (Figure 14).

Hence a network of AWESOME and SID monitors has now been set up in Algeria, Tunisia, Libyan Arab Jamahiriya, Nigeria, Morocco, Egypt, Ethiopia, South Africa, India, Uzbekistan, Serbia, Fiji, Malaysia, Indonesia, Kazakhstan, and Azerbaijan. These VLF receivers can also look for earthquake-associated disturbances due to the regions of high earthquake density in the Persian Gulf, Southwestern China, Northeast India Ocean, and Oceania.

5.4.2. Indian Institute of Astrophysics (IIA) educational spectrographs and radio interferometers

A similar program has transpired in India, whereby scientists participate with schools throughout India, consisting of simple experiments to study the Sun in the visible and radio wavelengths. These are:

1. Box spectroscope to view the spectrum of the Sun and common terrestrial light sources.
2. A simple two-element radio interferometer to observe Sun and other strong cosmic radio sources.

The IIA has taken the initiative of arranging an adequate number of these instruments to be fabricated for distribution in schools and colleges throughout the country. Prototypes of these instruments were on display during the IIA Open House held 9–10 August, 2007. More details can be found *at http://www.iiap.res.in/outreach/ihyoutreach*. Posters on the Sun, Space Weather and Solar–terrestrial relationships were displayed as a theme exhibition celebrating IHY 2007. There was also a demonstration by Dr. Navnirmiti of Mumbai using low-cost, no-cost tools to understand the Sun.

With the two-element radio interferometer kit, the team provides hands-on astronomical observing experience to the interested science and engineering graduate student community by donating radio antennas and receiver systems to their institutions. The host institution is expected to provide a personal computer for data acquisition. The students are be trained to: (i) carry out observations of radio emission from Sun and other strong cosmic radio sources with the above set up, and (ii) develop software for deriving quantitative information from the data acquired. More information on this project can be found in Sect. 7.

5.5. IHY schools and teacher training

5.5.1. IHY schools

5.5.1.1. Overview

IHY's focus on developing new and exciting EPO programs provided unique opportunities for the global community to increase the visibility and accessibility of heliophysics outreach programs. To address this focus, the IHY Secretariat developed an IHY Schools Program that developed and promoted a series of schools in 2007 and 2008, with the purpose of educating students about Universal Processes and providing them with an opportunity to view their own research interests in a new context. Universal Processes are the organizational principles and universal laws that underlie our understanding of the universe (see Sect. 3). The establishment of heliophysics as a new discipline has required a cross-disciplinary comparative approach to understanding these underlying principles.

The IHY Schools Program was organized and operated by the IHY Schools Committee (ISC), which was chaired by David Webb and consisted of select members of the IHY Secretariat, Steering and Advisory Committees, the IHY Regional Coordinators, and coordination teams of the individual schools.

By enhancing the scope, impact, and outreach of existing space physics programs at various host locations, the IHY schools have enabled student cultural exchanges to establish innovative new models for graduate education and training in heliophysics. Moreover, the schools helped provide fertile grounds for collaborative research in this field, and facilitated the development of a diverse, globally engaged scientific community that recognizes the increasing relevance of heliophysics.

The ISC developed a general curriculum as a model for all the IHY schools. It included seminars and hands-on sessions with databases acquired particularly through the IHY CIP and U.N. Basic Space Science programs, as well as collaborative efforts with other affiliated groups. The overall scope of the schools was heliophysics, Universal Processes, Sun–Earth interactions as well as those at other planets, and the outer heliosphere. The goal of the lectures and data laboratory activities was to cover the cross-disciplinary studies of Universal Processes, responses to external drivers including lectures covering the main IHY science topic areas, achieving international scientific cooperation, preserving the history and legacy of the IGY on its 50th Anniversary, public outreach, and global studies with an emphasis on science in developing countries.

Other lecture topics included science discipline areas, such as:

- Solar Physics;
- Solar sources of heliospheric variability;
- Emission of energetic particles, and their acceleration and propagation through the Heliosphere;
- Atmospheric studies of the planets, involving ionospheres, thermospheres, and mesospheres;
- Interaction of energetic particles with magnetospheres (Earth and planets), and their effects in the planetary atmosphere;
- Coronal Mass Ejections and Space Weather; and
- Solar irradiance and climate.

The topics were approached from both the theoretical and modeling aspects as well as using observations focusing on new instrumentation and missions such as STEREO, Hinode, THEMIS and IBEX. The structure of the schools usually consisted of 8–10 full days of class time, with brief morning and afternoon breaks and longer day and evening breaks for socializing and discussions with classmates and the instructors. Typically, there were about four sessions per day with each session lasting one to two hours. The amount of time devoted to data analysis labs depended on the local availability of computers and networking, the number of students and the amount of pertinent, analyzable data available by the time of the school. Each student was given individual attention and encouraged to focus on their own individual research areas.

The IHY Schools have helped support students associated with the two key elements of the IHY science program: the CIPs and the UNBSS IHY instrument program. The schools were able to provide these students with two unique opportunities not available through any current program: the chance to develop their studies as part of a new "basic" research discipline, and the ability to work with researchers active in the field during the school as a means of broadening their research. It is expected that upon completion of the program, graduates are familiar with several new topics, have made fundamental connections with and undergone personal mentoring by a number of leading researchers, and have identified new arenas and pathways to apply their own knowledge in the future. To enhance the experience of each student and help guide them after the school, CDs or DVDs containing the presentations and additional material were made available to the participants at the end of the each School. Finally, continuing contacts between the lecturers/mentors and the students has been encouraged to help them solidify their first exposure to this new scientific discipline.

5.5.1.2. Reports on each school

An attempt was made to have one or more schools hosted in each of the IHY Region. The resulting schools covered the regions of North America (U.S.), Latin America (Brazil), Asia-Pacific (India and China), and Europe–Africa (Nigeria). Unfortunately a sixth school planned for Malaysia in 2009 had to be cancelled due to local conflicts. However, two IHY-related Polar Aeronomy and Radio Science Summer Schools were held in Alaska in 2007 and 2008, and there were a number of IHY-affiliated schools, which are summarized below.

5.5.1.2.1. The North America school

The NASA Living With a Star Heliophysics Summer Schools occur over three years: 2007, 2008 and 2009. Each school has a separate curriculum in heliophysics, and yields a book covering the topics of that school (see "Publications" appendix). The first school in 2007 was on "Plasma Physics of the Local Cosmos" and is the only one officially affiliated with IHY. It was held at NCAR in Boulder CO, USA, from 30 July to 7 August 2007 and was organized by George Siscoe, Karel Schrijver, Don Hassler and David Webb. The school was held over 8 days and had 34 students in attendance. There were about 25 lecturers and computer lab

coordinators participating. The schedule, faculty and students and other details are listed on the schools website. CDs of the lectures and labs were made available after the school, so that students may reinforce the lessons.

5.5.1.2.2. The First Asia-Pacific school: India

This school was organized by the Indian Institute of Astrophysics (IIA), Bangalore and held at Kodaikanal Solar Observatory, 10–22 December 2007. The School Directors were Ashok Ambastha, R. Ramesh, K. E. Rangarajan, and Nat Gopalswamy, and the local organization was led by R. Ramesh. About 50 students attended. This international school offered an intensive 2-week course on topics related to heliophysics. The course was aimed at Ph.D. students and post-doctoral research associates. A similar, pre-IHY school on Solar Physics was held at Kodaikanal in December 2006 (Figure 15).

Fig. 15. *Participants and lecturers of the 2007 IHY Asia-Pacific school.*

5.5.1.2.3. The Latin America schools

A major Latin American IHY school was held 25–29 September 2006 in Buenos Aires, Argentina (IAFE) (see Sect. 7). The next IHY school was organized by the Centro de Radioastronomia e Astrofisica Mackenzie (CRAAM) and held at Presbyterian Mackenzie University in Sao Paulo, Brazil, 14–20 February 2008. The school was organized by Alisson Dal Lago, Jean-Pierre Raulin, Adriana V. R. Silva and Cristina Mandrini. Nearly 20 lecturers and 80 students attended. Lectures were in the mornings, and computer labs and other activities, such as visits to local institutes, were in the afternoons. A preliminary IHY School was held 25–29 September 2006 in Buenos Aires, Argentina. The presentations from each school were made available as a CD.

5.5.1.2.4. The Second Asia-Pacific school: China

The Second Asia-Pacific IHY School was held at the Center for Space Science and Applied Research (CSSAR), Chinese Academy of Sciences (CAS), in Beijing, China, 20–31 October 2008. The organizers were L. W. Ren and J. Liang. The school was sponsored by the State Key Laboratory of Space Weather, the Center for Space Science and Applied Research (CSSAR), the Department of Earth Sciences, the National Natural Science Foundation of China, and the Chinese Academy of Sciences (CAS). This IHY school was open to students and to researchers from the Asia-Pacific region and abroad, and about 40 students and 25 lecturers attended. The purpose of the school was to bring a greater understanding of underlying Universal Processes and to promote international collaboration and exchange of ideas and views in heliophysical studies. This two-week autumn school, focused on a broad range of extended heliophysics topics, such as solar physics, heliospheric physics, geospace physics and space weather, related to the

IHY science topic areas. The school consisted of about 50 lectures, as well as a series of laboratory exercises and observatory visits.

5.5.1.2.5. The Europe and Africa school

IHY-Nigeria hosted the African Regional IHY School to be held at the Centre for Basic Space Science, National Space Research and Development Agency

(NASRDA), the University of Nigeria, Nsukka., Nigeria from 10 to 22 November 2008. The school was organized by Dr. A. Babatunde Rabiu and was supported by the following agencies: NASA, the Center for Basic Space Sciences (CBSS), the Abdus Salam International Center for Theoretical Physics, Northwestern University (South Africa), the University of Bergen (Norway), and the space environment research Center, Kyushu University (Japan). It was declared open with a ceremony led by Dr. Joseph Akinyede, the Deputy Director General of the National Space Research and Development Agency (NARSDA) on 10 November 2008. The school featured nearly two dozen internationally recognized researchers from many branches of heliophysics, whose lectures covered topics in all areas of heliophysics. This international school offered an intensive 2-week course aimed at postgraduate students and new postdoctoral associates. About 60 students participated, including students from 14 different African nations.

5.5.1.3. IHY-affiliated schools

5.5.1.3.1. North America

Polar Aeronomy and Radio Science IHY Summer Schools, 27 July–2 August 2007 and 20–31 July 2008

The US Air Force Research Laboratory, the Office of Naval Research (AFRL/ONR) and the National Science Foundation (NSF) jointly funded a summer school with International Heliophysical Year sponsorship based at Fairbanks Alaska from 23 July to 2 August 2007 and 20–31 July 2008. These schools focused on solar terrestrial relationships and their influence on the ionosphere and upper atmosphere. Typical attendance was 27 graduate students drawn broadly from the United States, along with their faculty mentors. The school program began with lectures explaining the physical processes at work in regions of interest followed up by hands-on observations and experiments using radio and optical devices. The experimental part of the school was conducted in partnerships between students and their mentors.

Seventeen scientific lectures spanned topics from the neutral atmosphere and ionosphere through magnetosphere–ionosphere coupling, magnetosphere–solar wind coupling to the origins and propagation of the solar wind. Included in the set were technical lectures on radar and optical techniques applied to the ionosphere and upper atmosphere and on ionospheric modification with a high power radio transmitter. Specific instruments included in the hands-on sections of the school were the AFRL/ONR High Altitude Auroral Research Program (HAARP) 3.5 MW transmitter at Gakona, Alaska, the NSF Advanced Modular Incoherent Scatter Radar (AMISR) at Poker Flat, Alaska. Also, optical imagers and a LIDAR were available for the observation of noctilucent clouds. Students provided oral and written reports of their experiments.

5.5.1.3.2. Latin America

A Latin American school was in Lima, Peru, in 17–22 April 2006. Twenty-five students attended. Another Latin American school was held in San Jose dos Campos, Brazil (INPE), 23–26 October 2006. One hundred and ten students attended for 1 week.

Another school happened in conjunction with the 8th Latin American Conference on Geophysics (COLAGE) in Merida, Mexico, 12–17 July 2007. Contact: Jean-Pierre Raulin and Alisson Dal Lago.

5.5.1.3.3. Western Europe

The ISSS Advanced School in Space Environment – ASSE 2006: Solar–Terrestrial Physics, 9–14 September 2006, L'Aquila, Italy: "Turbulence and Waves in Space Plasmas"

The scientific program titled "Series of Events on Relations in the Sun–Earth System and Space Weather" (SERSES) offers a coherent series of courses for young researchers in the years 2006–2009 in different disciplines which concur to the Sun–Earth relationships, and presents the solar–terrestrial environment as a global system in order to allow a significant improvement in the global knowledge of the entire system. SERSES courses are organized at L'Aquila, in close cooperation by the Consorzio Area di Ricerca in Astrogeofisica and the International School of Space Science of the Consorzio Interuniversitario di Fisica Spaziale:

1. Spring 2006: The Physics of the Sun (i.e. The Active Sun on your Active Desktop);
2. Fall 2006: Solar–Terrestrial Physics;
3. Spring 2007: Magnetospheric Dynamics (9–15 April 2007);
4. Fall 2007: Turbulence and Waves in Space Plasmas (9–14 September 2007);
5. 7–12 April 2008: Geomagnetism and Ionosphere;
6. Fall 2008: Solar–terrestrial relations in Antarctica.

5.5.1.3.4. Balkan/Black/Caspian Sea region

A number of schools and student workshops brought new heliophysics knowledge to future scientists:

- Tusi 3rd Summer Astronomical School, 2006, held at Baku, The Republic of Azerbaijan.
- 4th Tusi Summer Regional Astronomical School "Sun and Geosphere" and Young Scientists Conference – 2007, Azerbaijan.

- Young Scientists International School on "Heliosphere and Galaxy", May 3–5, 2007 in Bucharest, Romania.

5.5.1.3.5. Africa

IHY-Africa Space Weather Science and Education Workshop, 11–16 November 2007, Addis Ababa, Ethiopia

See Sect. 4 for a report on this workshop.

5.5.2. Teacher workshops

5.5.2.1. Addis Ababa, Ethiopia, 2007

The IHY recognizes that, in order to develop space science research infrastructure in Africa, space science education must also be developed to support the long-term operation and use of scientific instrumentation. In response to these needs, the USA IHY Education Advisory Committee organized a Geophysical Information For Teachers (GIFT) workshop for Ethiopian high school physics educators on 10 November 2007 in the Ethiopian capital of Addis Ababa. The workshop, held in conjunction with the IHY Africa Space Weather Science and Education Workshop, gathered 70 teachers from around Ethiopia for a 1-day intensive professional development program that focused on fundamental physics concepts relevant to space weather.

The program discussed fundamental physics concepts relevant to space weather; examples of inquiry-based lessons plans; access to hands-on activities relating to magnetism, spectroscopy, and the electromagnetic spectrum, solar viewing through an H-alpha telescope; and an overview of the IHY and why it is relevant for their students. Time was also spent in comparing science education in the USA with that in Ethiopia. The teachers went away from the workshop with packets of resources including CDs and DVDs on space weather and the NASA THEMIS mission, Magnaprobes® and magnetic teaching supplies, a collection of 40 simple toys and games used to demonstrate microgravity, and class sets of spectrometers for students to build. A post-workshop survey showed that the vast majority of teachers learned much from, and enjoyed, the workshop and found ample material to use in their classrooms (Figure 16).

Thanks to Rice University and the Houston Museum of Natural Science, Ethiopia, became only the fourth country in Africa to have its own digital planetarium. Scientists had brought to the workshop a Discovery Dome®, an inflatable planetarium and programs developed to teach Earth and space sciences. The 5-m Discovery Dome, which can seat about 50 people, was demonstrated to

Fig. 16. *Attendees and presenters of the Ethiopian Teacher Workshop.*

Fig. 17. *(a) UCLA researcher Mark Moldwin demonstrates how to use an H-alpha telescope to observe features in the solar atmosphere. (b) Menelik High School students with the Discovery Dome®. (c) Zemenu Mekonnen's Science and Technology Club at Higher Seven Senior Secondary School in Addis Ababa using spectroscopes provided to Mr. Mekonnen at the teacher workshop.*

the GIFT teachers, later to the IHY conference attendees, and also to a Menelik high school class before being donated to the Ethiopian National Museum in Addis Ababa (Figure 17).

5.5.2.2. Acapulco, Mexico, 2006

The USA IHY Advisory Committee for EPO hosted two science outreach events at the November 2006 American Geophysical Union (AGU) conference in Acapulco, Mexico. Working with the IHY EPO coordinator for Mexico and with AGU Education Director Ines Cifuentes, the committee presented a Geophysical Information for Teachers (GIFT) workshop to 74 very enthusiastic teachers from all over Mexico. Many teachers had to travel on buses for over 8 h to attend this day-long workshop. Presentations and activities for the workshop were given by Mark Moldwin, Richard Schope, Blanca Mendoza, Roberta Johnson, and Pat Reiff with translation by Marina LaGrave.

Afterwards the Committee hosted a Family Science Event, providing resources and hands-on activities for both the teachers involved with the workshop and local families. The portable planetarium programs provided by Pat Reiff of Rice University and Carolyn Sumners of the Houston Museum of Natural Sciences were a big hit, as were the numerous giveaways including the Stanford Solar Center spectrographs and UV bead bookmarks, and Cherilynn Morrow's Sun Puzzle Page, all translated into Spanish for the occasion.

5.5.2.3. Baltimore, MD, USA, 2006

Nineteen K-12 teachers from around the US attended the Geophysical Information for Teachers (GIFT) Workshop entitled "Sun–Earth Relationships and the International Heliophysical Year" sponsored by the American Geophysical Union and held in Baltimore, MD on May 23–24, 2006. In the evaluation, the participants reported that the workshop was very useful with a great combination of experts.

5.6. Web and Internet resources

5.6.1. SunEarthPlan website

IHY-UK led the development of IHY's primary educational web resource, SunEarthPlan.net. A consortium of scientists spanning all of the IHY scientific disciplines designed the site so that the website would have comprehensive information on all areas of IHY science from the solar interior to the Earth's atmosphere. Modules are written for readers at all educational levels, with regular

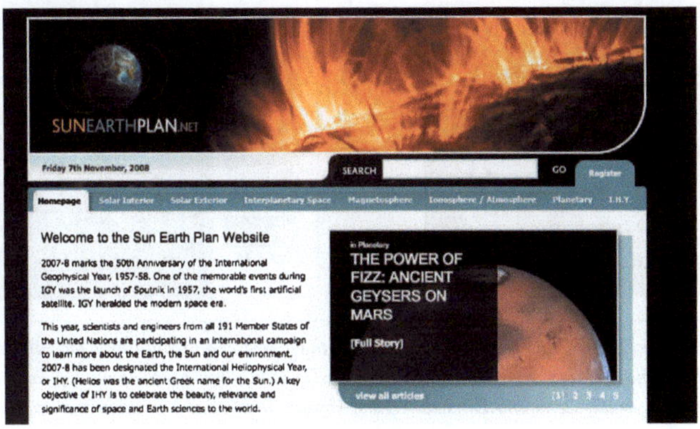

updates, news articles, and activities to engage readers on a regular basis. It features video feeds, interviews with scientists, and a "Your Questions" feature that allows readers to ask direct questions to scientists.

One of the hallmarks of this website is its inviting character. All of the contributing scientists are featured with photos and mini-biographies, with many opportunities to engage the scientists and learn about their research. Regular updates and news articles (see "The Power of Fizz" on the left side of the figure) give the reader a more dynamic interface, and the site offers many resources such as classroom activities and "build your own" science experiments.

5.7. Multilingual and multicultural resources

5.7.1. Translations and multilingual materials

Although English is the universal language in scientific research, this is not true for public outreach and education (especially at pre-college levels). Multi-lingual

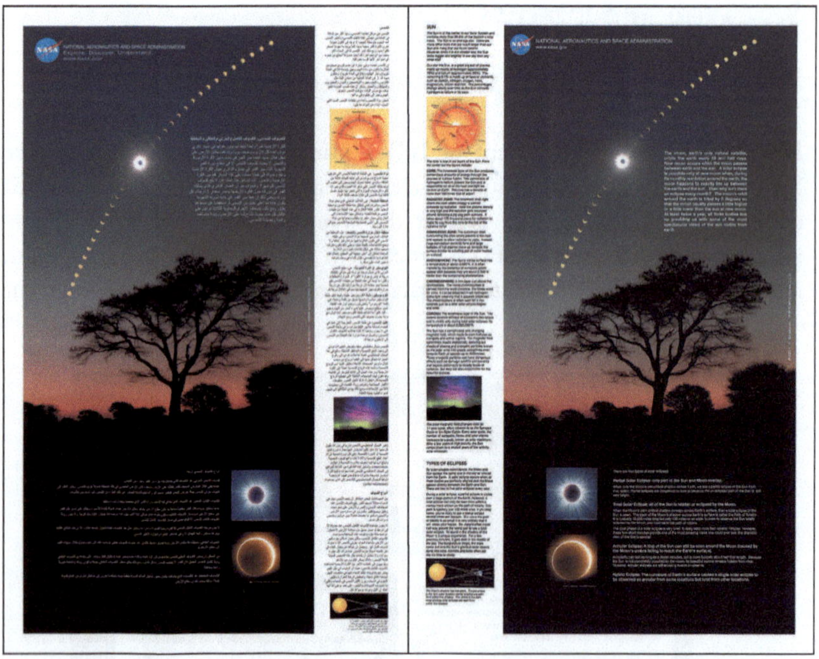

Fig. 18. *A poster describing the total solar eclipse path and geometry and containing facts about the Sun, space weather effects, and eclipse viewing safety was developed in four languages commonly spoken along the eclipse path. Shown are Arabic (left) and English (right). French and Spanish versions were also developed.*

adaptations and translations of educational resources are important for the globalization of heliophysics outreach and making the materials and activities accessible worldwide. IHY resources were made available for translation, and institutions and individuals were encouraged to help in this effort. The resulting resources were accessible to all from the IHY website, and many additional materials translated into Romanian are available on the IHY-Romania website. Among the languages translated from and to English were: Arabic, Armenian, Azerbaijani, Chinese, Czech, Dutch, Esperanto, French, Gaelic, German, Greek, Italian, Japanese, Norwegian, Polish, Portuguese, Romanian, Russian, Slovak, Spanish, Thai, and Turkish (Figure 18).

The AWESOME SID Space Weather Monitor (described earlier in this section) project undertook the goal to translate all accompanying training and educational material for the instrument distribution into the six official languages of the United Nations. Through the use of Stanford Alumni volunteers, most translations were accomplished. In addition, Dr. Sharon Murrel, Professor of Mathematics at Peking University, had her students translate the entire set of SID materials into Chinese, to support the creation of a large SID network in China.

Many other activities and resources have undergone translation or have been developed multilingually. Throughout this report there are many examples of this.

5.7.2. Data sonification for blind students

Christiana Dumitrache of IHY-Romania collaborated with musicians to create the "Music of the Spheres" project (Muzica Sferelor in Romanian). Scientific phenomena, such as the Venus transit of the Sun, geomagnetic indices, and the "Big Bang" were transposed into music. *http://aira.astro.ro/~crisd/muz/.*

The US IHY National Outreach Committee has supported activities in partnership with Marty Quinn, an expert in data sonification. Quinn, along with NASA's THEMIS team, is providing a tool to sonify the SID monitor data, primarily to make it accessible to blind students. Unfortunately, we were not able to obtain funding to pursue translating the package to Braille.

5.7.3. Multicultural communication

Cherilynn Morrow has spearheaded discussions of multicultural linkages between different cultures and adaptations of IHY outreach materials to make them relevant to other cultures. At the 2006 Committee on Space Research (COSPAR) meeting in Beijing China, a special session on outreach discussed international education initiatives, and a special "Globalizing Space Science Education and Outreach" splinter meeting allowed outreach leaders from around the world to discuss the challenges and results of efforts to multiculturalize education products. A special issue of the journal *Advances in Space Research* documents the discussions and results of the session, as well as descriptions and results of the participating programs (see Appendix III). Additionally, special sessions on "Global Geoscience Education and Outreach" and "Perspectives on Cross-Cultural Collaboration and Communication in Science Education and Public Outreach" also drew a great deal of interest at the 2005 and 2006 American Geophysical Union Fall Assembly meetings.

6 "IGY Gold" History initiative

6.1. Goals

The International Geophysical Year 1957–1958 was one of the greatest scientific events of the modern era. Although a great deal of historical effort has been dedicated the IGY, there is still a large amount of valuable historical information in danger of being lost in time. Therefore, an important part of the IHY 2007 activities was preserving the history and memory of IGY (for a brief history of IGY and International Science Years, please see Appendix V). The "IGY Gold" History initiative (gold symbolizing the 50th anniversary) has several goals:

- Identifying and recognizing planners of and participants in the first IGY;
- Preserving memoirs, articles, photographs, and all items of historical significance for the IGY;
- Making these items available to historians, researchers, etc.;
- Serving as a contact service for these activities;
- Spreading awareness of the history of geophysics; and
- Planning special events and "reunions".

The IGY Gold program, launched in 2004, is accomplishing these goals by creating a network that identifies participants in the first IGY, awards them a special certificate and commemorative pin, and encourages them to collect items of historical interest and make them available to historical archives. Support for this program is provided by the International Union of Geophysics and Geodesy, while organizations such as the AGU and IAGA History Committees have played a significant leadership role in the development of this activity.

6.2. The IGY "Gold" Club

The IGY "Gold" Club identifies and recognizes the exceptional accomplishment of participants from the first IGY. To be inducted into the IGY Gold Club, one *must meet three requirements*:

- You must have been a participant in the first IGY.

- You must contribute an item of historical significance to the initiative. It can be a letter, a recollection, an article, a photograph, etc. Unusual items are often the most significant to historians.
- You must be willing to have that item made publicly available to historians, librarians, and other people interested in investigating and preserving the history of geophysics.

"Gold club" participants will be rewarded with and a special commemorative "IGY Gold" lapel pin. They also receive a special "IGY Gold Anniversary" certificate of recognition from one of the IGY legacy sponsors (IHY, IYPE, eGY or IPY) and one of the associated geophysical organizations (currently IUGG, AGU and IAGA, other organizations are welcome to join). *The "IGY Gold" commemorative lapel pin has been specially designed for the 50th anniversary of IGY, and only members of the IGY Gold club will be issued this pin* (Figure 1).

There have been hundreds of people, from all of the continents, inducted into the IGY Gold Club, and a great deal of valuable historical information has been obtained and preserved for posterity. The first award was presented at a special IHY ceremony to Alan Shapley, who served as Co-Chair of the U.S. IGY Committee. Additionally, the University of Iowa had a special presentation for James Van Allen, who was the Explorer I lead project scientist and one of the original proposers of the International Geophysical Year, as well as His Royal Highness Prince Philip, Duke of Edinburgh (see Figure 2). The IGY Gold History Initiative proudly continues to award certificates and pins worldwide.

Fig. 1. *All members of the IGY Gold Club receive a special certificate (left) and IGY Gold commemorative pin (right). Shown is the award granted to James A. Van Allen.*

Fig. 2. *His Royal Highness Prince Philip, Duke of Edinburgh, receives his IGY Gold award from IHY team members Danielle Bewsher and Richard Stamper.*

6.3. Historical commemoration events

All IHY nations were invited to celebrate the 50th anniversary of IGY. Many events not only commemorated IGY, as many institutes, organizations and scientific activities were also celebrating their 50th anniversary.

The World Space Week and Yuri's Night World Space Parties (described in Sect. 5) are both IHY partner programs inspired by the dawn of the space age. These events were celebrated annually throughout IHY, and many individual IHY nations supported these programs with their own events.

The international science community was invited to join the "Joint CAWSES/ IHY Virtual Conference: The State of the Sun–Earth System During Extreme Space Weather", to "Return to the Auroral Oval for the Anniversary of the IGY". The conference was held online from November 13–17, 2006, and virtual sessions on heliophysical topics from the Sun to the ionosphere allowed scientists from around the world to participate. The spirit of the virtual conference was the advancement of the "next era" of scientific interaction, where researchers from around the globe can easily and efficiently collaborate via the internet. The meeting produced many interesting scientific results, and provided the basis for virtual workshops in the future.

The International Union of Geophysics and Geodesy meeting, in Perugia, Italy, July 2–13, 2007, was the site of a major "IGY + 50" celebration and public exhibition. The meeting also hosted several special sessions on the history of IGY and the space age, attracting historians of science from around the world.

Many nations hosted meetings with widespread international participation. Russian scientists played a major role in IGY, and IHY team members hosted two major commemorative meetings:

- International Conference: "50th Anniversary of International Geophysical Year (IGY) and Electronic Geophysical Year (eGY)", Suzdal, Russian Federation, September 16–19, 2007.
- "International IHY Symposium and Sputnik 50th Anniversary Celebration", Zvenigorod, Moscow, November 5–10, 2007.

Japanese scientists also celebrated many significant achievements, with "IGY + 50" sessions and exhibitions at the 2007 and 2008 Japan Geoscience Union Meeting. An international "Fifty Years after IGY – Modern Technology and Earth and Solar Sciences", was held November 10–13 in Ibaraki, Japan, and attracted scientists and historians from around the world.

In the United States, there were special sessions at many of the American Geophysical Union Meetings, along with history committee meetings, talks on IGY history, and even a special "drop-in" room for the IGY Gold Club.

The major celebration in the United States was the "Sputnik Golden Anniversary Gala" co-hosted by the Russian Embassy in Washington, DC. The event attracted many participants from around the world, including cosmonauts, government representatives, historians, scientists, and diplomats. The keynote

Fig. 3. *The "Sputnik Golden Anniversary Gala", co-sponsored by IHY the Russian Embassy in Washington, DC, was the primary Sputnik commemorative event in the United States.*

speaker was to be Valentina Tereshkova, who became the first woman in space in June 1963, but a brief illness forced her to cancel her visit (Figure 3).

Many national scientific organizations also celebrated their anniversary, as the IGY inspired the foundation of many scientific support structures, organizations and committees. This includes NASA, which celebrated its 50th anniversary in 2008. There were also an international conferences on the "50th Anniversary of the Founding of the Geophysical Observatory of Addis Ababa University", in Ethiopia, and the "Fifty Years of Romanian Astrophysics" conference in Bucharest, Romania.

7 IHY National and Regional Committee Reports

This section contains reports from all of the IHY geographical regions as well as reports filed by the IHY National Committees from many participating countries. They highlight advances in all aspects of IHY, and demonstrate how the goals of IHY were adapted to meet the needs and priorities of each individual region.

These reports are by no means comprehensive, as this publication is too brief for a thorough summary detailing the many years of effort in scores of nations. However, each report provides examples of how the efforts of IHY have been refined to suit the needs of each particular area, and how each nation played a significant role in this global initiative.

7.1. Africa

Participating nations: Algeria, Cameroon, Cape Verde, Cote D'Ivoire, Egypt, Ethiopia, Libya, Kenya, Morocco, Nigeria, South Africa, Tunisia, Zambia. The Democratic Republic of Congo, Ghana, Mozambique, Uganda, Malawi, Sudan, Tanzania are participating in IHY but do not have national committees to include in this report.

Many nations within Africa have organized IHY initiatives, and have participated in continent-wide activities taking place throughout IHY. The following are objectives of the IHY-Africa planning team. These objectives are clearly interwoven with the educational priorities with scientific priorities:

- Education and training in satellite and ground-based instrumentation
- Education and training in basic and universal physical processes and numerical modeling
- Topical lectures on advances in heliophysical phenomena
- Identify common priorities in space science research and education in participating countries that may lead to regional space science centers and annual space science schools at post graduate level
- Become a conduit for future generations of African space scientists and educators to gain professional experience
- Adoption and development of space science curricula based on the K12 curricula in the USA

- Develop a comprehensive space education strategy and contribute to the improvement of science education at all levels
- Conduct informal education to bring about public awareness of the field of space science
- Encourage the participation of women in space science education and research
- Stimulate quantity and quality of space research and education at undergraduate and graduate level at African universities and institutions
- Enable African scientists and institutions to build rewarding partnerships with the global space science community

African space scientists and institutions have participated in most of the IHY observatory development projects (see Sect. 4), and hosted workshops and schools centered around the science of the instrumentation.

IHY workshops were hosted in many nations, along with several international conferences and IHY schools. Sixty-three African scientists, representing 19 African countries, and 40 scientists from other nations met in Addis Ababa, Ethiopia for the IHY-Africa Space Weather Science and Education Workshop on 12–16 November 2007 (described in Sects. 4 and 5) to discuss space science research and education issues. The 6-day workshop focused primarily on observation of the ionosphere, the tenuous layer connecting Earth's atmosphere to the space environment. The attendees passed resolutions recognizing Africa's unique position along the equator, that makes it ideal for this type of observation, and a second resolution urging the improvement of the internet in Africa as a tool for scientific research. The workshop also had a special teacher training activity (see Sect. 5), and an additional SCINDA workshop (see Sect. 4). The final IHY-Africa workshop is to be held in Livingstone, Zambia in June 2009.

The following is a series of reports from nations all over Africa. They describe a wealth of activities designed to accomplish the scientific and educational objectives of the IHY-Africa planning team. It is clear that the dedicated effort of the national organizing committees and IHY scientists in Africa have resulted in many success stories.

7.1.1. Algeria

Submitted by Samir NAIT AMOR, IHY-Algeria National Coordinator, Centre de Recherche en Astronomie Astrophysique et Géophysique, Algiers

Astronomy and Astrophysics research is mainly done at the CRAAG (Centre de Recherche en Astronomy Astrophysique et Géophysique) in Algiers, previously known as the Bouzareah Observatory. 14 researchers and 22 engineers are working on solar physics, three of them are currently Ph.D. students at the Max Planck Institute in Germany. Related projects include Helioseismology, Solar activity, and Solar astrophysics. Additionally, some researchers are working on atmospheric turbulence for daytime measurements, which can connect to processes in space.

Astrophysics is also done at some universities inconjunction with the physics research teams. This includes nuclear and plasma astrophysics at Bab Ezzouar and Batna universities, General Relativity and astroparticle physics at Constantine University, Cosmic Ray physics at Annaba University. Space Research is done mainly at the CNTS (Centre National des Techniques Spatiales) at Arzew which is an engineering school and a research center in Geodesy, Teledetection and space techniques. It is also in charge of the conception and development of the Algerian microsatellite program (The AlSat series) in coordination with the ASAL (Agence Spatiale Algerienne) whose seat is at Algiers. The INCT (Institute National de Geodesie et de Teledetection) at Algiers is also a dedicated center for teledection and related activities. Some research activities in geodesy and teledetection are also carried out at the Earth Science departments of various Algerian universities.

7.1.1.1. IHY UNBSS instrument array report for the nation of Algeria

Introduction

After participating in the first IHY/UN Basic Space Science Workshop in 2005 in the United Arab Emirates, researchers from CRAAG made contact with Professor Umran Inan from Stanford University who agreed to establish one of the AWESOME Very Low Frequency (VLF) receiver experiments (see Sect. 4) in Algiers. The instrument was installed in August 2006, and has recorded data every day since then. Some of the data were recorded in conjunction with the EuroSprite 2005 and 2007 campaigns for atmospheric observations. Different results and analyses were presented in several major scientific meetings. Another IHY instrument, the AMBER magnetometer (see Sect. 4), will be hosted by the Department of Geophysics in CRAAG and will be installed at regional station in Medea (Algeria).

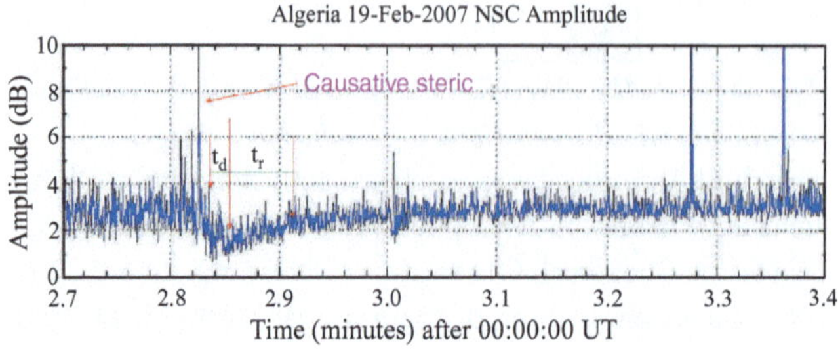

Fig. 1. *Data showing LEP perturbation observed by the VLF receiver in Algiers.*

IHY Research performed in Algeria

Lightning-induced electron precipitation events. The AWESOME VLF receiver's research topics are related to the earth's atmospheric perturbations due to local activity (lightning, sprites and elves, TGFs), solar events (solar wind, solar flares, solar radiation, CMEs) and extragalactic sources (gamma ray bursts). At the beginning of the data analysis our effort was focused on the search of Lightning-Induced Electron Precipitations (LEP) observed during thunderstorms. These events are observed in the narrowband signal amplitudes where their characteristic parameters were studied according to the transmitters' locations. In Figure 1 an example of LEP is presented with its characteristic parameters (onset times, duration time and the recovery time).

Two transmitter signals were used for this comparative study (DHO, and NSC), the results show that the number of events is more important in the NSC signal than DHO. There were 73 NSC events and 28 DHO events. The mean perturbation amplitude was greater in the DHO signal than NSC, while the mean time onset is smaller in NSC events than DHO. Additionally, the time duration is longer in DHO events than NSC, and the recovery time is greater in DHO events than NSC. These results were presented at IHY2007 workshop in Japan.

Solar flare effects on VLF propagating signals. Solar flare eruptions are also sources of ionospheric disturbances. Electromagnetic radiation (especially X-ray emission) during flare eruption penetrates down to the lower altitude region of the ionosphere (the D-region) and cause ionization of atmospheric atoms and modify the ionospheric conductivity which affect the signals of propagating VLF waves. During the period from 1 to 5 June 2007, a new sunspot (960) was a source of many flares of different classes ranging from B to M class. Figure 2 shows X-ray emissions associated to the solar flares occurred during this period as recorded by the GOES satellites.

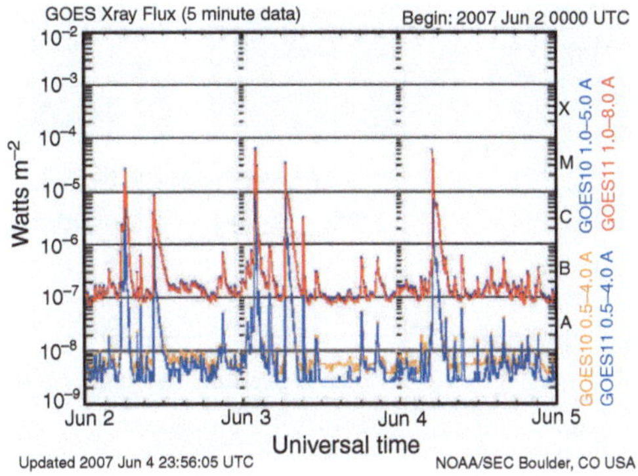

Fig. 2. *X-rays associated with solar flares for the 2 June 2007 data.*

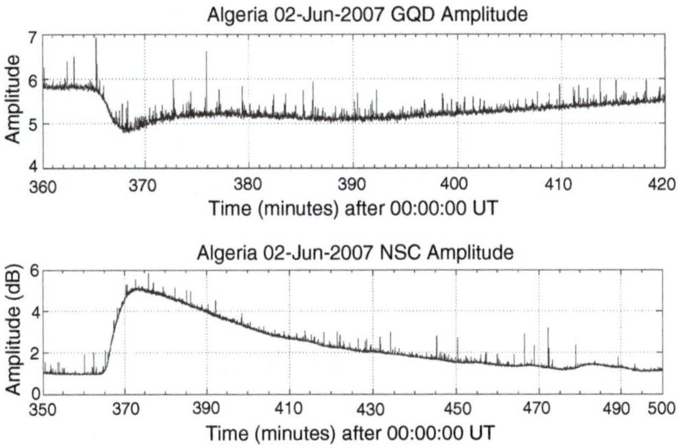

Fig. 3. *VLF signal perturbations due to solar flares.*

The VLF signal response to the flare eruption is characterized by a long duration time (few minutes) to the maximum perturbation amplitude followed by a longer recovery time (several minutes). Also, the perturbation is positive for some frequencies and negative (decreasing in signal amplitude) for GQD frequency, as shown in Figure 3.

VLF perturbation observed in conjunction with EuroSprite2007 campaign. Transient Luminous Events (TLEs) are also sources of perturbations in the VLF signals. They occur at approximately 80 km altitude, which is during the VLF reflexion high at night time. Wave signals which propagate near TLE events are backscattered by the TLE body and show an abrupt increase in signal

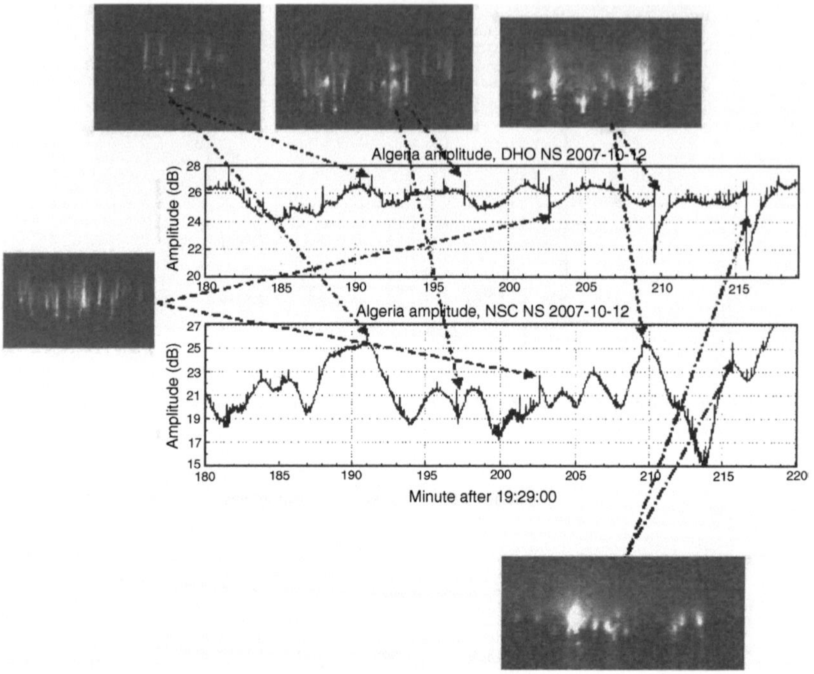

Fig. 4. *Sprite events associated with early events and LEP observed at NSC and DHO signals.*

amplitude ranging between 20 ms and 1 s; these early perturbations are well known.

Between July and December 2007 EuroSprite was launched to observe the TLE events occurring in central and southern Europe. We got a chance to participate with our instrument to this observation campaign and contribute to the scientific discussion on the TLE events and their association with early events observed with VLF receiver. In October 2007 many TLE events were observed and were associated with early and LEP events, as shown in Figure 4.

Unlike new observations made by Haldoupis *et al.* on the one-to-one association between TLE and early VLF perturbations where they concluded that early VLF events occur only when TLE are observed, many disturbances were recorded with our VLF receiver but no TLE events were observed by EuroSprite2007 cameras. The reason is due to the long distance between Haldoupis receiver and the thunderstorm locations which were near HWU transmitter in central France (~2000 km). So small early VLF perturbations can occur but cannot survive the long distance to distant receiver, especially when the thunderstorms occur far from the transmitter. In our case the thunderstorms were located in the Mediterranean

Sea (~600 km from Algiers receiver), and even the small perturbations which were due to lightning discharges were recorded.

Parts of these results were presented in IHY/SCINDA workshop is Addis Ababa, and will presented at AWESOME VLF workshop. A paper on the association between TLE and early VLF perturbations is in progress and will be sent to the Journal of Geophysical Research.

The scientific results presented here are proof that capacities exist in the developing countries and can do research and development. Activities like IHY are an important and useful way to support and help scientists in developing countries to launch new research areas in their own institutions.

7.1.2. Cameroon

Submitted by Emmanuel Guemene Dountio, IHY-Cameroon National Coordinator, Ministry of Scientific Research & Innovation, Energy Research Laboratory, Yaounde-Longkak

7.1.2.1. Report on IHY activities in Cameroon

The IHY program began in Cameroon since 2005, when the Office of Outer Space Affairs (OOSA) sent a letter of intention to invite the country to join the initiative. The letter was received by the Minister of Scientific Research and Innovation, who mandated that the Energy Research Laboratory in Yaounde-Longkak represent the country at those important events.

The purpose of the IHY program in Cameroon was to take part of the ambitious new IHY programs and evaluate the possibility to participate and contribute to the advancement of international scientific knowledge, concerning specially the heliophysical domain. The purpose was also to get in touch with the scientists of other countries and be aware of the state of art in that domain, as well as the new findings, and finally to put on the bases of an international collaboration, which could lead to the establishment of national research networks in space sciences.

The aim of this participation was to contribute in advancing the understanding of the heliophysical processes that govern the Sun, Earth and heliosphere, especially in connection with life on Earth. Following this extended plan, the IHY program has been very useful in many Cameroonian national and international initiatives. The after effects of the program will continue beyond the official end of IHY activities.

The IHY program promoted national collaboration between researchers, students, non-governmental organizations (NGOs), universities and research institutions. The activities focused on the sharing of data from ground-based measurements of atmospheric parameters, in particular, solar data and satellite images. Resources that were available in different institutions were shared, for the benefit of students, researchers, teachers and NGOs. This dissemination process involved the participation of the Institute for Geological and Mining Research (IRGM), the Energy Research Laboratory (LRE), the University of Yaounde 1 (UY1), the National Meteorological Centre, many local NGOs, and many Cameroonians who participated at the National Exhibition JERSIC2007 organized by the Ministry of Scientific Research and Innovation (MINRESI).

The IHY also promoted an international collaboration with external and international institutions and scientists, including universities, research institutions, researchers, teachers and engineers. Among those institutions, we can list the United Nations Office of Outer Space Affair (UN-OOSA), the North Carolina A&T State University, the University of California Los Angeles (UCLA), the University of Colorado, the Indian Institute of Astronomy, the Àir Force Research Laboratory (AFRL), the Physical Society of Ethiopia, the International Union of Astronomy and the whole IHY scientific family.

Under the IHY program, a Cameroonian researcher received from the OOSA mobility facilities, and funding support to participate and present his works at two international workshops. The first one, UN/NASA/ESA Workshop, was held in the United Arab Emirates in 2005. At this important workshop, many contacts were established. It was the beginning of collaboration with instrument providers for the expanding possibilities of measurement networks around the world. It also enabled us to get involved with the scientific state-of-the-art research in the heliophysical sciences and subjects.

The second one, the UN/NASA Workshop was held in 2006 at the Indian Institute of Astrophysics (IIA). This workshop gave us the opportunity to get in touch with other instrument providers, to strengthen the relationship established previously, to coordinate our effort, and to be aware of the new scientific developments in the domain of heliosphere. Many educational opportunities were also discovered there, with the effectiveness of IHY schools. Those opportunities were disseminated in Cameroon through national exhibitions, printed material distribution, and Yuri's Night World Space Party postings, with the help of both the UN-OOSA and the IIA which enabled us to develop printed materials to support the effort. Unfortunately, the IHY schools did not get the expected participation of Cameroonians students, because of the lack of astronomical courses in Cameroonians universities.

Under the auspices of the IHY program, two Cameroonian scientists also received funding support from the SCINDA2007 and the IHY-Africa 2007 workshop cosponsors, to participate and present their works at the mentioned workshops, held in Addis Ababa, Ethiopia. At those workshops, the final arrangements were found on the deployment of ground-based instruments, and the development of research infrastructural capabilities, such as internet availability and the recognition of important space science in developing countries. Following those resolutions, and the opportunity given by the International Astronomical Union, a discussion is ongoing to join the IAU and develop astronomical courses in Cameroon. The internet facilities have been progressively improving in the country and research institutions.

Two ground-based instruments were installed in the country under the IHY program. The SCINDA system has been installed at the Yaounde University as well as the AMBER system. The first one is a GPS system and the second one is a magnetometer (see Sect. 4). Other GPS systems are under negotiation.

As a response to these important initiatives, the government, on the base of a national funding, is onstructing ten solar data measurement stations, to signify its interest in space science and climate changes studies. The construction of a national observatory of climate changes has been announced by the Head of State, and studies for its construction have already begin. Even if this last initiative is not directly linked to IHY program, it can be seen as a proof of the government interest in such research. It can also reorient and develop new research opportunities. Therefore, if there is an experiment or plan for a ground based instrument or observatory that need new observing locations in Cameroon, local institutions are eager to negotiate and provide those locations.

7.1.3. Egypt

The following organizations are involved with IHY and Basic Space Science in Egypt:

(1) Research and postgraduate studies are carried out at the National Research Institute of Astronomy and Geophysics (in Helwan). Kottamia Observatory operates a 2-m optical telescope with photometers, a spectrograph and a charge-coupled device camera. The observatory was built in 1963. This telescope is the largest optical/infrared telescope in North Africa and the Middle East. The Carl Zeiss Company is involved in modernizing the optical system of this telescope. The Institute plans to build a radio telescope at Abu Simbel in the south of Egypt as part of the European Very Long Baseline Interferometry (VLBI) Network, or EVN, to bridge the gap between the radio telescope in Western Europe and the radio telescope at Hartebeesthoek in South Africa.

(2) Egyptian scientists have hosted several international meetings and workshops during IHY, including a World Space Environment Forum in October 2007, an International Astronomical Union Regional meeting in April 2008, and an International Association for Geomagnetism and Aeronomy meeting in October 2008.

(3) The teaching of and training in Basic Space Science and solar physics are provided in a few universities, particularly at Cairo University, which has a good astronomy department and about 30 astronomers and solar physicists. It offers two Bachelor of Science degrees in Physics/Astronomy (interdisciplinary) and Astronomy (single discipline), in addition to Master of Science and doctoral degrees in solar physics, astronomy and space science. The department of astronomy at Cairo University contributes to different fields of solar physics & BSS, such as astrophysics, astro-chemistry, celestial mechanics, solar physics, and cosmology and space physics. In schools, astronomy is a part of the general science course at the primary and preparatory levels and a part of the physics course at the secondary level.

7.1.4. Ethiopia

The most significant IHY activities transpiring in Ethiopia was the hosting of three events:

- A special Geophysical Information for Teachers (GIFT) workshop held on 10 November 2007 in Addis Ababa. Teachers from 70 schools from all around Ethiopia participated in this intensive workshop and professional development program (a more complete description can be found in Sect. 5).
- The 2007 SCINDA Workshop, where scientists participating in this international research and observation program could gather and discuss science analysis and results, the operation of the SCINDA array of instruments (see Sect. 4), and the establishment of future SCINDA sites.
- The IHY-Africa Space Weather Science and Education Workshop, held 12–16 November in Addis Ababa.

A special IHY-Africa focus group identified a need for a workshop to support African research activities and to further develop research partnerships in conjunction with the instrumentation program. In response to this need, more than one hundred international scientists met in Addis Ababa for a special IHY workshop focused on Space Weather Science and Education. The meeting was the culmination of more than 2 years of planning, spurred on by the IHY-United

Nations Basic Space Science (UNBSS) initiative to promote African space weather science and further deployment of instrument arrays in developing countries.

The IHY-Africa workshop had two overall objectives:

- To study space weather science at mid and low latitudes in the African longitude sector
- To support African space science and education

The science focus was built around three major themes:

1. Ionospheric Irregularities
2. Ionospheric Total Electron Content
3. Equatorial Electrodynamics and Plasmasphere/Ionosphere Coupling

Many individual scientists participated in other IHY programs, such as campaigns and CIPs. In addition, and active undergraduate physics club at the University of Addis Ababa held several events for the public promotion of heliophysics.

7.1.5. Kenya

Submitted by Maurice Odondi K'Orowe, IHY-Kenya National Coordinator, Jomo Kenyatta University of Agriculture and Technology, Nairobi

The Kenya Chapter of IHY is quite young. However, since the time we joined the IHY community, in 2006 or thereabouts, we have actively participated in most of the activities.

7.1.5.1. Science cooperation

The workshops attended by Kenyans are namely:

1. 2nd UN/NASA/ESA Workshop on the IHY and BSS, Bangalore, India (27 November–1 December 2006) Jared Ndeda of Jomo Kenyatta University of Agriculture and Technology presented a paper
2. 3rd UN/ESA/NASA/JAXA Workshop on IHY and BSS. Tokyo, Japan (18–22 June 2007)
3. 2nd SCINDA Workshop and IHY-Africa Space Weather Science and Education Workshop, Addis Ababa, Ethiopia (11–16 November 2007). Four Kenyans participated and presented papers.

7.1.5.2. Equipment

Two sites were established for the SID VLF experiments (see Sect. 4). One was given to the University of Nairobi Physics Department and the second one to the Jomo Kenyatta University of Agriculture and Technology (JKUAT). MAGDAS and SCINDA are being hosted at the University of Nairobi, and SCINDA equipment to JKUAT is still pending.

7.1.5.3. Training

One candidate for a Ph.D., Jared Ndeda is about to submit his Thesis on "Solar radiative forcing of climate change on seasonal to decadal scale" in Kenya. He has immensely benefitted from the IHY program. Two more students are also at different levels in their Post-graduate work.

7.1.5.4. Outreach

The JKUAT Physics Department is participating in IHY outreach activities, and plans are on the way to initiate basic undergraduate physics projects using the equipment we already have. We hope to continue as active members in the next phase of IHY.

7.1.6. Libya

Prepared by Joseph M. Davila, IHY Executive Director, NASA Goddard Space Flight Center, Greenbelt, Maryland

The Libyan Space Centre started building a 2-m robotic optical telescope in 2005. Observations and research are being carried out for choosing a suitable site out of three being studied. A training program for astronomers and technicians is also in process, in addition to the telescope manufacturing process.

With the cooperation of the U.S. Department of State, Dr. Joe Davila led a NASA team to Libya to observe the 2006 total solar eclipse from the Libyan desert. He was accompanied by several scientists, and in addition NASA provided a film crew to document the trip. The science team gave lectures on the significance of Space Weather at several institutions in Libya including the Remote Sensing Institute, and Al Fatah University in Tripoli and at Sebha University in Sebha.

Fig. 5. *The Libyan and US science teams at the Waw Namous observing site.*

Fig. 6. *Dr. Davila sets up the eclipse experiment in the Libyan desert.*

Observations were performed near Waw Namous, in southern Libya. Transportation was provided by the Libyan military, and a desert camp was provided by the Libyan government. The trip resulted in the formation of the first scientific collaboration between U.S. and Libyan scientists for research purposes since at least the early 1980s. This collaboration continues today with a VLF radio receiver operating at Sebha University by several faculty members, and additional planned installations at Tripoli and in Benghazi (Figures 5 and 6).

A full 1-h documentary of the trip was prepared and thousands of copies were distributed around the world, and although the narration was translated into Arabic, the English version is used in Libyan schools to help with the teaching of English as a foreign language.

7.1.7. Nigeria

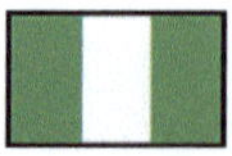

Submitted by A. Babatunde Rabiu, IHY-Nigeria National Coordinator, Space Physics Laboratory, Department of Physics, Federal University of Technology, Akure
http://ihy2007.org/ihy_nigeria.shtml

7.1.7.1. Report on the activities of International Heliophysical Year IHY in Nigeria

Preamble

Nigeria joined the International Heliophysical Year project as a nation during the preparatory stage in 2004 when Dr. A. Babatunde Rabiu became the National Coordinator of IHY in Nigeria. Professor Ekundayo E. Balogun, who was at the time the Director of the African Regional Center for Space Science and Technology for English speaking countries (ARCSSTEE), chaired the National Organizing Committee of IHY in Nigeria. The co-chair was Professor P. N. Okeke of the Center for Basic Space Science in Nigeria. The IHY-Nigeria group championed the course of IHY in Nigeria. We worked in collaboration with the International Coordinating team as well as the African regional team and ensured the achievement of the objectives of IHY in Nigeria and Africa as a whole.

Conferences and workshops

A total of three National Workshops were held as follows:

i. 1st National Workshop on IHY, Federal University of Technology, Akure, FUTA, Nigeria. Attendance: 33
ii. 2nd National Workshop on IHY, Tai Solarin University of Education TASUED, Ijagun, Ijebu Ode, Nigeria, 19–21 July 2006. Attendance: 75
iii. 3rd National Workshop on IHY, the Centre for Basic Space Science CBSS, National Space Research and Development Agency (NASRDA), University of Nigeria, Nsukka. Attendance: 35

Conferences in Nigeria where sessions on IHY were held are:

i. Twenty-eightth Annual Conference of the Nigerian Institute of Physics (NIP)/World Year of Physics. August 17–20, 2005, Obafemi Awolowo University, Ile-ife, Nigeria.
ii. Twenty-ninth Annual Conference of the Nigerian Institute of Physics (NIP). August 16–19, 2006, University of Nigeria, Nsukka, Nigeria.
iii. First National Conference of the Nigerian Astronomical Society (NAS), August 5–7, 2007. Centre for Basic Space Science (CBSS), Nsukka, Nigeria.
iv. Thirtieth Annual Conference of the Nigerian Institute of Physics (NIP). August 15–18, 2007, Lagos State University, Ojo, Lagos, Nigeria.
v. 2007 African Regional Conference of the International Academy of Astronautics; December 3–5, 2007, Abuja, Nigeria.

Nigerian delegates also presented papers relevant to IHY activities in at least 12 international meetings.

IHY/UNBSS instrument array

The following are the IHY facilities functional in Nigeria (see Sect. 4 for further descriptions of instrumentation):

i. MAGDAS: a real-time magnetometer; one unit is located along the dip equator at the University of Ilorin, while another is located at the edge of the equatorial electrojet at the Redeemer's University, Lagos, Nigeria.
ii. SCINDA: a real-time, data driven, communication outage forecast and alert system. SCINDA is located at the Federal University of Technology, Akure.
iii. IGS GPS: scintillation detection system acquired in 2007 and operational at University of Lagos.
iv. AWESOME: Atmospheric Weather Electromagnetic System for Observation Modeling and Education (AWESOME) is a Very Low Frequency (VLF)

receiver capable of monitoring radio waves from 24 transmitters distributed around the globe. AWESOME is located at the Federal University of Technology, Akure.

v. SID MONITOR: about five units of Sudden Ionospheric Disturbance SID monitors were deployed to Nigeria in 2007. They are located at Centre for Basic Space Science, Nsukka; Tai Solarin University of Education, Ijebu Ode; Adekunle Ajasin Univeristy, Akungba Akoko; University of Ibadan; and Redeemer's University, Redemption City.

vi. ASTRONOMICAL TELESCOPE: two astronomical telescopes obtained from National Astronomical Observatory of Japan NAOJ are functional at the Centre for Basic Space Science, Nsukka.

vii. The Nigerian Environmental and Climatic Observing Program (NECOP): at the national level, Center for Basic Space Science CBSS, under the Directorship of Prof. P. N. Okeke set up a network of real-time observations for meteorological variables across the nation. NECOP presently has seven observation sites.

Figures 7 and 8 are relevant to the installation of some of the above listed facilities.

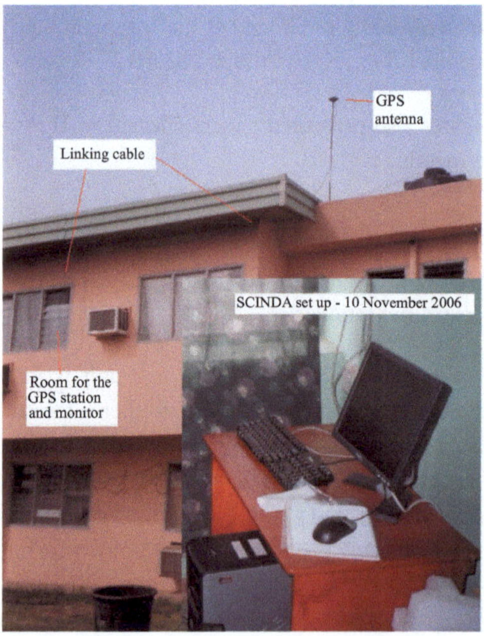

Fig. 7. *SCINDA GPS facility at Akure.*

Fig. 8. *Installation of the MAGDAS Instrument at Ilorin, Nigeria.*

Outreach

Public lectures: a number of public lectures on IHY and its activities were delivered at many secondary (high) schools and universities across the country. Some of the institutions included the Model Secondary School, Akure; the Federal University of Technology, Akure, FUTA; the Tai Solarin University of Education TASUED, Ijebu Ode; the University of Ilorin, UNILORIN; and Redeemer's University, among others (Figures 9 and 10).

Fig. 9. *Prof. K. Yumoto (the Principle Investigator of the MAGDAS project) on Nigerian Television in August 2006.*

Fig. 10. *Prof. K. Yumoto Delivering a public Lecture on IHY at Ilorin, Nigeria, August 2006.*

Total Solar Eclipse of 29 March 2006, Outreach: on Wednesday, 29 March 2006, a total eclipse of the Sun was visible from within a narrow corridor which traversed accessible part of Nigeria. The solar eclipse was widely publicized and viewed in Nigeria. Two major viewing centers were coordinated by IHY volunteers within the country. The IHY International Secretariat of and the Education and Public Outreach office of IHY made available a number of eclipse kits consisting of educational posters, eclipse viewing glasses, eclipse literature and IHY literature at no cost for public outreach. More than 100 solar viewing glasses were also made available. The IHY volunteers organized various outreach activities centering on public enlightenment through various media. A number of high schools were visited and public seminars were held at various tertiary institutions. The educational posters were massively used in public enlightenment campaign in preparation for the eclipse. The Federal Ministry of Information paraded a lot of jingles all gearing at demystifying solar eclipses and at encouraging people to go about their "normal" activities. The climax of the IHY – inspired public enlightenment was the public lectures organized by the University of Ilorin, Nigeria on 28 March 2006, which was transmitted live through the state television services. There was a demonstration on how to use the solar viewing glasses at the lecture and a number of them were distributed to people.

Special viewing centers were created at the major towns where the total eclipse path was expected, including Saki and Igboho, the IHY team at Igboho did a video recording of the eclipse proceedings. The town of Saki was invaded by unprecedented thousands of viewers, which included tourists, scientists and school children who came to view the scientific event. The viewing centre OYSADEP Stadium was turned into a carnival ground. The moon began to cast its shadow on the Sun at about 8:20 UT, this marks the onset of the partial eclipse at Saki town. By 9:10 UT, the big round Sun had been reduced to a mere crescent. At exactly

9:20 UT there was a total eclipse which brought about total darkness in a broad day light. That was a great experience. This lasted for about 3.5 min. The experience at Igboho was quite similar. Within the short duration, darkness fell like a thick blanket over the ancient town of Saki and its environs, as it is in the night. By 10:45 UT the moon had retreated away from the Sun, and the moon could no longer be seen. Figures 11 and 12 show participants and leaders of the IHY eclipse viewing station in Saki.

Other States in Nigeria which experienced the partial eclipse include Lagos, Abuja, Ogun, Oyo, Kogi, Edo, Niger, Katsina, Kano, Kaduna, Osun, Kwara,

Fig. 11. *Waiting for the Eclipse at Saki (Inset is the Total Eclipse as seen at Saki).*

Fig. 12. *IHY volunteers with Model secondary school students viewing the eclipse at Saki, Nigeria.*

Lagos, and Ondo. More information on the 2006 total solar eclipse activities and IHY viewing stations can be found in Sect. 5.

IHY African Regional School in Nigeria: Nigeria is hosting the first IHY African Regional School scheduled for November 10–22, 2008 at the Centre for Basic Space Science, National Space Research and Development Agency (NASRDA), University of Nigeria in Nsukka (see Sect. 5).

Nigerian Union of Radio Science: Nigerian Union of Radio Science (NURS) – the national committee of International Union of Radio Science URSI – was reactivated in March 2008 by a group of scientists as a spinoff of IHY activity in the country.

IGY Gold history activities

The following people were received IHY Gold award (see Sect. 6 for a description of this program):

1. Professor Ekundayo E. Balogun
2. Professor Cyril Agodi Onwumechili
3. Professor Mrs Ebun Oluwafunmilayo Oni

Professor Balogun worked at a meteorological station located in Minna, central Nigeria during the IGY. He used to take routine measurements and to be submitted to the officer in-charge of the station on daily basis for onward transmission to UK.

Professors Onwumechili and Oni were with the Department of Physics, University of Ibadan where ionospheric Physics was pioneered during IGY. The three, though still alive, have retired from active service. They have made tremendous contributions to our science.

Gains of IHY in Nigeria

In addition to the set objectives of IHY, the following have been identified as gains of IHY in Nigeria:

- Knowledge & technological transfer.
- Positive collaboration.
- Availability of Research facilities for internationally competitive research.
- Windows of postgraduate opportunities.
- Control of brain drain (i.e. retention of talented students).
- Development of Research in Basic Space Science.
- Capacity building.

7.1.8. South Africa

Submitted by Prof. Marius S. Potgieter, IHY-South Africa National Coordinator, Unit for Space Physics, North-West University, Potchefstroom and Dr. Lee-Anne McKinnell, IHY-South Africa Outreach Coordinator, Hermanus Magnetic Observatory, Hermanus

7.1.8.1. Report on the International Heliophysical Year for South Africa

The activities for the International Heliophysical Year 2007 (IHY) and the International Polar Year (IPY) in South Africa were linked and they were in most instances planned together under the auspices of the South African funding agency, the National Research Foundation. The SA National IPY & IHY Committee was established in 2005 and had three meetings, in August 2005, January 2006, and June 2007. Prof. Dave Walker, a space physicist from the University of Kwazulu Natal (UKZN), was the chairperson of this NRF-sponsored committee. As a recommendation of the mentioned national committee, a special IPY & IHY regional workshop was held in Cape Town from 24–27 October 2006. It was fully sponsored by the NRF and the ICSU Africa Regional Office. Several stakeholders from the rest of Africa were invited, in particular the IHY coordinators. Figure 13 is the poster announcement of the IPY-IHY Regional Workshop that was held in Cape Town in October 2006. Other key events include:

- A special African Skies edition, edited by Dr. Peter Martinez, on IHY projects was published in November 2005 (see Appendix III).
- The Hermanus Magnetic Observatory (HMO) in South Africa became officially a national facility of the NRF, dedicated to Space Physics. A stakeholder workshop involving all space physicists in South Africa was held at the HMO on 24–25 April 2006. A session was devoted to IHY and IPY activities in South Africa.
- Sessions on IHY-related research were held during the annual conferences of the South African Institute of Physics (www.saip.org.za) in July 2006 and July 2007. Researchers from Africa could apply to the NRF for travel grants to attend these conferences.
- A list of all IHY related research projects in South Africa was compiled. These projects were submitted to the SA National IPY & IHY Committee for recognition as formal IHY 2007 projects. Unfortunately, despite a peer review

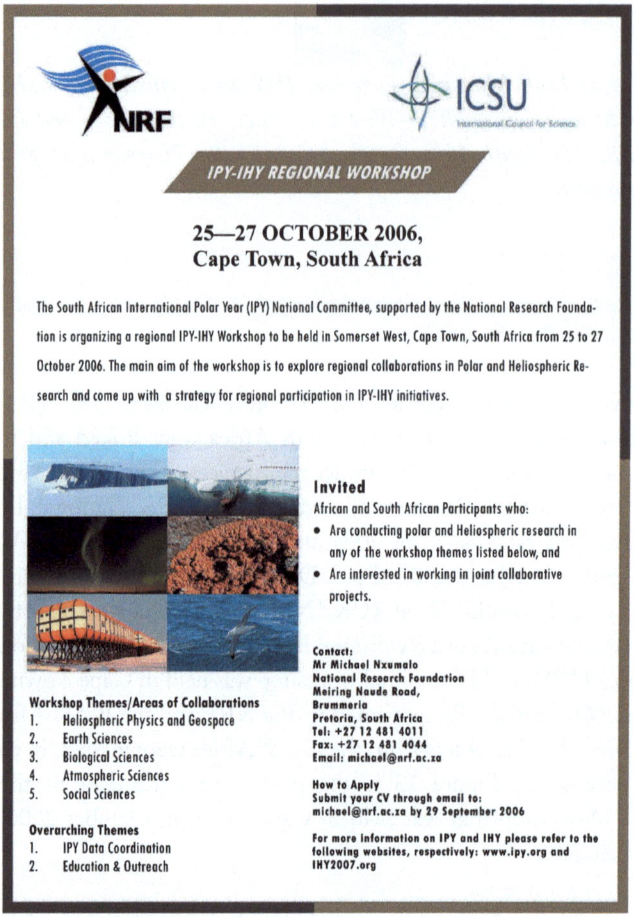

Fig. 13. *The poster announcement of the IPY-IHY Regional Workshop that was held in Cape Town in 2006.*

process and endorsement from the international heliophysical officers, none of these IHY projects eventually got funded.

- Prof. Pieter Stoker, a space physics pioneer in South Africa was one of the South Africans receiving a member certificate of the International Geophysical Year (IGY) Gold Club. The certificate and lapel pin were presented to him on his 80th birthday in November 2006. Figure 14 shows an envelope issued during the Geophysical Year in 1957–1958, postmarked on Marion Island, one of South African Antarctic Islands with a research base.
- A South African Space Agency Stakeholder Workshop was held at the Institute of Satellite and Software Systems in Grabouw in December 2006 and in November 2007.

Fig. 14. *A postmarked envelope issued during the Geophysical Year in 1957–1958.*

- During 2007 the summer and winter schools, entitled "Introduction to Space Physics" and run by the HMO, included an IHY colloquium. The colloquium was given by guest speakers with the topics being IHY related and was open to (and widely supported by) the public. Students attending the school as well as permanently based HMO students were made aware of IHY through the colloquium as well as lectures at the schools, and informal discussions with researchers.
- In October 2007, HMO held an open day where the public, learners and students could visit the HMO and learn about the different aspects of Space Physics. Tours of the facility were on offer as well as scheduled presentations and classroom activities.
- The Science Centre at the Potchefstroom Campus of the North-West University hosted a display of eight IHY related posters, focusing on Space Physics, during 2007.
- South African space scientists and institutions participated in the following IHY projects: AWESOME, the VLF Receivers and Educational System for Observation and Modeling of Space Weather Effects; MAGDAS, the Japanese Magnetic Data Acquisition System; and SCINDA, the Scintillation Network Decision Aid – Ionospheric Scintillation (see Sect. 4 for a description of these programs).
- The 4th Annual AFRICA ARRAY Workshop was held from 17 to 20 June 2008 at the University of Witwatersrand in Johannesburg. Members of the staff of the HMO are participating in the distribution of experiments in space physics all over Africa.
- A formal collaboration between South African space physicists and physics staff at the University of Zambia in Lusaka has been introduced. One result of this

collaboration is the final IHY Africa Workshop to be held in Livingstone, Zambia in June 2009.

7.1.9. Tunisia

Submitted by Hassen Ghalila, IHY-Tunisia National Coordinator, Laboratoire LSAMA, Départment de Physique, Université de Tunis El Manar I, Tunis

7.1.9.1. 2005 annular eclipse

In April 2005 the IHY Executive Director Joseph Davila, accompanied by astrophysicist Dr. Mehdi Benna (a Tunisian native) was invited to Tunisia for a series of lectures and to participate in the observation of an annular eclipse on 3 October 2005. Dr. Davila presented a series of lectures at the Museum of Science in Tunis, and at the American Corner discussing the science of stereo observations from the STEREO spacecraft. The trip resulted in many newspaper stories, several radio interviews, and an appearance by Dr. Benna on Al Jazeera television (Figure 15).

A 30-min documentary was filmed by a Tunisian company for Tunisian television. This documentary was intended for teenage viewers in the Arab world, and the narration was provided in English, French, and Arabic.

Also on this trip, Dr. Davila was able to facilitate the collaboration between Stanford University and Hassen Ghalila, University of Tunis (see discussion below) for the installation and operation of an AWESOME low-frequency radio receiver.

Fig. 15. *Group photo of Junes Science group and Dr. Davila taken at the annular eclipse observing site.*

7.1.9.2. Report of the IHY AWESOME observatory in Tunisia

This report is the first one about the Extremely Low Frequency/Very Low Frequency (ELF/VLF) activity in the LSAMA Laboratory (Laboratoire de Spectroscopie Atomique, Moléculaire et Applications). This field of research begun in October 2005 with the acquisition of a new equipment from Stanford University to contribute to the collection of data through the records of the ELF/VLF activity in the Earth's atmosphere. This work forms part of an international wide-area remote sensing network known as the AWESOME project involving many countries all over the world. See Sect. 4 for a more complete description of the AWESOME program.

The local VLF team in LSAMA consists of the following: Nadra Tounsi, Master student at LSAMA; Ines Bensaïd, Master student at ENIT; Rym Amara, Professor at ENIT; Atef Khadhraoui, Master student at ENIT; Omar Chebly, engineering student at Jeune Sciences (Tunis); and Hassen Ghalila, Professor at LSAMA.

The principal topic of the LSAMA laboratory is theoretical and experimental spectroscopy. The theoretical calculations consist in ab initio investigations of the electronic structure and the spectroscopy of the molecular systems. The experimental researches use Laser Induced Plasma Spectroscopy (LIPS) technique or Laser Induced Spectroscopy (LIF) or Cavity Ring Down Spectroscopy (CRDS) for the detection of pollutants at the state of trace. The main applications of these experimental studies are biology and environmental issues. Most of the colleagues work on ab initio calculations and a smaller part works on experiments.

Many experimental studies in spectroscopy involve plasma states, like LIPS technique and more generally plasma gas used to study the molecular spectroscopy, and these skills establish the link with ionospheric plasmas. Indeed, the ELF/VLF range is far from the visible spectrum and spectroscopic issues but the sources of these ELF/VLF are plasmas. Thunderstorms, sprites, and Lightning-Induced Electron Precipitation (LEP) all act like gas discharges.

The equipment consists of two antennas, GPS, amplifier, receiver and computer. The LSAMA laboratory is located at the Department of Physics of the University of Tunis El Manar, in Tunis. The antennas are installed on the roof of the department (longitude 10.08, latitude 36.5) as shown in Figure 16. The receiver and computer are installed on the flat just below the roof.

One research focus has been the identification and analysis of "tweeks." Tweeks are the manifestations of the dispersion of the waveguide's mode near the cutoff frequency. Their name is due to the sound which they produce. They appear in the spectrogram (Figure 17) as vertical lines which end by hooks. We show here, through successive zooms, the difficulty to observe them and how it is easy to pass close without seeing them. Analysis of tweeks leads to several information such as

Fig. 16. *Left Panel: University of Tunis El Manar: The antenna site is indicated with the black cross-ring. Right panel: Omar Chelby of LSAMA with the antennas.*

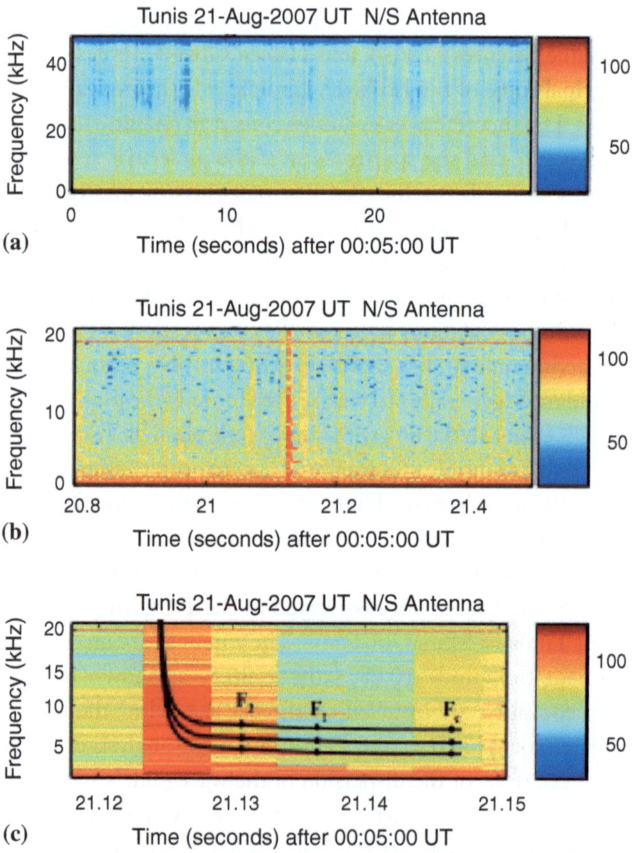

Fig. 17. *(a) Spectrogram showing several spherics, among them some have tweeks structure. (b) Zoom in the region (20.8, 21.5) sec of the spectrogram above. (c) Zoom in (21.12, 21.15) sec to be able to estimate the dispersion.*

the ionosphere height, cutoff frequencies, group velocity and also distance and location of the cause of the tweeks.

7.1.10. Zambia

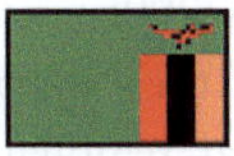

The IHY-Zambia program was headed by Dr. Geoffrey Munyeme, Dr. Adrian Habanyama and Dr. Habatwa Mweene from the University of Zambia in Lusaka. The University of Zambia has been active in collaboration with other African scientists, and results have been presented at several IHY workshops.

The upcoming IHY-Africa workshop will be hosted by IHY-Zambia. This workshop is a continuance of the tremendously successful IHY Space Weather Science and Education workshop held in Addis Ababa, Ethiopia in November, 2007. The upcoiming Zambia workshop is aimed at scientists and students from African countries, and those international scientists interested in working together with African scientists, who are involved in all aspects of Space Physics, and will include at least four main Heliophysics science sessions:

(1) Total electron content,
(2) Ionospheric irregularities and scintillation,
(3) Sun–heliosphere connection, and
(4) Magnetosphere–ionosphere coupling.

In addition to these four main science sessions, the meeting will include a Space Science Education and Public Outreach (EPO) section. Moreover, the meeting will be expanded and include panel discussions on the collaborative common database over the continent and about the infrastructure and communications in Africa.

The Scintillation Network Decision Aid (SCINDA) 2009 meeting will take part in conjunction with the IHY-Africa 2009 meeting, and sessions concerning the SCINDA participants will take place on the Sunday (7 June 2009) and Monday (8 June 2009) preceding the IHY-Africa workshop.

7.2. Eastern Asia and the Pacific

Participating Nations: Australia, China, India, Indonesia, Japan, Nepal, Malaysia, Mongolia, Philippines, Singapore, South Korea, Taiwan Province of China, Thailand, Viet Nam

The Asia-Pacific region began activities in 2004, starting with a planning meeting during 10–12 July in India. This was followed by another meeting in

Kiyosato, Japan held 26–29 October, 2004. The Asia-Pacific region consists of a large number of countries, with primary activities concentrated in Australia, China, India, Japan and Thailand. The Asia and Oceanic Geophysical Society (AOGS) meeting in Singapore during July 2005 was the first instance where many countries came together and presented their activities and laid the plans for the future. Further planning took place at regional and international IHY workshops.

In addition to the CIP activities, a large number of countries from Southeast Asia and Southern Asia are participating in the U.N. Basic Space Science instrument deployment program. The IHY/UNBSS workshops had participation from nearly all of the nations in the region. Many of these countries participated in the UNBSS instrumentation program. Additionally, educational innovators from the Asia-Pacific region are provided leadership for a number of IHY Education and Public Outreach (EPO) activities, including the IHY Asia-Pacific School held 20 November–1 December, 2008.

The following reports briefly summarize activities happening in individual nations in this region.

7.2.1. Australia

Submitted by Brian Fraser, IHY-Australia National Coordinator, Department of Physics, University of Newcastle, Newcastle

7.2.1.1. International Heliophysical Year activities in Australia 2007–2008

Introduction

Heliophysical research in Australia is organized around existing research groups which study various regions of our space environment, from the surface of the Earth to the Sun. The National Committee for Space Science under the Australian Academy of Science is the focus for international activities, including the IHY. However, in the absence of a dedicated IHY national program results of activities have been reported at appropriate national and international forums where on some occasions dedicated IHY sessions have been organized.

IHY related research in Australia

The majority of IHY related activities in Australia are undertaken by the following research groups. This combination of groups covers IHY research from the atmosphere, the ionosphere, the magnetosphere and Sun–Earth connection.

University of Adelaide: studies of the mesosphere lower thermosphere (MLT) include determination of mean winds, planetary waves, tides, gravity waves and turbulence. Other studies relate to electron densities, meteors and momentum fluxes. Instrumentation includes radar, lidar and passive optical techniques. To investigate spatial variability additional radars are operated at Katherine and Darwin (Northern Territory), Christmas Island, Pontianak (Indonesia – with University of Kyoto and LAPAN) and Davis (Antarctica – with Australian Antarctic Division).

La Trobe University: this group, located in Melbourne conducts research related to the behaviour of the thermosphere and mesosphere, ionosphere and magnetosphere. It also studies the properties of radio wave propagation, where the results are of practical importance in communications and surveillance systems. Research activities include leadership in the Tasman International Geospace Environment Radar (TIGER) consortium which operates two HF SuperDARN radars, one located at Bruny Island, Tasmania, and the other (Unwin) located near Invercargill, New Zealand. The radars look south towards Antarctica and have overlapping beams in order to map ionospheric motions by observing ionospheric scatter. These radars can also be used to study mesospheric winds through meteor observations, and sea-state from backscatter echoes. La Trobe also operates a digital ionosonde near Melbourne and Fabry-Perot Spectrometers at Davis and Mawson (Antarctica), and near Melbourne.

University of Newcastle: the Centre for Space Physics' research interests are primarily concerned with plasma waves, current systems, and boundary regions in the Earth's magnetosphere and ionosphere. These include particular studies of ultra-low frequency (ULF) hydromagnetic and ion cyclotron waves in the magnetosphere and ionosphere using observations from satellites, the TIGER radars, and ground magnetometer arrays located at low and high latitudes. Also of interest are the transmission of ULF waves through the ionosphere, and the use of ULF waves as diagnostic probes to monitor the state of the magnetospheric plasma. High latitude studies of magnetosphere–ionosphere coupling are undertaken using magnetic data from the Iriduim low altitude orbiting satellite constellation of over 60 spacecraft and ground magnetometers, HF SuperDARN radars and standard and imaging riometers.

University of Sydney: research in the Centre for Wave Science is focusing on the detailed plasma physics of the growth of plasma waves and radiation, particle acceleration at shocks and in reconnection regions, solar system radar emissions associated with shocks, and the effects of dust and generation of complexity in plasmas. Applications include solar and interplanetary radio bursts, solar and magnetospheric magnetic reconnection regions and space weather. Interests of a second team within the Centre include the statistics of solar X-ray bursts, development of magnetic structures and magnetic reconnection.

Ionospheric Prediction Service: the Ionosphere Prediction, Radio and Space Services (IPRSS), a unit of the Australian Government Bureau of Meteorology, includes the World Data Centre for Solar Terrestrial Science. IPRSS operates a ground-based network of 17 ionospheric monitoring sites located in Australia, the Pacific region and Antarctica to provide near real time and archival data. Magnetometers at five sites in Australia provide local K-indices and a "Pc3 index", a measure of geomagnetic field variability due to ULF waves. Solar radio emissions are recorded at the IPRSS Culgoora Solar Observatory, near Narrabri, western NSW. The Learmonth Solar Observatory provides continuous H-alpha monitoring and radio observations from a wide frequency range spectrograph. IPRSS, in cooperation with DSTO collects oblique propagation data, including polar paths and scintillation and TEC observations at three sites (low, mid and high latitude). The WDC for Solar–Terrestrial Science archives ionogram and magnetic data along with TIGER and Australian Antarctic data, and four (all-sky) riometer data sets.

Australian Antarctic Division: a division with the Australian Government Department of the Environment, the AAD operates four scientific stations in Antarctica. These include Casey, Davis, Mawson, and Macquarie Island, a sub-Antarctic station. Each station operates magnetometers, riometers, ionosondes and photometers and some have radars and other experiments operated in conjunction with Adelaide, La Trobe and Newcastle universities and IPRSS Radio and Space Services. A Lidar is operational at Davis and a cosmic ray monitor at Mawson.

IHY-Associated International Programs

The Australian research groups noted above are variously participating in the following parallel programs of interest to the IHY. Interhemispheric Conjugacy Effects in Solar Terrestrial and Aeronomy Research (ICESTAR) is a SCAR-IPY program (see Appendix V). Institutes involved included University of Newcastle, Australian Antarctic division, IPRSS.

Ultra Large Terrestrial International Magnetic Array (ULTIMA) is a world wide magnetometer consortium (see Sect. 4) including the Australian Antarctic Magnetometer network. Institutes involved include University of Newcastle, Australian Antarctic Division.

Super Dual Auroral Radar Network (SuperDARN) is worldwide network of identical radars operating in the high latitudes and polar regions of both hemispheres. More recently radars have been deployed at middle latitudes. These systems study the Earth's upper atmosphere, ionosphere and connection to the magnetosphere and near-Earth space. Institutes involved include La Trobe University, University of Newcastle, Australian Antarctic Division, IPRSS, Monash University.

IHY-Related Conferences

Australian researchers have communicated scientific results at a number of conferences with IHY interests and connections. Within Australia these include:

- AGU Western Pacific Geophysics Meeting, Cairns, July 2008. A variety of IHY research related papers were presented.
- Australian Institute of Physics Congress, Adelaide, December 2008. A special session on the IHY in Australia included six leaders in their fields providing an overview of current research in Australia, including international linkages.
- Australian Space Science Conference. A joint meeting of the National Committee on Space Science with the National Space Society of Australia to present papers on all aspects of space research, including IHY related activities.

IGY Gold Club

Australia, New Zealand members elected to the Gold Club include Prof. Valeria Troitskaya, La Trobe University (retired), Prof. Keith Cole, La Trobe University (retired), Prof. Richard Dowden, University of Otago, New Zealand (retired), and Mr. George Goldstone, IPRSS Radio and Space Services (retired).

The Australian Decadal Plan for Space Science

The Australian Academy of Science National Committee for Space Science is preparing its first decadal plan for building a national presence in space and a draft was released in February 2008 with the final presentation to the Australian Government through the Australian Academy of Science anticipated in mid-2009. The plan addresses innovative space science and technology, strong education and outlook and international collaborations that will build Australia a long term and productive presence in space.

7.2.2. China

Submitted by Fan Quanlin, State Key Laboratory of Space Weather, Chinese Academy of Sciences, Beijing

7.2.2.1. Major IHY Accomplishments in China

China has been playing an active role in all the related IHY activities, as well as the coordination efforts of the Asia/Pacific region programs. We report here three major activities, namely the IHY Asia/Pacific Regional Planning Meeting, the Meridian Project, and the IHY Asia-Pacific School in 2008.

2006 IHY Beijing meeting

The 2006 IHY Asia/Pacific Regional Planning Meeting & International Space Weather Meridian Circle Program Workshop was a successful foundation for many IHY plans, which was held in Beijing during 10–12 October, 2006, and was hosted by the State Key Laboratory of Space Weather and Chinese Meridian Project Management Office, with the sponsorship from the Center for Space Science and Applied Research, Chinese Academy

Fig. 18. *Photo of participants of the 2006 IHY Asia/Pacific Regional Planning Meeting in Beijing.*

Fig. 19. *Chinese national television CCTV interviewed Dr. Nat Gopalswamy on IHY.*

Fig. 20. *Prof. Wu Ji, head of CSSAR, delivered the welcome message for the 2006 IHY/Asia Pacific Regional Planning Meeting.*

of Sciences, National Natural Science Foundation of China, and the China National Space Administration. Representatives from IHY headquarters, China, Japan, Korea, Malaysia, Russia, and USA participated in the meeting. Representatives from Canada and Mongolia also submitted their reports to the meeting (Figures 18–20).

After a comprehensive exchange of information and discussion, all parties expressed their strong interest in multilateral cooperation within the framework of IHY as follows:

1. The multilateral activities included:

   (a) Interplanetary Scintillation (IPS) network to study the propagation of geoeffective solar disturbances
   (b) Asia-Pacific H-alpha network to study the early life of solar eruptions
   (c) Possible collaboration with CALLISTO (see Sect. 4) to monitor shock waves near the Sun
   (d) Data Center for Solar and geospace data to promote world-wide collaboration
   (e) Magnetometer Data Acquisition System (MAGDAS, see Sect. 4) to study the geospace impact of solar disturbances
   (f) Particle Detector network to study cosmic ray modulation by magnetic structures in the heliosphere

2. The International Meridian Circle Project is an important IHY activity in the Asia-Pacific region and all countries in the region agreed to support the project.
3. A committee was established to promote cooperative modeling of the Sun–Heliosphere system. Developing a framework for data-driven integrated simulation models for the study of the essential physics involved in the initiation and propagation of solar disturbances from the Sun to Earth and beyond to the edge of the heliosphere.
4. A team to plan the IHY schools in the Asia/Pacific region was established.

In the second part of the 2006 IHY Beijing meeting, the workshop discussed the issues of space weather monitoring. Space weather refers to the conditions in space that affect our space assets and ground based technological systems. The Meridian Space Weather Monitoring Project (Meridian Project) is a Chinese ground-based multi-station chain mainly along 120° E longitude to monitor space environment. There is a need to extend it the full meridian circle. The International Space Weather Meridian Circle Program (ISWMCP) to connect the ground-based monitor chains along the 120° E and 60° W meridians of around the globe will significantly enhance the capability to monitor the space environment. The 2006 International Space Weather Meridian Circle Program Workshop is the first international workshop focusing on the ISWMCP.

After comprehensive exchange of information and discussion, all parties expressed their strong interest in participation of the International Meridian

Circle initiative. The main points reached in this workshop are summarized below:

1. All parties regard the International Meridian Circle Project to be of great scientific value to monitor the space environment worldwide, and are willing to participate in the project.
2. The multilateral activities to be promoted include:

   (a) Extending the International Meridian Circle Project to include stations along existing parallel meridian circles in order to get the most scientific benefit;
   (b) Establishing a close collaboration among the existing magnetometer chains run by USA, Canada, Japan, Russia, China and other possible countries in the region. Establishment of new stations is suggested to fill the gaps along the full meridian circle, such as Siberia, South East Asia;
   (c) Encouraging collaboration among various ionospheric, magnetospheric and interplanetary observatories; and
   (d) Forming the Incoherent Scatter Radar chain in the framework of EISCAT along the Meridian Circle.

3. The main functions of the International Meridian Circle Project should include:

   (a) Coordinating observational campaigns and joint research;
   (b) Making data available to the scientific community;
   (c) Encouraging collaboration on scientific research and observations; and
   (d) Promoting education and public outreach on space science and technology.

The Meridian project

Meridian Space Weather Monitoring Project, a Chinese national mega-project of science research and engineering, has been formally approved and funded by the Chinese government. The construction phase was started on 5 January 2008, and is expected to begin operation in 2010. Since the ground-based observatory chain of this project is mainly located in the neighborhood of 120° E meridian, it is thus named Meridian Project for short. Meridian Project is an integrated scientific project to assemble the monitoring, studying, predicting with applications related to space environment. It will build three main systems, namely space environment monitoring system, data and communication system, and research and forecast system.

In the observational scheme, new or upgraded geomagnetic, radio, optical instruments, and sounding rocket are carefully collocated in 15 observatories along the 120° E longitude plus the latitude of 30° N, forming an integrated continent-scale meridian monitoring chain. Employing the advanced ground-based techniques comprehensive multi-layered and inter-disciplinary survey and exploration of space environment will be conducted, continuously monitoring the middle-upper atmosphere, ionosphere, magnetosphere and the interplanetary space, covering the spatial domain from an altitude higher than 20–30 km up to tens of earth radii far away, aiming at measurements of the geomagnetic & electric field and cosmic rays on Earth's surface, wind field, air density, temperature and its composition in the middle and high layers of the atmosphere, density of electron and proton, temperature, irregular structures, and electric current in ionosphere, solar wind plasma speed in interplanetary space. The observations will be implemented with a variety of equipment, such as magnetometers, ionosondes and digisondes, Incoherent Scattering Radar (ISR), HF back-scattering radar, LIDARs, Fabry-Perot interferometer, Interplanetary Scintillation (IPS), and sounding rockets, in a bid to probe space environment in geospace (Figure 21).

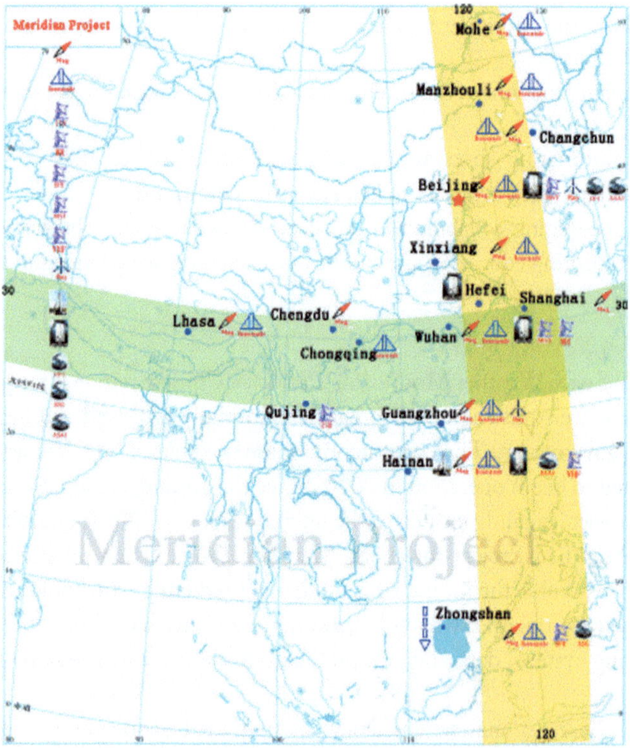

Fig. 21. *Observatory distribution and apparatus allocation of Meridian Project.*

China's Meridian Project adopts the unique geographical advantage and is one of the most meaningful and widely recognized Chinese contributions to the international solar-terrestrial scientific missions including IHY. Based on the coordinated ground-based observations implemented by Meridian Project, an International Space Weather Meridian Circle Program has been proposed to connect the other space-monitoring stations in Russia, Australia and other countries or regions running through the east longitude 120° E as well as in related countries whose territories are traversed by the west longitude 60° W. Through such appropriately distributed ground-based observation arrangements, a circum-terrestrial system for monitoring the spatial environment along the complete meridian line will be coming into being, which is the only one standing on most continents of Earth. Coordinated research, data exchange on the circle and short visit to each other among the contributed counties is the best way for international cooperation based on the Program. In this way, a worldwide survey on the near-Earth space environment by stations located at the same meridian line will be realized during the time interval of the Earth's single rotation, greatly enhancing the ability of global space environment monitoring and reducing/avoiding the hazards to human activities and technical system caused by adverse space weather events.

In addition, as the first step of the International Space Weather Meridian Circle, the extension of the Meridian Project into Russia has been included into the bilateral space science cooperation agreements between China and Russia governments.

The second International Space Weather Meridian Circle Program Workshop is under preparation.

The 2008 Beijing IHY School

The Second Asia-Pacific IHY School (see Sect. 5) was held in Beijing, China, 20–31 October 2008. The school was sponsored by the State Key Laboratory of Space Weather, the Center for Space Science and Applied Research (CSSAR), the Department of Earth Sciences, the National Natural Science Foundation of China, and the Chinese Academy of Sciences (CAS). This IHY school was open to students and to researchers from Asia-Pacific and abroad, and about 40 students and 25 lecturers attended. The purpose of the school was to bring a greater understanding of underlying universal processes and to promote international collaboration and exchange of ideas and views in heliophysical studies. This 2-week autumn school, focused on a broad range of extended heliophysics topics, such as solar physics, heliospheric physics, geo-space physics and space weather, related to the IHY science topic areas. The school consisted of about 50 lectures, as well as a series of laboratory exercises and observatory visits.

7.2.3. India

Submitted by Siraj Hasan, Indian Institute of Astrophysics, Bangalore and Arvind Paraynjpye, Inter-University Centre for Astronomy and Astrophysics, Pune
http://www.prl.res.in/~ihyindia/

7.2.3.1. IHY activities in India

Several institutions in India have been working on various aspects of solar and solar-terrestrial physics. Many instruments are in operation for these studies. The IHY-India program provides an opportunity for national and international collaborative studies of the Sun–Earth system and the heliosphere. The IHY-India program has the following broad composition:

- Indian representatives on the IHY International Steering Committee: S.S. Hasan, S.M. Chitre & A.R. Choudhuri
- Chair National Advisory committee: G. Madhavan Nair
- IHY-India National Coordinator: P.K. Manoharan
- IHY-India Outreach Coordinator: Arvind Paranjpye
- Working groups: Sun, Space Weather, Heliosphere and Solar wind, Climate and Earth atmosphere, Instrumentation, Education & Public Outreach

IHY/UNBSS program

One of the major foci of IHY is the United Nations Basic Space Science initiative, whereby a wide distribution of small instrument arrays like magnetometers, radio spectrometers, GPS receivers, all-sky cameras and so on, locating them at different longitudes to get continuous data (see Sect. 4). The following instruments were acquired through the above program:

- CALLISTO, a radio spectrograph, provided by ETH-Zentrum, Zurich. Two such instruments these have been installed at the Radio Astronomy Centre (Tata Institute of Fundamental Research), Udhagamandalam (Ooty) and at the Gauribidanur radio observatory of the Indian Institute of Astrophysics (IIA).

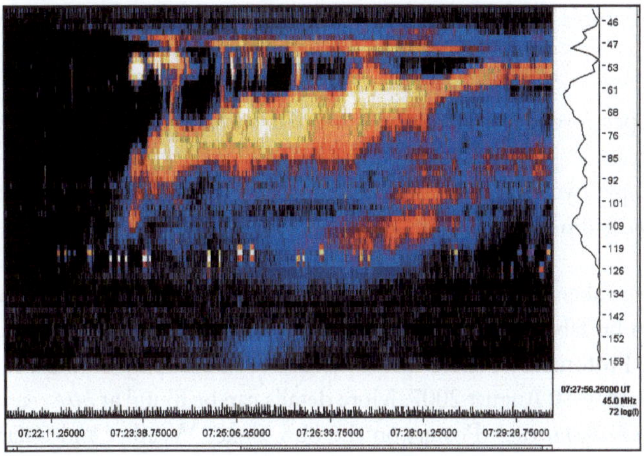

Fig. 22. *Dynamic spectrum of the type II solar radio burst observed with the CALLISTO-Gauribidanur setup on 23 May 2007.*

- AWESOME receiver for VLF remote sensing of the lower atmosphere, which has been set up at Nainital under the aegis of Indian Institute of Geomagnetism, Mumbai. Stanford University provided this instrument.

All the above instruments are in operation and data is being obtained regularly.

Figure 22 shows the dynamic spectrum of a type II radio burst (considered to be because of electrons accelerated by flare and/or CME related magneto-hydrodynamic shocks) from the solar corona observed with the CALLISTO-Gauribidanur instrument on 23 May 2007. Interestingly, this is the first type II radio burst event to be observed during the "official" IHY, which started on 19 February 2007. The abscissa and ordinate in the figure are time and frequency (MHz), respectively. The slowly drifting path of intense emission from 96 MHz at 07:23 UT to 58 MHz at 07:28 UT, is the aforementioned type II burst. The two horizontal patches of intense emission around 101 and 46 MHz are due to radio frequency interference. The above instrument, along with other CALLISTOs at different longitudes around the world is expected to provide round-the-clock coverage of the Sun.

IHY Education and public outreach activities

A 1-day meeting was held at the Indian Institute of Astrophysics (IIA) in Bangalore on 13 January 2007 to discuss various public outreach (PO) related activities. Several scientists from various national centers attended the meeting. Scientists at the IIA (Indian Institute of Astrophysics) came up with a concept design of simple experiments to study the Sun in the visible and radio wavelengths. These are:

1. Box spectroscope to view the spectrum of the Sun and common terrestrial light sources.
2. A simple two-element radio interferometer to observe Sun and other strong cosmic radio sources.

IIA has taken the initiative of arranging an adequate number of these instruments to be fabricated for distribution in schools and colleges throughout the country. Prototypes of these instruments were on display during the IIA Open House held 9–10 August 2007. More details can be found at *http://www.iiap.res.in/outreach/ihyoutreach*. Posters on the Sun, Space Weather and Solar-terrestrial relationships were displayed as a theme exhibition celebrating IHY 2007. There was also a demonstration by Dr. Navnirmiti of Mumbai using low-cost, no-cost tools to understand the Sun.

With the two-element radio interferometer kit, the team will provide hands-on astronomical observing experience to the interested science and engineering graduate student community by donating radio antennas and receiver systems to their institutions. The host institution is expected to provide a personal computer for data acquisition. The students are be trained to: (i) carry out observations of radio emission from Sun and other strong cosmic radio sources with the above set up, and (ii) develop software for deriving quantitative information from the data acquired. The radio frequency (R.F.) signal reception setup is a

Fig. 23. *Two-element radio interferometer antenna system (the antennas in the foreground) fabricated at the Gauribidanur observatory for donation to colleges/universities in India as a part of the IHY-PO program.*

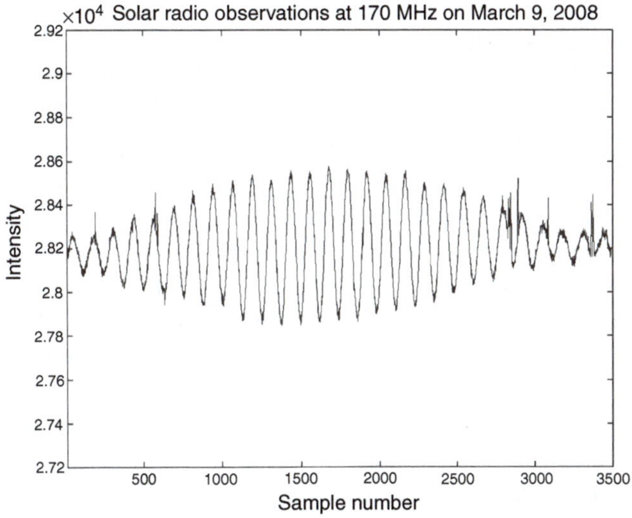

Fig. 24. *Interference fringes obtained from the Sun at 170 MHz on 9 March 2008 with the above two-element radio interferometer.*

simple radio interferometer consisting of two half-wave dipole antennas (tuned to receive R.F. signal at 170 MHz), separated by ~50 m. A prototype system has been designed and is presently in operation at the Gauribidanur radio observatory. Figures 23 and 24 show the prototype antenna set-up (the two Yagi antennas in the foreground) and the interference fringes obtained with them on Sun in the meridian transit mode. The overall variation in the observed fringe amplitude

Fig. 25. *Prof. Siraj Hasan handing over the first IHY radio kit to the students from Udaipur.*

is due to the gain variation of the response pattern of the individual antennas as a function of hour angle (half-power width $\sim 60°$).

The first IHY radio kit was handed over by the Director, Professor Siraj Hasan to the students of Mohanlal Sukhadia University, Udaipur on 27 August 2008 (Figure 25). The Head of the Department of Physics supported by the Dean, Faculty of Sciences of the MLS University had written to Professor Hasan, requesting for the equipment for use in their M.Sc. program. The MLS University was joined by the Rajasthan Technical University also in Udaipur as a partner, to share the equipment for their engineering students to also learn radio astronomy. The students of both the universities were with us 24–29 August 2008 and were trained by Dr. R. Ramesh and his team at Gauribidanur Observatory on how to install and use the equipment. They learned the basics of radio interferometry in the process. They also carried out observations of Sun and Cygnus A with the equipment during their stay at Gauribidanur. We wish them all the very best for the successful installation of the antenna and the receiver system at their institution and carry out radio observations.

Requests for the radio kit have been received from another 10 colleges and universities and fabrication work is in progress at the Gauribidanur observatory of IIA.

The Inter-University Centre for Astronomy and Astrophysics (IUCAA) is also actively involved in the PO program. These activities include making small telescopes (refractor telescopes with 40 mm lens) and demonstrating to children how to observe the sun by projecting the solar image. Public talks have been arranged on IHY themes at school and college levels. IHY related public talks and training programs have been conducted during mid 2006–mid 2007 by the Radio Astronomy Centre (Ooty) in various schools and colleges where students were provided training to use the radio telescope and carry out data analysis.

Muktangan Vidnyan Shodhika, the Science Centre of IUCAA, celebrated a Science Festival during November 11–14, 2007. This was to commemorate the birth anniversary of late Shri Pu.La. Deshpande, which fell on November 8. Fondly called "PuLa", Shri Deshpande was one of the visionaries behind the IUCAA Science Centre. The centre building is named "Pulastya" after him.

The first event, on November 11, was poster competition on the "Sun–Earth Relationship". This was planned as a part of IUCAA's participation in the International Heliophysical Year. The participants were science-minded high school students. A brief lecture on the topic "Sun and Us" was given by Samir Dhurde. The ideas gathered by the kids from it were very adeptly put onto paper. The posters were displayed on the rest of the days.

On 22 December 2007, the day of winter solstice (for the northern hemisphere), an observing experiment to measure latitude and longitude of a place was carried using EDUSAT. Ten stations from different parts of India participated in this program.

Meetings held so far

- 1st IHY-India planning meeting was held at the Radio Astronomy Centre – Tata Institute of Fundamental Research, Ootacamund during 10–12 July 2004. About 30 participants from various research organizations and universities across the country attended the meeting;
- The International Living with the Star Conference, February 2006, Goa, India, organized by the Indian Institute of Geomagnetism which included IHY-themed talks;
- 2nd UN/NASA workshop on IHY and BSS was held IIA, Bangalore during 26 November–1 December 2006. About 150 participants from 30 member states of the UN attended the workshop;
- IHY Science and IGY Gold meeting at Aryabhatta Research Institute of Observational Sciences (ARIES), Nainital during 7–10 May 2007;
- The First Asia-Pacific School on IHY (see Sect. 5) was held at the Kodaikanal Observatory of IIA during 10–22 December 2007. The school was co-sponsored by EOARD-AFOSR. Speakers were drawn from both India and abroad and there were 45 lectures in total. There were also lab exercises for the students to get experience in observations and data analysis.

7.2.4. Japan

Submitted by Kiyohumi Yumoto, IHY-Japan National Coordinator, Space Environment Research Center, Kyushu University, Fukuoka and the STPP Sub-Committee of the International Subgroup of the Earth and Planetary Science Committee, Science Council of Japan

http://www2.nict.go.jp/y/y223/sept/IHY/IHY.htm

7.2.4.1. IHY activities in Japan

The IHY is an extensive international program to study the universal physical processes in heliospace for a better understanding of the Sun-heliosphere system. In particular, the neutral and ionized matter in heliospace and the interaction between them are studied, from the atmospheres of Earth and planets to the interplanetary medium. The IHY will continue the legacy of the IGY activities in Japan during 1957–1958 by extending the geophysical studies of 50 years ago to the combined system of the Sun and the planets. IHY also extends the physical realm from geospace to heliospace, recognizing the enormous progress made over the past 50 years.

In June of 2006, the Science Council of Japan formally recognized the STPP (Solar Terrestrial Physics Program) Subcommittee for International Affairs, Committee on Earth and Planetary Sciences, as the IHY National Steering Committee for Japan. The STPP subcommittee is chaired by K. Yumoto, IHY-Japan National coordinator. The IHY National Steering Committee promotes

(1) The international heliophysical observing network,
(2) Public outreach,
(3) International workshops, and
(4) the nomination of IGY Gold Club members.

Scientific investigations

1. MAGDAS Network (PI: Prof. K. Yumoto, Space Environment Research Center, Kyushu Univ.)
As Japan's leading contribution to the IHY, the *Space Environment Research Center* (SERC) of Kyushu University, Fukuoka, Japan is in the process of deploying globally 50 state-of-the-art magnetometers. The *MAGnetic Data Acquisition System* (MAGDAS) Group (see Sect. 4) is deploying around the world in a strategic fashion a new generation of tri-axial fluxgate magnetometers (called MAGDAS) that transfer the digitized data to a central SERC server in real-time for space weather study and application during the IHY period (2007–2009). The project aids the study of the dynamics of geospace plasma changes during magnetic storms and auroral substorms.

2. Muon Detection Network (PI: Prof. K. Munakata, Shinshu Univ.)
This network of muon detection instrumentation consists of nine institutes from seven counties, and performs cosmic ray monitoring for space weather study. This system, the Muon Detector Network (see Sect. 4), performs space weather monitoring through the observation of the directional intensity of high-energy cosmic rays. In March 2006, this world-wide network of muon detectors was upgraded with both an enlargement of a detector in Brazil and the installation of a

new detector in Kuwait-City, Kuwait. This enlargement vastly improved the coverage of cosmic ray pitch angle.

3. Interplanetary scintillation (IPS) Network (PI: Prof. M. Kojima, Solar-Terrestrial Environment Laboratory, Nagoya Univ.)

Interplanetary scintillation (IPS) is a remote sensing technique to observe the solar wind, which has advantages over some in situ spacecraft measurements: It can observe three-dimensional solar wind in a short time, and the observations can be carried out consistently over a solar cycle. They are planning coordinate IPS observations among IPS facilities, so that they can observe the solar wind in the full distance range from near sun region to the earth orbit and monitor the solar wind 24 h a day. The IPS network does synergistic collaboration with the Solar Mass Ejection Imager (SMEI), which gives measurements of bulk density changes with much higher spatial resolution than the IPS. They also carry out a complementary collaborative project between the muon network and the IPS network, both of which can derive 3D CME structure; The IPS observe the density compressed region ahead of the ICME, while the muon network observes flux rope structure in the ICME. The IPS observations can be carried out consistently over a solar cycle. They are planning coordinate IPS observations among IPS facilities, so that the solar wind can be monitored in the full distance range from near sun region to the earth orbit and in 24 h a day. The coordinated network does synergistic collaboration with the Solar Mass Ejection Imager (SMEI), which gives measurements of bulk density changes with much higher spatial resolution than the IPS. The IPS observes the density compressed region ahead of the ICME, while the muon network observes flux rope structure in the ICME (Figure 26).

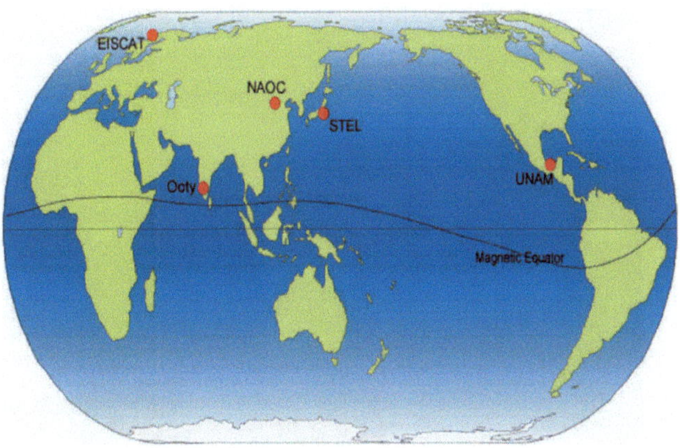

Fig. 26. *Sites in the IPS observation network.*

4. CHAIN Network (PI: Dr. S. Ueno and Prof. K. Shibata, Kwasan and Hida Observatories, Kyoto Univ.)

Continuous H-alpha Imaging Network (CHAIN)-project was planned to monitor solar flares and erupting filaments continuously by using several of characteristic telescopes. As part of CHAIN-project, they selected Peru as the country where the first oversea Flare Monitoring Telescope will be installed. They are investigating various items, aiming to start the operation of the FMT in Peru by the end of 2009, such as the seeing condition and the size of the turbulence due to the heat haze, the best structure of the housing of the telescope, the efficient method of remodeling the telescope with corresponding to the latitude of Peru, the best combination of the observing wavelengths, the appropriate software for data processing under the computer environment at Ica University, the human environment and the way of training of the local staffs, etc.

Public outreach

Public outreach is carried out through the Network of International Space Environment Services of the National Institute of Information and Communication Technology (NICT, PI: Dr. S. Watari). Pictured below is the NICT ("National Institute of Information and Communication Technology" of Japan) Space Weather Information Center of International Space Environment Services (ISES). Here, real-time data from satellites and ground-based observatories are monitored, and a forecast is issued everyday at 6:00 UT. The exploitation of space requires that we have a better understanding of space weather. At Space Weather Information Center, real-time data from satellites and ground observatories are monitored, and a forecast is issued everyday at 6:00 UT as shown in Figure 27.

Fig. 27. *Space weather now- and fore-casting at NICT.*

The exploitation of space requires that we have a better understanding of space weather.

The U.N./IHY Basic Space Science Tokyo Workshop

Japan hosted an UN/ESA/NASA Workshop on Basic Space Science and IHY 2007 at the National Astronomical Observatory of Japan (NAOJ), Mitaka, Tokyo, on 18–22 June, 2007. Many workshops are being held in various countries in conjunction with IHY, with an aim to benefit scientists and engineers from developing nations. To make the IHY Tokyo Workshop a reality, considerable financial support was provided by NAOJ and SERC. For example, SERC paid for the travel expenses of five Japanese persons and about 10 foreign scientists.

IGY Gold Club

Part of IHY is to celebrate the accomplishments of the International Geophysical Year of 1957. With this in mind, the "IGY Gold Club" was initiated (see Sect. 6 for more information). Members are limited to those individuals who participated in IGY. To date, the following 12 persons from Japan have been selected as Gold Club Members (all nominated by SERC): Dr. Kaichi Maeda, Dr. Hiroshi Maeda, Dr. Masahisa Sugiura, Dr. Noboru Wakai, Dr. Mutsumi Ishitsuka, Dr. Hiroyoshi Tanabe, Dr. Keizou Nishi, Dr. Eijiro Hiei, Dr. Masami Wada, Dr. Tai-ichi Kitamura, Dr. Takashi Oguti, and Dr. Ichiro Kondo.

More members from Japan are expected, and are currently being nominated.

Acknowledgments

The IHY Organizer in Japan, K. Yumoto (SERC, Kyushu Univ.) would like to appreciate the following Japanese IHY Coordinators: Secretary of Japan IHY Dr. S. Watari (NICT), Prof. T. Sakurai (Natl. Astron. Obs. of Japan), Prof. M. Kojima (STE Lab., Nagoya Univ.), Prof. K. Shibata (Kwasan/Hida Obs., Kyoto Univ.), Prof. M. Fujimoto (JAXA/ISAS), and Assistant Secretary of Japan IHY Mr. G. Maeda (SERC, Kyushu Univ.) for their tireless effort and contributions to the IHY program in Japan.

7.2.5. Nepal

Submitted by Jayanta Acharya, IHY-Nepal National Coordinator, Mahendra Sanskrit University, Kathmandu

http://www.aimnepal.com.np/content/view/25/41

7.2.5.1. IHY 2007 Activities in Nepal

Note from the IHY-Nepal National Coordinator

Nepal is a developing country. Only about 60% of the people in this country are literate. People of the older generation in this country have a deep belief in the stories of legends and they consider the planets and stars as the Gods and Goddesses. These beliefs have influenced Nepalese society to some extent. When we started to show planets through our telescope in public areas, many people including children and housewives gathered, showing great excitement. They were eager to answer their queries about planets and stars. Although we could not answer all the people in the small program, it was very much exciting and really informative for the general public (Figure 28).

Talk Programs

On the occasion of IHY 2007 we organized various talk programs on different subjects of astronomy. Renowned astronomers, scientists, professors and amateurs astronomers were the speakers on the talk programs. A talk program on 24 April 2008 at Nepal Academy of Science and Technology (NAST) where Prof. Dr. Michel Gerbaldi from Institut d' Astrophysique de Paris, gave very lively speech on astronomy imaging was one of the remarkable talk events conducted by us. The following is a list of talk programs held on different subjects (Figure 29):

- Talk on astronomy imaging, on 24 April 2008 given by Prof. Dr. Michel Gerbaldi.

Fig. 28. *IHY-Nepal Planning Team.*

Fig. 29. *Collage of photos from Talk Programs at various locations in Nepal.*

- Talk cum Colloquium on Astronomy, on 14 August 2007 Tuesday, at NAST Meeting Hall, Speakers are Er. Rishi Shah, Academician, NAST, Mr. Kedar Badu, GASPO, by Jayanta Acharya, Balmiki Campus.
- Talk programs at Balmeeki Campus, on 26 May 2007, on the subject General Astronomy. Speaker Jayanta Acharya, Er. Rishi Shah.
- Talk programs at Takshashila Academy, on 21 July 2007 on the subject Basic Astronomy by Basanta Acharya.
- Talk programs at the Acharya Intstitute of Mathematics (AIM) Nepal, on 10 August 2007, on the subject "Awareness on Astronomy" by senior students of the AIM Nepal institute.

Star Parties

We found the "Star Party" as the most effective and lively program among students and the general public. They were very much interested in viewing the craters of moon, ring of Saturn and crescent of Venus through telescope, so the parties attracted more participants if the scope was broadened to include all of planetary astronomy. Because of the small power of the telescope, we could not reach beyond the planets of our solar system. Star parties on the occasion of meteor showers and Moon-Mars-occultation were really impressive and unforgettable.

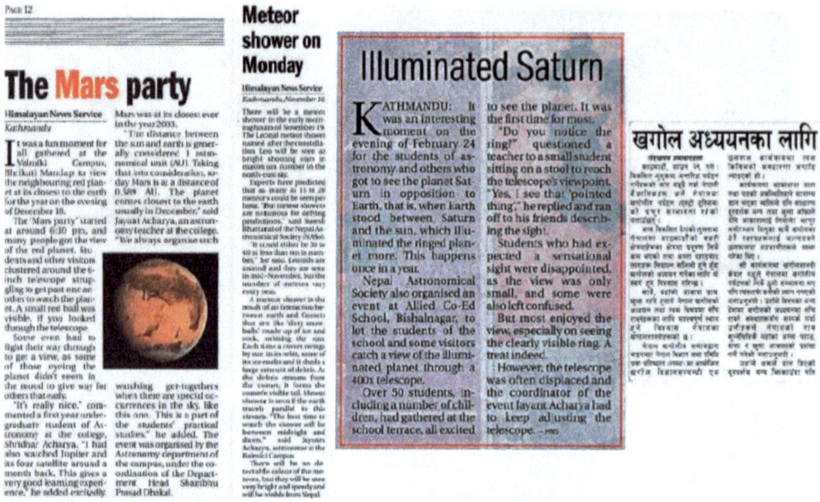
Page 12

The Mars party

Himalayan News Service
Kathmandu

It was a fun moment for all gathered at the Valmiki Campus, Bhrikuti Mandap to view the neighbouring red planet at its closest to the earth for the year on the evening of December 18.

The 'Mars party' started at around 6:30 pm, and many people got the view of the red planet. Students and other visitors clustered around the 6-inch telescope struggling to get past one another to watch the planet. A small red ball was visible, if you looked through the telescope.

Some even had to fight their way through to get a view, as some of those eyeing the planet didn't seem in the mood to give way for others that early.

"It's really nice," commented a first year undergraduate student of Astronomy at the college, Shridhar Acharya. "I had also watched Jupiter and its four satellite around a month back. This gives a very good learning experience," he added excitedly.

Mars was at its closest ever in the year 2003.

"The distance between the sun and earth is generally considered 1 astronomical unit (AU). Taking that into consideration, today Mars is at a distance of 0.589 AU. The planet comes closest to the earth usually in December," said Jayant Acharya, an astronomy teacher at the college. "We always organise such watching get-togethers when there are special occurrences in the sky, like this one. This is a part of the students' practical studies," he added. The event was organised by the Astronomy department of the campus, under the co-ordination of the Department Head Shambhu Prasad Dhakal.

Meteor shower on Monday

Himalayan News Service
Kathmandu, November 16

Illuminated Saturn

KATHMANDU: It was an interesting moment on the evening of February 24 for the students of astronomy and others who got to see the planet Saturn in opposition to Earth, that is, when Earth stood between Saturn and the sun, which illuminated the ringed planet more. This happens once in a year.

Nepal Astronomical Society also organised an event at Allied Co-Ed School, Bishalnagar, to let the students of the school and some visitors catch a view of the illuminated planet through a 400x telescope.

Over 50 students, including a number of children, had gathered at the school terrace, all excited to see the planet. It was the first time for most.

"Do you notice the ring?" questioned a teacher to a small student sitting on a stool to reach the telescope's viewpoint. "Yes, I see that 'pointed thing'," he replied and ran off to his friends describing the sight.

Students who had expected a sensational sight were disappointed, as the view was only small, and some were also left confused.

But most enjoyed the view, especially on seeing the clearly visible ring. A treat indeed.

However, the telescope was often displaced and the coordinator of the event Jayant Acharya had to keep adjusting the telescope. – HNS

खगोल अध्ययनका लागि

Fig. 30. *Media coverage of Nepali Star Parties and Talk Programs.*

Our parties:

- Star party on a Road. Date: 7 April 2007;
- Star party on the occasion of the meteor shower. Date: 17 November 2007;
- Mars party at Valmeeki campus. Date: 17 December 2007;
- Saturn party at Takshashila Academy. Date: 14 February 2008.

Public Awareness Programs

On the occasion of IHY 2007, we published 2000 copies of a poster which contained general information on the planets. The poster was designed and edited by a B.Sc student Nishu Karna of NASO. The poster was published by Jagadamba press. We found the poster very useful for school students. We have also launched the website *http://www.astronepal.org.np*. This site aims to give basic idea about astronomy to laymen. This site is in both English and local language. The coordinator of this site is Mr. Basanta Acharya (Figure 30).

7.2.6. Thailand

Submitted by Busaba KRAMER, IHY-Thailand National Coordinator for Education and Outreach, and Boonrucksar SOONTHORNTHUM, IHY-Thailand National Coordinator, National Astronomical Research Institute of Thailand (NARIT), Chiang Mai

http://www.narit.or.th/ihy2007/

Astronomy development in Thailand has improved significantly during the last few years. The government has approved the establishment of the National Astronomical Research Institute of Thailand (NARIT). The roles of NARIT in the development of astronomical research and astronomy education in Thailand include a national framework, national facilities, collaborative research networks, teacher training and public outreach programs. The new 2.4-m reflecting telescope (expected completion in late 2008) will serve not only the astronomy community in Thailand but also throughout Southeast Asia.

NARIT joined the International Heliophysical Year (IHY) program in August 2006, and both scientific and public outreach activities were planned to promote IHY in Thailand from 2006 to 2009. Joining IHY not only helps promote scientific awareness and understanding for the public, but also promotes astrophysical research and collaboration at the national and international levels.

7.2.6.1. IHY activities in Thailand

For public outreach, NARIT coordinated several nationwide activities in collaboration with the Geo-Informatics and Space Technology Development Agency (GISTDA) and other institutions in Thailand. The website providing basic knowledge in Thai language about physics of the Sun and space weather were established in August 2007, *http://www.thaispaceweather.com/IHY/*, in collaboration with Mahidol University.

NARIT proposed the celebrations of IHY activities to the Ministry of Science and Technology as a highlight for the annual National Science and Technology Fair 2007. The event was held at the Bangkok International Trade & Exhibition Centre (BITEC) during 8–19 August 2007 and had almost 800,000 participants

Fig. 31. *Exhibits at the National Science & Technology Fair in Bangkok, Thailand.*

during the 12-day period. On 10 August 2007, H. R. H. Princess Maha Chakri Sirindhorn presided over the opening ceremony and visited the IHY Pavilion (Figure 31).

Dates	Activities
Nov.–Dec. 2006	IHY Website and IHY Thai-logo
	Monthly AstroRadio Shows
	Productions and distribution of outreach materials (e.g. AstroCalendar 2007 – featuring the Sun)
9–13 Jan. 2007	Science Boulevard & National Children's Day, Bangkok
13 Jan. 2007	National Children's Day, AstroFun, Chiang Mai
12–23 Feb. 2007	Exhibitions at the 44th Session of UN-STSC and IHY official opening ceremony, Vienna, Austria
19 Mar. 2007	Partial solar eclipse, nationwide eclipse observations and radio/TV interviews
22–24 Mar. 2007	Thai National Astronomy Meeting (TNAM2007) and the 1st South-East Asian Astronomical Network Meeting (SEAAN2007) – highlighting IHY Science
30 July–3 Aug. 2007	The 4th Asia Oceania Geosciences Society (AOGS) Meeting, Bangkok – Thailand-IHY Exhibition
8–19 Aug. 2007	National Science & Technology Fair 2007, BITEC, Bangkok – IHY Pavilion
4–10 Oct. 2007	World Space Week 2007, GISTDA & NARIT coordinated nationwide activities

For IHY science activities, NARIT has promoted IHY science sessions in the Thai National Astronomy Meeting (TNAM2007), which had 10 oral presentations, including the review talk, "IHY Science: Ground-Based Instruments for Space Physics", by Prof. David Ruffolo, Mahidol University. During TNAM2007, NARIT organized the First Southeast Asia Astronomy Network (SEAAN2007) Meeting, which had participants from five countries in Southeast Asia, including Indonesia, Malaysia, Philippines and Laos.

At the level of school research activities, the Solar Sudden Ionospheric Disturbance (SID) monitoring program (see Sect. 5), under the IHY/United Nations Basic Space Science Initiative, has been introduced in Thailand. A Team of scientists from NARIT, Stanford's Solar Center and Mahidol University help supervise a team of talented students from the Mahidol Wittayanusorn School as a pilot project. Stanford's Solar Center, in conjunction with the Electrical Engineering Department, Stanford University, USA, have developed inexpensive SID monitors that students can install and use at their local high schools. Students joined the project by building their own low cost antenna, and collecting and analyzing the data on a local computer. Students were able to exchange and discuss the data with other students in global network via the SID Blog and SID Database website. The SID pilot project was displayed at the National Science and Technology Fair 2007. It is aimed that this program will be expanded to other schools throughout Thailand.

7.2.6.2. Summary

The International Heliophysical Year (IHY) is an international program of global research to understand the external drivers of the space environment and climate, continuing the tradition of previous International Science Years. For Thailand, joining IHY not only helps to promote scientific awareness and understanding for the public, but also helps to promote the astronomical research and collaborations both at the national and international levels.

7.3. Eastern Europe and Central Asia

Participating Nations: Kazakhstan, Russia, Ukraine, Uzbekistan

Please note that many nations from Eastern Europe are included in the report for the Balkan, Black Sea and Caspian Sea Regional Network. There was a great deal of coordination and collaboration between these two regions, and thus region's report continues into the next section on the Balkan, Black Sea and Caspian Sea Regional Network.

7.3.1. Russia

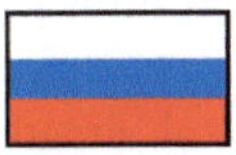

Submitted for the IHY-Russia National Organizing Committee, by *G. Zherebtsov, Institute of Solar-Terrestrial Physics, Irkutsk, A.V. Stepanov, Pulkovo Observatory, Saint Petersburg and V. Obridko, Izmiran, Moscow*

7.3.1.1. IHY activity in Russia

About 40 institutes, universities, observatories, solar and geophysical stations from Far East (Kamchatka, Ussurijsk) to the West (Kaliningrad) e.g. in 10 time zones in Russia are involved in IHY. A number of ground-based observation programs and solar-terrestrial space missions were realized in 2006–2008 and more are planned in the future.

Ground-based instrument contributions

- Pulkovo Observatory: Horizontal solar telescope ATSU-5, Horizontal solar telescope, Photoheliograph for the chromosphere flares, Big Pulkovo Radiotelescope.
- Kislovodsk Solar Mountain Station of Pulkovo Observatory (2100 m, since 1948): Nikolsky Big Coronagraph, Lyot coronagraph, OPTON Telescope, Small radio telescopes at $\lambda = 2$, 3 and 5 cm (I), photoheliograph, spectral heliograph, helioseismograph.
- Baikal Astrophysical Observatory (ISTP, Irkutsk) is located on the south shore of Lake Baikal. Large Solar Vacuum Telescope (LSVT) provides here observations of a thin structure of active solar formations and registration of solar flares and other non-stationary phenomena in solar atmosphere. In 2007 the LSVT was equipped with adaptive optical system.
- Siberian Solar Radio Telescope (SSRT) is a special-purpose solar radio telescope (600 m cross) designed for studying solar activity in the microwave range (5.7 GHz) where the processes occurring in the solar corona are accessible to observation over the entire solar disk.
- Sayan Solar Observatory (ISTP, 2100 m, since 1962) has three unique solar instruments: Horizontal solar telescope with magnetograph, Solar telescope of operative forecasts, and Non-eclipse coronagraph. Measurements of solar magnetic fields and spectral observations of solar active features and dynamic processes in the solar atmosphere are carried out at the observatory – within the framework of a special IHY program.
- IZMIRAN Forecasting Center of geophysical situation: New IZMIRAN magnetograph (magnetic maps of the solar surface).
- Special Astrophysical Observatory (SAO), North Caucasus: RATAN-600 was upgraded recently for solar observations. New multi-wavelength receiver with simultaneous analysis in frequency range 3–18 GHz was created. Observations the Sun with RATAN-600 is possible now from 0700 to 1100 UT.
- Joint Geomagnetic Observatory (ISTP, Irkutsk) consists of several observation stations: INTERMAGNET magnetic observatory Irkutsk in Patrony village, Baikal Magneto-Telluric Observatory at Olkhon Island, ULF observation station at Sayan Solar observatory, and magnetic station at Norilsk observatory.

- Irkutsk incoherent scatter radar (ISTP) for ionosphere plasma parameter measuring by incoherent scatter technique is the member of worldwide radar net which consists of nine radars. During IHY radar measurements are being made with enhanced mode of operation.

Space-based observations

- *Microsatellite "Universitetski-Tatyana"*: altitude 970 km, Inclination 82 (launched in 2005). Electrons: 1 keV, >40 keV, 0.1 ÷ 0.9 MeV, 3 ÷ 10 MeV, Ions: 2–180 MeV, UV flux 300–400 nm, + educational functions
- *Meteor-M*: sun-synchronous orbit, apogee 1000 km Inclination 99 (launched in 2008). Four instruments measuring electrons, ions, high-energy ions and a radio frequency mass spectrometer
- *Electro*-L: geostationary meteosat (launch in 2009). Five instruments measuring electrons, soft X-rays, a solar photometer, and a magnetometer
- *Compas-2*: altitude 600 km, inclination 82 (launched in 2006). Two experiments measuring electromagnetic waves, ions, electrons, UV flux, a radio spectrometer, a GPS occultation receiver, and a radio sounding transmitter
- *Canopus-Vulkan*: altitude 500 km, inclination 82 (launch in 2009). Measures electromagnetic waves, radiospectrometer, GPS occultation receiver, and a radio sounding transmitter
- *Electromagnetic Environment Experiment "OBSTANOVKA" at International Space Station*: multipoint multi-component wave and field experiment. Deployment in 2009
- *CORONAS-Photon*: altitude 500 km, inclination 82.5 (launch in 2009). Nine instruments performing gamma ray spectroscopy, neutron measurements, low-energy gamma rays, X-ray and gamma ray spectro-polarimetry, UV monitor, solar XUV telescope, cosmic ray, electrons, proton and alpha particle measurements.

Meetings

There were several national and international meetings pertaining to IHY hosted by IHY-Russia. Russian delegations also participated in many IHY meetings around the world.

- IHY Meeting in Zvenigorod "New Insights into Solar–Terrestrial Physics" (5–11 November 2007)
- Annual International Baikal School in Astrophysics, Space Physics, and Geophysics (Irkutsk, 2006–2008)

- Russian-Chinese Workshop on Space Weather (September 2008, Space Res. Institute, Moscow)
- Russian-Polish School for Solar Terrestrial Physics (Pulkovo Observatory, October 2008).

Main scientific results

- *Reconstruction of key parameters of the Space Weather over 400-year time scale, by Yu. Nagovitsyn, E. Miletsky, V. Ivanov (Pulkovo)*: For the period 1610–2005 the following time series of different parameters of the Space Weather were constructed: absolute total sunspot magnetic flux, open magnetic flux, dipole-octupole (A–) index of the solar large-scale magnetic field, IDV-index of geomagnetic activity (prolonged version), aa-index of geomagnetic activity, interplanetary magnetic field strength. The obtained data are a new information resource to study the solar-terrestrial relations overlong time scale including the Maunder minimum epoch.
- *On Prediction of the Strength of the 11-Year Solar Cycle No. 24 by Obridko V.N. and Shelting B.D., Izmiran*: Various forecast techniques have been analyzed with reference to the solar activity cycle 24. Three prediction indices have been proposed: the intensity of the polar field, mean field at the source surface, and recurrence index of geomagnetic disturbances. As a rule, the forecast based on the polar field and extrapolation of local fields gives the height of cycle 24 smaller than the height of cycle 23. The use of the recurrence index and the global field value leads us to the conclusion that cycle 24 will be medium high: the same as or somewhat higher than cycle 23.
- *Jet-like events above null points of coronal magnetic field by Filippov B.P., Golub L., Koutchmy S.*: Jet-like events are observed on different scales in the solar corona in EUV and X-rays. Polar coronal holes show a significant activity manifested in small loop brightening and jetlet formation. The geometrical shape of the jetlets and their position indicate that they appear near the singular points of the magnetic field, namely, null points or X-points, which arise due to the interaction between new emerging small dipoles and large-scale magnetic field of the coronal hole. The jetlets could serve as injectors to the process of the solar fast wind acceleration.
- *Photospheric and chromospheric oscillations in the solar faculae by Kobanov N., Pulyaev V., ISTP, Irkutsk*: Altitude inversion in localization of 3-minute oscillations of the radial velocity in the solar spots is found. Location of the oscillation power maximum in the chromosphere corresponds to the minimum in the photosphere. In the same time the chromosphere faculae are characterized

by the enhancement of low frequency components of oscillations in the period range of 8–15 minutes and indications of wave propagation. The point of this propagation origin coincides with the maximum of the magnetic field strength.

- *Coronal seismology & flare plasma diagnostics (Team: Pulkovo, Institute of Applied Physics, Crimean Astrophysical Observatory)*: Two approximations were used: flare loop as MHD-resonator and as an equivalent electric (RLC) circuit. With these approaches the main flare plasma parameters (n, B, T) were obtained and electric currents in flares were determined ($\sim 10^{11}$–10^{12} A).
- *Studies of the Earth's magnetopause shape (Space Research Institute)*: It is revealed that the magnetopause of the Earth is ~5% compressed in the directions perpendicular to the plane formed by solar wind velocity and interplanetary magnetic field vectors. This compression can results from the stress of the magnetic field lines draping the magnetopause. The influence of B_z component of the interplanetary magnetic field on the subsolar magnetopause position was not found in Prognoz's and Interball 1 data.
- *Total solar eclipses (March 29, 2006 and August 1, 2008)*: Eclipses are a very good opportunity for the upgrading of the observatories and stations. Quite a few solar telescopes in the Russia (Kislovods Mountain Solar Station, Russian-Ukrainian Observatory at Terskol-Peak and Niznnyi Arkhyz) will be upgraded due to the eclipses. There is a special program in the Russian Academy just for these events.

Educational outreach

There is quite low level of education in astronomy today in the Russian secondary schools, unfortunately. To improve the situation observatories, planetaria and astronomical institutes in Russia try to keep up astronomical knowledge among youth via lectures and excursions. For example, about 15,000 school children visit Pulkovo Observatory per year. The Planetarium in Nizhny Novgorod was open recently and very active now due to high level of scientific management. Russian scientists are doing public lectures in General Astronomy and Solar-–Terrestrial Physics and publishing the papers in popular journals like "Zemlya i Vselennaya" (Earth & Universe). There are examples:

- *Educational program of Russia in Space Physics (Moscow University)*: The program includes multimedia lectures, interactive demonstrations, special computerized lab exercises. Special projects including launching of mini satellites "Universitetskiy" and "Kompas-2" with the scientific equipment

"Tatyana" are realized by Skobeltsyn Institute of Nuclear Physics of Moscow University (Headed by Prof. M. Panasyuk) *http://cosmos.msu.ru*
- *Multimedia lectures "Life of the Earth in the Solar Atmosphere"*
- *Textbook "Plasma Helio-Geophysics" (L. Zelenyi, Ed.) in Russian* will be published in December 2008.

IGY Gold history

There are many scientists in Russia who took a part in IGY-1957. Many Russian scientists received IGY Gold Awards, including Prof. G. Ivanov-Kholodny and Prof. E. Mogilevsky (Izmiran), Prof. E. Ponomarev (ISTP, Irkutsk), Prof. G. Gelfreikh (Pulkovo), Dr. N. Soboleva and Dr. V. Ikhsanova (Special Astrophysical Observatory, N. Caucasus), and Prof. I. Podgorny (Institute of Astronomy, Moscow).

7.4. Balkan, Black Sea and Caspian Sea Regional Network

Participating Nations: Armenia, Azerbaijan, Bosnia-Herzegovina, Bulgaria, Croatia, Georgia, Greece, Romania, Russia, Serbia and Montenegro, Slovakia, Turkey, Ukraine

Submitted by Katya Georgieva, Coordinator of the Balkan, Black Sea and Caspian Sea Regional Network

http://www.stil.bas.bg/IHY

The Balkan, Black Sea and Caspian Sea Regional Network for Space Weather Studies was created in June 2005 to coordinate the IHY activities in the region. Initially it consisted of 11 countries (Armenia, Azerbaijan, Bulgaria, Croatia, Georgia, Greece, Romania, Russia, Serbia and Montenegro, Turkey, Ukraine), later Bosnia and Herzegovina also joined. The main impetus behind establishing the network was to coordinate the observational, scientific and educational activities related to the IHY, and to strengthen activities and collaboration in the region in the field of space science.

The planning activities for this network began in June 2005, at the IHY Balkan, Caspian and Black Sea Regional Planning Meeting held in Sozopol, Bulgaria. The second meeting of the Balkan, Black Sea and Caspian Sea Regional Network on Space Weather Studies was organized by the Kandilli Observatory and Earthquake Research Institute, Boğaziçi University, Istanbul, Turkey. It was held in Antalya, Turkey between March

31 and April 2, 2006, right after the Total Solar Eclipse (TSE) on March 29, 2006. Most of the participants witnessed this exceptional, natural event for the first time in their lifetimes while other scientists have made relevant experiments on eclipse observations. Eighty-one scientists and leading experts in various research areas of Space Weather from many countries, including Armenia, Austria, Azerbaijan, Belgium, Bulgaria, Croatia, France, Georgia, Germany, India, Japan, Poland, Romania, Russia, Slovakia, Turkey, Ukraine, and USA attended this Meeting and presented 59 presentations. Most of these presentations were published in "Sun and Geosphere" (Figure 32).

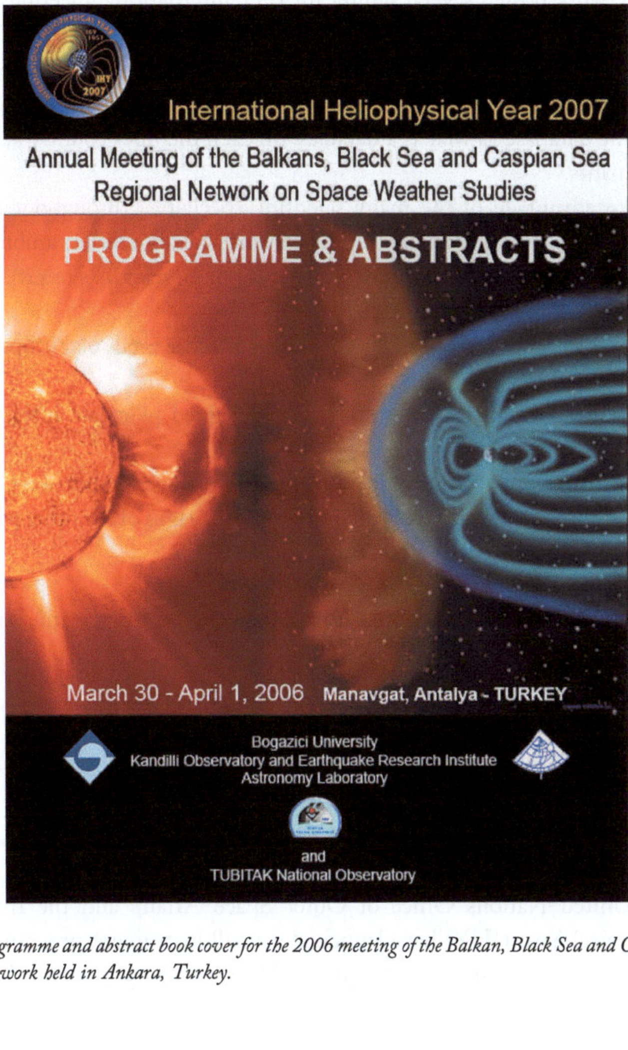

Fig. 32. *Programme and abstract book cover for the 2006 meeting of the Balkan, Black Sea and Caspian Sea Regional Network held in Ankara, Turkey.*

The following list itemizes the priorities and accomplishments of the network:

- A number of scientific problems were addressed in the framework of IHY and scientists from this network led nearly 20% of the IHY CIPs.
- The Regional Network established an International refereed scientific journal “Sun and Geosphere,” with an international editorial board (see Appendix III).
- Instrument activities that are relevant to IHY science were identified, particularly the Space Environment Viewing and Analysis Network (SEVAN) led by IHY-Armenia (see Sect. 4)
- The network hosted several significant IHY international meetings, including the 4th UN/ESA/JAXA/NASA IHY workshop.
- The network web site provides coordination support for the region, with information about current and past regional and worldwide activities, and a scientific database as well as a database on each member country activities and participants.
- To take advantage of the many scientific specialties throughout the region, bilateral and multilateral joint projects and collaborations among member countries were identified, and the support of young scientists received particular attention.
- An expedition was organized for observations of the total solar eclipse in March 2007.
- Many outreach activities were initiated, designed to create interest in IHY, heliophysics and space weather.
- A series of popular articles on the Sun and space weather were written by members of the network, translated into the languages of the other participating countries and published in local periodicals.

The following national reports show the level of activity in the participating nations.

7.4.1. Armenia

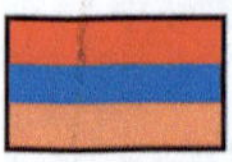

Submitted by Ashot Chilingarian, IHY-Armenia National Coordinator, Alikhanyan Physics Institute, Yerevan

The United Nations Office of Outer Space Affairs and the International Heliophysical Year (IHY) have launched a small instrument programme as one of United Nations Basic Space Science (UNBSS) activity. A network of particle detectors located at the middle to low latitudes, SEVAN (Space Environmental Viewing and Analysis Network) aims to improve the fundamental research on particle acceleration in vicinity of sun and space environment conditions. The key issue of the SEVAN network research is better understanding of the physics of most energetic processes on the sun, on the propagation of the ICME in the interplanetary space and its interactions with magnetosphere. The excellent facilities of the ongoing and planned space missions will bring direct information about processes on the sun and in vicinity of Earth. Secondary fluxes of elementary particles measured on the Earth surface can provide compatib le information on highest energy solar cosmic rays and on the modulation effects solar activity poses on the ambient population of galactic cosmic rays. Particle detectors located at Aragats Space Environmental Center (ASEC), measure charged and neutral secondary fluxes and access wide energy domain of the primary cosmic rays. Please see Sect. 4 for more information on the SEVAN network.

IHY-Armenia hosted the international workshop: "Forecasting of Radiation and Geomagnetic Storms by networks of particle detectors" (FORGES-2008), held on September 29–October 3, 2008 in the International Conference Center, Nor Amberd, Armenia, 40 km from Armenia's capital Yerevan. The focus of the symposium was pointed on the space weather drivers and on possibilities of the networks of particle detectors measuring changing fluxes of neutral and charged particles to forewarn on upcoming severe radiation and geomagnetic storms, which was central to many of the IHY instrumentation programs. Forty scientists and students from Germany, Italy, Great Britain, Croatia, Greece, Ukraine, Russia, USA, Costa Rica and Armenia listen to 8 invited lectures and 25 original papers. One of topics of symposium was training of the SEVAN groups. Research groups from Croatia and Costa Rica were introduced to the SEVAN detector operation and data analysis.

We also launched a new educational site, which will serve as a major resource for Armenian students in space weather and cosmic rays, and of course for IHY.

7.4.2. Azerbaijan

Submitted by Elchin S. Babayev, IHY-Azerbaijan National Coordinator, Shamakhy Astrophysical Observatory, Baku
http://www.shao.az/IHY/

Azerbaijan, which is often called a gateway between Asia and Europe, has an astronomical glory thanks to the presence of the famous Maragha Observatory in southern Azerbaijan established by Middle Ages Azerbaijani astronomer Nasiraddin Tusi (XIII century). A new remarkable stage in the development of the national astronomy started in 20th century, and now there are two big observatories in Azerbaijan with a number of unique optical telescopes: Shamakhy Astrophysical Observatory (ShAO, in Shamakhy) and Batabat Astrophysical Observatory (BAO, in Nakhchivan). There are following major traditional scientific trends in astronomical researches: physics of stars and nebulae, solar physics, solar-terrestrial relations, investigation of the planets and solar system small bodies.

Expanding research activities in the field of Space Weather and its effects, throughout the whole world during the last several decades, has undoubtedly influenced the Azerbaijani scientific community, who have also initiated investigations on these problems. There is a well-established group on study of solar–terrestrial relations in the Azerbaijan National Academy of Sciences, joining efforts of highly skilled scientists and specialists from different fields of science and technology – astrophysicists, geophysicists, physiologists, doctors, engineers, etc. which conduct experimental, theoretical and statistical studies on Space Weather effects and influences on the different technological systems and human life and health (Figure 33).

Fig. 33. *Space weather plays a central role in IHY-Azerbaijan activities.*

The Azerbaijani group pays close attention to the public awareness and publication of scientific information about IHY activities, particularly, Space Weather, impending hazards from space, especially the prediction of solar and geomagnetic storms, etc. It is carried out, in particular, in domestic media, newspapers, the IHY-Armenia website, television, and radio broadcasts, as daily space weather information, weekly interviews, monthly newspaper columns, which are addressed mainly to public, medical and technical specialists. Scientific-popular articles are usually published in special journals, such as ELM ve HEYAT ("Science and Life"), YOL ("Way"), IRS ("Heritage"), and others.

Research on solar and solar-terrestrial physics as well as education and public outreach activities in these areas in Azerbaijan are becoming very popular. Throughout IHY, attention was given to the education process of astronomy and space sciences, particularly on solar and solar-terrestrial physics and space weather, carried out in high-secondary schools, gymnasiums, universities, academic institutions through lessons, lectures, seminars, domestic conferences, etc. There were prepared souvenirs, including hand-made small carpets, decorated with IHY logos. There also was a special corner established for IHY in the Astronomy Museum in ShAO.

Azerbaijan is also active in regional coordination for IHY, as Azerbaijan is one of founders and active members of the Balkan, Black Sea and Caspian Sea Regional Network for Space Weather Studies. The IHY Azerbaijani National Committee was established in January 2006 for the coordination all of IHY-related activities. This Committee includes not only scientists but also representatives of famous education organizations, journalists, TV and radio broadcasts. Since 2001 we have organized a Yuri's Night World Space Party every year on 12 April, and there has been a Tusi Summer Astronomical School every summer. They were continued in 2006–2008 as part of IHY activities. Sun–Earth Day "NOVRUZ" is celebrated in Azerbaijan every year. It is a National Spring Holiday; "NOVRUZ" means a "New Day". It comes from ancient times. Azerbaijanis believe that New Year starts from this day, which corresponds to the Vernal Equinox (Figure 34).

Two domestic conferences were organized in 2007 and 2008 in Azerbaijan with special sessions on Space Weather and Solar–Terrestrial Physics. Unfortunately, Azerbaijan could not organize International IHY meetings due to financial difficulties. However, representatives of Azerbaijan participated in many regional and international IHY-related meetings, symposia and conferences, either as a member of the Scientific Organizing Commitee, invited speaker or oral/poster presentation contributor. There is also a well-established "Working Group on Study of Solar-Terrestrial Relations and Space Weather Effects". Weekly Baku City Astro-Seminar discusses contributions from all organizations of Azerbaijan dealing with astronomy and space science, particularly, space weather, solar and solar–terrestrial physics.

Fig. 34. *Artwork created for the Yuri's Night World Space Party.*

Within IHY activities there were established bilateral and multilateral collaborations with the Institute of Solar–Terrestrial Influences of the Bulgarian Academy of Sciences (Bulgaria), the Pushkov Institute of Terrestrial Magnetism, the Ionosphere and Radio Wave Propagation (IZMIRAN) and the Space Research Institute of the Russian Academy of Sciences (Russia), Athens University (Greece), Kandilli Observatory and ERI, Bogazici University (Turkey), the Belgian Institute for Space Aeronomy (Belgium), the Rabin Medical Center and Tel-Aviv University (Israel), and others. These collaborations have resulted in number of joint scientific papers in refereed journals. One of these collaborative (Azerbaijani-Ukrainian) papers, studying space weather's possible impacts on power distribution systems in middle latitudes, was published in Greece and has been awarded by National and

Fig. 35. *Public lectures helped educate people about heliophysics and IHY.*

Kapodistrian University of Athens in 2008 as on one of outstanding papers (Figure 35).

As part of IHY activities and a joint INTAS grant, 2 Ph.D. students from Azerbaijan completed long-term training in solar and solar–terrestrial physics in Bulgaria, Turkey and Russia. IHY science also provided motivation for the "Laboratory of Heliobiology," which was established in 2007 in the Azerbaijan National Academy of Sciences as a joint laboratory of ShAO and Medical Center INAM with participation and consultancy of scientists from Russia, Bulgaria, Israel, and Belgium. IHY's impact also resulted in the establishment of Department of Physics of Solar–Terrestrial Relations as a new structural unit in ShAO in 2008. A special 1-h Report on IHY, Space Weather and Solar–Terrestrial Relations was done by E.S.Babayev in the Meeting Presidium of the Azerbaijan National Academy of Sciences among only 12 selected top-level reports per year.

In August 2008, an AWESOME Very Low Frequency (VLF) radio receiver (see Sect. 4) was installed. M. Cohen and S. Bijoor from Stanford University visited ShAO for these purposes. System, which functions fine, is used for remote sensing of the ionosphere and the magnetosphere and contacted to global net. Scientific collaboration is planned with Indian and Belgian colleagues (Figure 36).

Finally, an International Journal of Research and Applications called SUN & GEOSPHERE (see Appendix III) was established as the journal of IHY's Balkan, Black Sea and Caspian Sea Regional Network for Space Weather Studies. It is

Fig. 36. *A team sets up the new AWESOME VLF receiver.*

edited and published in Azerbaijan, and is a peer-reviewed journal that is abstracted in NASA ADS. The Journal website is: *http://www.shao.az/SG*

7.4.3. Bosnia and Herzegovina

Submitted by A. Andic, IHY National Coordinator for Bosnia and Herzegovina, University of Banjaluka, Republic of Srpska

The actions undertaken from IHY resulted in significant progress in Bosnia and Herzegovina. As a result of the actions, several new amateur Astronomer organizations were founded, one in Sarajevo and one in Tuzla. Together with the already existing one in Banjaluka they put a lot of effort into the popularization of

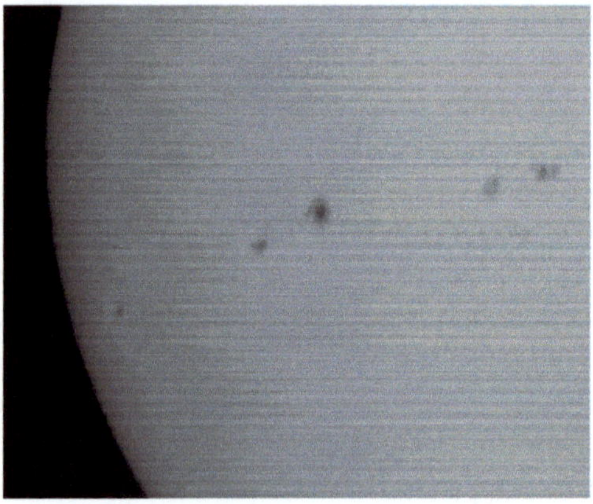

Fig. 37. *The image of the Sun taken during the observations in Sarajevo.*

Fig. 38. *Photo of the Yuri's Night planning team.*

astrophysics in general. Also, serious action and planning were conducted for educational outreach and popularization. The biggest activity in the last period was show by the Amateurs from Sarajevo. They organized the observation of the Sun for the general public and the celebration of the Yuri's Night (Figures 37 and 38).

They are very enthusiastic are already planning a plethora of new actions in the field of popularization of science.

7.4.4. Bulgaria 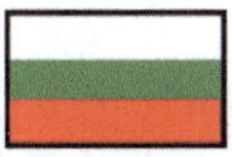

Submitted by Katya Georgieva, IHY-Bulgaria National Coordinator, and Penka STOEVA, IHY-Bulgaria Outreach Coordinator, Institute of Solar-Terrestrial Influences of the Bulgarian Academy of Sciences, Stara Zagora, and Alexey STOEV, Director of the Yuri Gagarin Public Astronomical Observatory and Planetarium in Stara Zagora

http://www.stil.bas.bg/IHY/

Bulgaria is the initiator of the Balkan, Black sea and Caspian Sea Regional Network for Space Weather Studies, created to coordinate the IHY activities in the region. The main scientific activities of Bulgaria are in the area of Heliophysical studies. Bulgarian scientists participate in several CIPs, and CIP54: "Inferring the magnetospheric global potential distribution from SuperDARN data through the IMech numerical model" is led by M. Kartalev from the Institute of Mechanics, Bulgaria. Four instruments have been designed and fabricated for space-borne measurements of heliophysical parameters, and a new SEVAN detector was established in Bulgaria as part of IHY.

Several IHY-related meetings were organized in Bulgaria, the most important ones being the Balkan, Black Sea and Caspian Sea Regional Planning Meeting in 2005, and the Fourth UNBSS-IHY meeting in 2008.

7.4.4.1. International Heliophysical Year – Education and public outreach activities in Bulgaria

In Bulgaria, joint initiatives of the Institute of Solar–Terrestrial Influences of the Bulgarian Academy of Sciences, and the Yuri Gagarin Public Astronomical Observatory and Planetarium, Stara Zagora are aimed to increase the visibility and accessibility of existing outreach programs and develop existing programs and activities to broaden their scope and impact. We have planned and coordinated activities – part of global outreach events. Many of them take place on a local scale according to the educational goals and resources.

Efforts were made with mass media – radio, TV, magazines, newspapers – and information and press conferences about IHY events have been regularly made available for journalists and special articles prepared for publishers. This period – March 2007–May 2008 was very successful. Due to all the activities, the discipline of heliophysics, we think, is already known as a basic science.

Lectures and Public Talks

"Astronomy for everybody" lectures are given and observations are conducted whole the year in the Astronomical Observatory and Planetarium on Monday.

Fig. 39. *Alexey Stoev gives an IHY lecture.*

At the day of the Vernal Equinox on March 21, 2007 a cycle of six lectures named "The Universe" began in the Art Gallery in Stara Zagora, devoted to the Sun–Earth Day and the IHY. The preparation of Yuri's Night 2008 began from the Sun–Earth day – March 20, 2008 – when the Spring arrives, with a cycle of six lectures of Dr. Alexey Stoev in the Stara Zagora Art Gallery – "Astronomy in Culture – an exciting journey through time and history of science" (Figure 39).
Additionally, public talks were given on the processes which govern the influence of the Sun on our solar system and its connection to the Earth and other planets. These talks were delivered in several locations in Bulgaria. Lectures devoted to prehistoric monuments and cults connected with the Sun considered as a God in the Antiquity were given in September.

Exhibitions

IHY-Bulgaria hosted a competition "We and the Sun" for paintings and photographs, for students in three age groups – 7–10, 11–15, and 16–20 years old – have been organized with about 1200 participants. The exhibition has been displayed in the star hall of the Planetarium and it has been visited by nearly 200 groups of students and adults. Additionally, the IHY "Open Doors Day" was held at STIL, Space Research Institute, National Astronomical Observatory and Public Observatories. The events featured videos, quizzes, space music, practical exercises, and observations (Figures 40 and 41).

Fig. 40. *Painted by Gergana Toteva – Kazanluk (18 years, 1st place).*

Fig. 41. *Painted by Irina Licheva – St. Zagora (11 years, 3rd place).*

Olympiad and Summer School

The National Astronomy Olympiad is an intellectual competition for students from Vth to XIIth class. The style of problems is aimed towards imagination development, cognitive and creative abilities and independent thinking of students. The National Summer School of Astronomy for students, solving problems for the International Astronomy Olympiad is usually in July. Many of the problems are connected with solar physics and space weather. Seven students

Fig. 42. *The Bulgarian olympiad team in Crimea, Ukraine.*

Fig. 43. *The Bulgarian team in Pune, India.*

participated in the International Astronomical Olympiad in Semeiz, Crimea, Ukraine in 2007 and visited the Astrophysical Observatory and solar telescopes (Figures 42 and 43).

The programme of identification and support of outstanding students on astronomy is realized in collaboration with The Ministry of Education and Science, Public Astronomical Observatories and Planetaria, Institute of Astronomy Bulgarian Academy of Sciences, Institute of Solar–Terrestrial Influences BAS

Fig. 44. *Penka Stoeva discusses with teachers and scientists the IHY EPO activities.*

and the Union of Astronomers in Bulgaria. Leaders: Dr. Alexey Stoev – Stara Zagora and Dr. Eva Bozhurova – Varna. Teachers from Astronomical Observatories and Planetaria have been actively involved in different world initiatives: Marlena Lyubenova – Silistra, Natasha Ivanova – Varna, Ioanna Kokotanekova – Haskovo, Boncho Bonev – Kurdzhali, Svetlana Tsekova – Yambol, Nadya Kiskinova – Stara Zagora, Iliya Iliev – Gabrovo (Figure 44).

Observatory Development

Stara Zagora is a host of a IHY SID Space Weather (see Sect. 5). Eng. Nikolay Stoyanov is the responsible person. We use it in the classroom to collect and analyze data for the space weather and its affect on the Earth's ionosphere, to exchange and discuss data in the web.

Solar Week 2007

We also participated in the 2007 Solar Week – Celebrating the Sun–Earth Connection in the classroom, organizing a Space Weather Action Centre around one of the computers in the Astronomical Observatory. Using materials from the STEREO observatories, we have made a 3D image of a rock sanctuary, which is very famous in our region. Students from the group of archaeoastronomy of the Yuri Gagarin Public Astronomical Observatory and Planetarium, Stara Zagora were very impressed and were happy to do this. More than 600 people visited the sanctuary at the day of Summer Solstice to observe the sunset in the opening of the equipment.

Sun–Earth Day 2007 and 2008

The theme in 2007 was "Living in the atmosphere of the Sun – IHY" and in 2008 – "Space Weather Around the World". We have given printed materials with the key understandings of the Sun–Earth day "Living in the atmosphere of the Sun – IHY" and annotations for the lectures. A lot of observatories and planetaria from Bulgaria, amateur astronomers, the European Centre for Education and Qualification "Europe Schools," Stara Zagora celebrated the Sun–Earth Day.

Yuri's Night World Space Party

Here, in Bulgaria, we usually have many celebrations on April 12. In 2007, along with the International Heliophysical Year, we organized 17 Yuri's Night parties at

Fig. 45. *Observations during Yuri's Night.*

15 towns in Bulgaria with more than 700 participants. We were very impressed from the idea to celebrate Yuri's historic flight and the first Space Shuttle flight in such free manner, everywhere. It was very attractive! Moreover, we, Bulgarians have two cosmonauts Georgi Ivanov (1979) and Alexander Alexandrov (1988), and traditions in space research (Figure 45).

In 2008 we organized four parties – in the Public Astronomical Observatories in four towns – Stara Zagora, Silistra, Haskovo and Varna, with about 500 participants. Yuri's Night Stara Zagora was the largest party where over 120 people attended the event. We had posters with the Yuri's Night 2008 programme posted everywhere in Stara Zagora to invite people to celebrate at the Public Astronomical Observatory and Planetarium. We made flyers and stickers and gave them to children and adults (Figure 46).

Fig. 46. *Students with the first Bulgarian cosmonaut, Georgi Ivanov.*

The party began with listening to the voice of Yuri Gagarin during the launch of VOSTOK spacecraft and video – a brief history of manned space flights. A special talk about Yuri Gagarin, Bulgarian cosmonauts and the Space Shuttle was given by Dr. Alexey Stoev. After that, students from the 3-year Astronomical school of the Observatory told about space missions and space stations – the cosmic home of mankind, showing many of slides. We had guest students from Holland and lecturers repeated the presentation in English for them. All the time people could see the exhibition "Beauty of the Planet Earth" of pictures painted from children in three age groups, devoted to the Year of the Planet Earth – 2008.

The celebration was followed by entertaining space games and quiz, space music, observations of celestial objects and the International Space Station. We had also food and drinks, and a lot of smiles – of the children and their parents or grandparents, of the students, teachers and scientists!

World Astronomy Day 2007–2008
Summer Solstice festival

A municipal competition "The Sun – our nearest star" has been organized and 150 children in four age groups participated in it. There were also lectures, films, questionary, competitions, broadcasts and astronomical observations.

World Space week – 4–10 October

Celebration of the 50th anniversary of the First launch of an artificial satellite of the Earth in Cosmos on November 4. A lecture about Bulgaria in space research. National competition of science fiction tale devoted to the Day of the National Leaders – 1st November.

Adaptations of educational resources and their translation in Bulgarian

Many printed materials were developed for distribution, including The Sun and Space Weather, Happy Equinox, What is Yuri's Night, UV detector, Parallax, Solar clock, and Solar biscuits. Also, printed materials with key understandings and annotations for the lectures were distributed to accompany the talks.

A series of educational comic books have been produced under the supervision and guidance of Prof. Y. Kamide. It is a collaborative project between the Solar–Terrestrial Environment Laboratory (STEL) at Nagoya University in Japan and the Climate and Weather of the Sun–Earth System (CAWSES) program. Their translation into Bulgarian allowed students throughout Bulgaria to learn about space weather.

7.4.5. Croatia

Submitted by D. Rosa, Zagreb Observatory, Zagreb, and B. VRSNAK, IHY-Croatia National Coordinator, Hvar Observatory, Faculty of Geodesy, Zagreb

7.4.5.1. Institutions participating in IHY activities

The main organizer of IHY-related activities in Croatia was the Zagreb Observatory, a public institution whose primary task is popularization of astronomy and other natural sciences. A significant contribution was provided also by Hvar Observatory, Croatian Astronomical Society, Technical Museum, and Astronomical Association of Zagreb.

7.4.5.2. Hosted instrumentation

In the framework of IHY program, two instruments are hosted in Croatia. In the scope of the Space Weather Monitor Program, a SID monitor (see Sect. 5) was installed and set to operation at the Zagreb Observatory in May 2008 (Figure 47).

Currently, preparations for installing The Hybrid Particle Detector from SEVAN project (Space Environmental Viewing and Analysis Network, PI Ashot Chilingarian from Alikhanyan Physics Institute, Armenia) are in progress. This

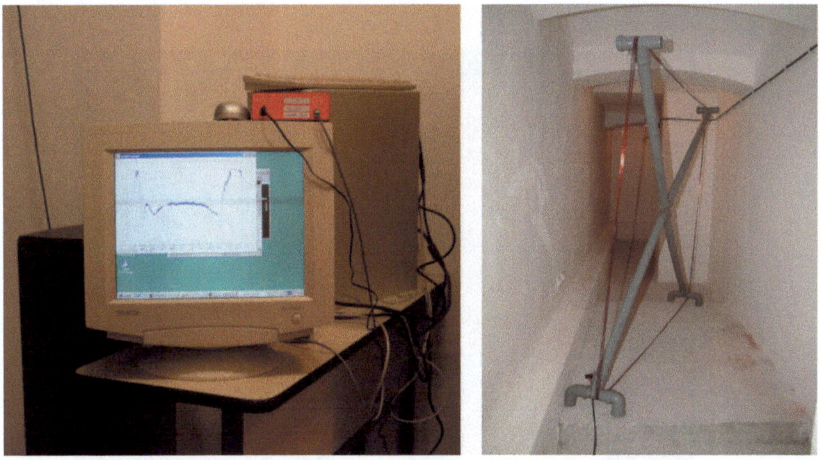

Fig. 47. *The SID monitor. Data recording (left) and the antenna (right).*

unit will be the first detector designed for measuring cosmic rays in Croatia and its installation will be finished in summer 2008.

Both instruments are installed at the Astronomical Observatory at the Astronomical Association of Zagreb.

7.4.5.3. Public outreach

Public outreach included several dozens of popular lectures held in Croatian primary and secondary schools, astronomical groups and societies and at public observatories. A number of contributions were presented in the media (newspapers, TV, radio).

In the program "Solar observation initiative – IHY," which includes white-light and H-alpha observations of the Sun at the Zagreb Observatory, thousands of school students and visitors participated. In collaboration with the Sonneobservatorium Kanzelhoehe (Austria), a poster "The Sun" (Figure 48), edited in three languages (Croatian, German, and English), was printed in 3000 copies and was/is distributed to pupils attending the IHY related public lectures.

There were also articles written by IHY scientists published in widely distributed magazines. The following articles on IHY were published in the Croatian astronomical popular-science magazine *Covjek I Svemir*: "International Heliophysical Year, IHY" by D. Rosa, "Sun's Magnetism" by M. Gigolashvili, and "Solar Rotation" by D. Rosa and D. Maricic.

The logo of IHY was printed in diplomas for pupils taking part in annual contests in astronomy 2007/2008.

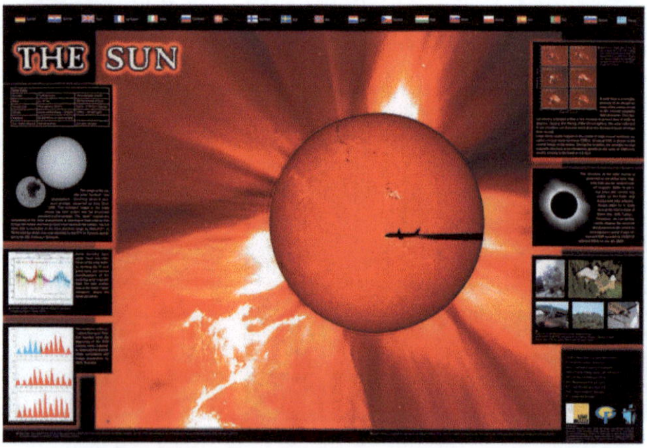

Fig. 48. *Poster "The Sun" printed in collaboration with the Sonneobservatorium Kanzelhoehe (Austria).*

7.4.5.4. Participation at IHY conferences

Croatian scientists actively participated in IHY conferences in Sozopol, Bulgaria, 2005, Nor Amberd, Armenia, 2006, and Sozopol, Bulgaria, 2008. Several contributions were presented, mainly related to the eruptive phenomena on the Sun. At 8th Hvar Astrophysical Colloquium held 24–29 September 2006, Hvar, Croatia, a session was devoted to IHY activities, including presentations several IHY international leaders. Similarly, the 9th Hvar Astrophysical Colloquium held 22–26 September 2008 also had a special session dedicated to IHY.

7.4.6. Georgia

Submitted by Prof. M. Gigolashvili, IHY-Georgia National Coordinator, Georgian National Astrophysical Observatory at Ilia Chavchavadze State University, Tbilisi, and Prof. G. Aburjania, Tbilisi State University, M. Nodia Institute of Geophysics, Tbilisi

7.4.6.1. IHY and related scientific activities in Georgia

In the framework of IHY in Georgia, research projects covered both the solar variations and their effects at Earth. Observations and theoretical studies were both performed, and recent results were achieved on the following topics:

Solar differential rotation (Prof. M. Gigolashvili and Dr. D. Japaridze): On the study of the solar differential rotation via the rates of H-alpha filaments and magnetic fields elements.

Investigation of total solar and spectral irradiance (Prof. M. Gigolashvili): On the variations of total and spectral solar irradiance and cosmic rays and their correlations with space weather effects during different phases of the sunspot cycle.

Lower and upper atmosphere investigation by optical methods in Abastumani (Prof. G. Didebulidze): On the observations of tropospheric aerosol distributions in the study of global physical ad chemical processes in the formation of structures in the atmosphere, and the variations of geomagnetic Ap index and 630 nm line intensity and their correlation with atmospheric variability.

Model of strong turbulence in the geospace environment (Prof. G. Aburjania): On the correlation between the space environment and nonlinear regular structures, and their excitation of Alfvénic waves and vortices, and the resultant turbulent and energetic effects.

7.4.6.2. Observatory development and hosted instruments

The two main observatories in Georgia performing IHY-related research are the Georgian National Astrophysical Observatory at Ilia Chavchavadze State

Fig. 49. *Abastumani Astrophysical Observatory, Mt. Kanobili, Georgia.*

Fig. 50. *The Lyot-type extra-eclipsing coronograph (D = 12 cm, F = 300 cm).*

Fig. 51. *The extra-eclipsing big coronagraph in Abastumani (D = 53 cm, F = 800 cm).*

University in Tbilisi, and the Abastumani Astrophysical Observatory at *Mt. Kanobili* (*Location*: 1700 m above sea-level). The Georgian National Astrophysical observatory (i.e. Abastumani Astrophysical Observatory) was integrated with Ilya Chavchavadze State University in 2007. With the help of the State University, the solar instrumentation was upgraded (see Figures 49–51).

In Georgia, several books for teaching astronomy in middle and higher level of education were printed during last 2 years:

1. Marina Gigolashvili – "Sun and Planets", The book on the Solar, Planetary and Solar–Terrestrial connections, printed in Tbilisi, Georgia, p. 68, 2007.
2. Marina Gigolashvili – Encyclopedia on Astronomy "Our Galaxy", is printed by "Elf's Printing House" Tbilisi, Georgia, p. 64, 2006.
3. George Aburjania – Self Organizationof Nonlinear Vortex Structures and the Vortex Turbulence in the Dispersed Media. Monography. Moscow: Publication Comp – "Kniga", URSS, p. 320, 2006.

7.4.7. Romania

Submitted by Cristiana Dumitrache, IHY-Romania National Coordinator, Astronomical Institute of Romanian Academy, Bucharest and Georgeta Maris, IHY-Romania National Coordinator, Institute of Geodynamics "Sabba S. Stefanescu" of the Romanian Academy, Bucharest

http://www.astro.ro/~ihy/

7.4.7.1. Introduction

Romanian scientists join the international effort in fulfillment of IHY objectives. We note that many of our activities (publications, conferences and the international school) were based on projects financially supported by the Romanian Ministry for Education and Research.

7.4.7.2. Historical perspective and scientific projects

Fifty years ago, on the occasion of the International Geophysical Year, two research teams started their activity in Romania: the solar group and the artificial Earth satellite group. This moment meant the initiation of the Romanian astrophysics research at Bucharest Observatory.

Solar observations have known three phases at Bucharest Observatory from the beginning until now:

(1) 1956–1997 – Photospheric and chromospheric survey – published in the own bulletin “Observations Solaires” and sent to main world data centers
(2) 1999 – the total solar eclipse projects
(3) 2001–2002 – CCD Hα observations Solar researches covered observational works and data interpretation, statistical studies and MHD numerical simulations of the coronal structures that started in 1995.

A comprehensive history of Romanian astrophysics, “Fifty Years of Romanian Astrophysics” by C. Dumitrache and N. A. Popescu, was published at the fiftieth anniversary of solar observations start in Bucharest. After 50 years a new team with young people started to work in solar physics, in Romania. The Bucharest Observatory solar physics group at present includes two senior researchers, one post-doc and five Ph.D. or master students, with research mainly on two topics: the dynamics of solar atmosphere and heliosphere, under the aegis of the Romanian Academy, and Romanian contributions to the Sun-Heliosphere Studies, a PECS-ESA project. The first project focuses on data analysis and interpretation, active regions and filament evolution, transient phenomena, numerical MHD simulations of solar phenomena. The second project has as objectives:

- Study of configurations observed in the solar atmosphere susceptible to give CMEs and other instabilities (SOHO and ground based observations)
- Heliospheric studies using Ulysses data

- Theoretical studies on magnetic topologies susceptible to give instabilities (CMEs) approached using the dynamical systems tools or numerical simulations; 3D simulations of heliospheric electric and magnetic fields.

Part of this second project is registered as CIP 71, which focuses on the study of CMEs/ICMEs using SOHO/Ulysses Data. Our main objectives are the study of solar sources of CMEs (polar prominences, active regions), and the follow up of CMEs from SOHO to Ulysses or track back of ICMEs from Ulysses to the Sun.

At the Institute of Geodynamics of the Romanian Academy, a project incorporating the physics of the Sun, the heliosphere, the magnetosphere and the terrestrial atmosphere was developed, called "Promoting Romanian Research Regarding the Study of Variations in the Sun–Heliosphere–Earth System in European and International Programmes". Additional programs, "Heliospheric Variabilities and their Impact on Some Components of the Terrestrial System" and "Long-Term Variability of the Geomagnetic Field in Relation with the Physical Processes in the Heliosphere," develop further studies into the relationship of the Sun, Earth and heliosphere, all topics fundamental to IHY.

Special attention was paid for international events organized in Romania. A conference was organized for an anniversary to bring together scientists from the main fields of astrophysics and space science. It was a celebration of the 50 years from the start of the solar observation at Bucharest Observatory and the beginning of astrophysics researches in Romania. Topics discussed included: solar atmosphere, magnetic structure on the Sun, solar and stellar winds, waves, oscillations and cyclicities, astroseismology, stellar astrophysics, and extragalactic astronomy and cosmology. Another event was an workshop divided in two important sections: (1) sun and heliosphere; (2) stars and galaxies. The first section treated solar magnetic structures such as active regions, flares and CMEs, prominences and coronal streamers. Both observational and models works were presented. Flows, boundaries and interfaces were approached in numerical simulations. Interface between streamers and coronal holes and also solar corona rotation were deduced from SOHO data. Coronal waves and Sun quakes were also treated in a few papers. Some papers came from dynamical systems or differential equations formalism to solar physics or heliospheric phenomena.

Additionally, the Second International Symposium on Space Climate "*Long-term Changes in the Sun andtheir Effects in the Heliosphere and Planet Earth*" was held in Sinaia, Romania, 13–16, September 2006, under the aegis of the Romanian Academy. It washosted by the Astronomical

Institute of the Romanian Academy and the Institute for Space Sciences, Bucharest, Romania.

7.4.7.3. Education and outreach

Since the solar group was renewed in last years, many young researchers join our team. Solar physics is not a topic taught in Romanian universities and for this reason we had to train the young researchers. A Young Scientists School on Heliosphere and Galaxy held in Sinaia in May 2007, consisted of twenty five students and seven professors. Both lectures and students presentations were published, and another book dedicated to researchers training is the first solar physics monograph in Romanian language ("Protuberantele Solare", edited by C. Dumitrache).

The Astronomical Institute of Romanian Academy (AIRA) organizes examinations for a doctoral degree. The solar group has one Ph.D. student at AIRA and one master student at Bucharest University. Their theses will approach a topological model for the electromagnetic heliospheric field and respectively halo CMEs solar sources and their effect on the interplanetary space and Earth. We also have several activities with high school students participanting in the national olympiad of astronomy.

We paid important attention to the outreach in our activities. We covered a large series of topics: publications, conferences, exhibitions, open doors (Figure 52) and on-line presentations in Romanian language. A booklet abut the Sun was published in Romanian and was freely distributed to the public.

The outreach section of the IHY-Romania website contains several projects: power point presentations in Romanian language, and a link to another site – a

Fig. 52. *Romanian students learn about heliophysics during "Open Doors Day."*

solar blog in Romanian (*http://blogsolar.blogspot.com/*). We also outline a project jointing science and music, called "Music of the spheres".

7.4.7.4. Conclusions

The 50th anniversary of IGY, in the frame of IHY, has been celebrated by Romanian astronomers as an anniversary of 50 years from the solar physics observations beginning at Bucharest Observatory. The International Heliophysical Year activities (2006–2008) lead smoothly into the International Year of Astronomy (2009). The Centenary of Bucharest Observatory has the privilege to be framed between these two important international events.

7.4.8. Slovakia

Submitted by Karel Kudela and Vojtech Rusin, IHY-Slovakia National Coordinators, Slovak Academy of Sciences, Koskice
http://ihy.saske.sk/

7.4.8.1. IHY-Slovakia Steering Committee

Aleš Kušera (Astronomical Institute SAS, Tatranská Lomnica) Karel Kudela (COSPAR NC, IEP SAS, Košice – coordinating) Alina Prigancová (Geophysical Institute SAS, Bratislava) Pavol Rapavý (Slovak Union of Astronomers Amateurs) Ján Rybák (Astronomical Institute SAS, Tatranská Lomnica) Vojtech Rušin (Slovak Astronomical Society) Július Sýkora (SCOSTEP NC).

7.4.8.2. Participating institutions (partial list)

Astronomical Institute, Slovak Academy of Sciences, Taranská Lomnica Geophysical Institute SAS, Slovak Academy of Sciences, Bratislava Institute of Experimental Physics, Slovak Academy of Sciences, Košice Slovak Union of Astronomers Amateurs Observatory in Rimavská Sobota High mountain observatory at Lomnický štít.

7.4.8.3. IHY Measurements in Slovakia

- Cosmic rays at high mountain laboratory at Lomnický štít
- Coronal emission spectral lines 530.3 and 637.4 nm and prominences at Lomnický štít
- Observations of the Sun at the Rimavská Sobota observatory

7.4.8.4. Instruments

1. Coudé refractor – daily drawings of the sunspots
2. Jensch type spectrohelioscope – patrol observationsof chromospheric flares
3. Lyot coronagraph – patrol observations of prominences
4. Horizontal Solar Telescope with Spectrograph (HSTS) – spectroscopic observations of the solar spectrum

7.4.8.5. IHY-Slovakia "Open Doors Day"

There were four sites in Slovakia participating in the IHY-Europe "Open Doors Day": High Mountain Observatories, Lomnicky Stit Cosmic Ray, the Slovak Central Observatories at Hurbanovo, and local schools at Tatranska Lomnica (Figure 53).

Fig. 53. *Participants in the IHY "Open Doors Day" learned how a telescope operates and participated in scientific activities.*

7.4.9. Turkey 

Submitted by Atila Ozguc, IHY-Turkey National Coordinator, Kandilli Observatory and E.R.I. Bogazici University, Istanbul and Yurdanur Tulunay, IHY-Turkey Outreach Coordinator, Department of Aerospace Engineering, Middle East Technical University, Ankara

http://www.ihy2007.boun.edu.tr

We can summarize our IHY activities in Turkey as follows:

National initiatives: Solar Physics, Planetary Ionospheres, and Climate studies.

International initiatives: Especially in the Balkans, Black Sea and Caspian Sea Region Planetary Magnetospheres, Heliosphere and Cosmic Rays cooperative projects.

Infrastructure: Build up new Excellency Centers and observing stations.

Outreach and dissemination: Organizing meetings on natural events and development some projects on about the undistinguished capabilities at schools.

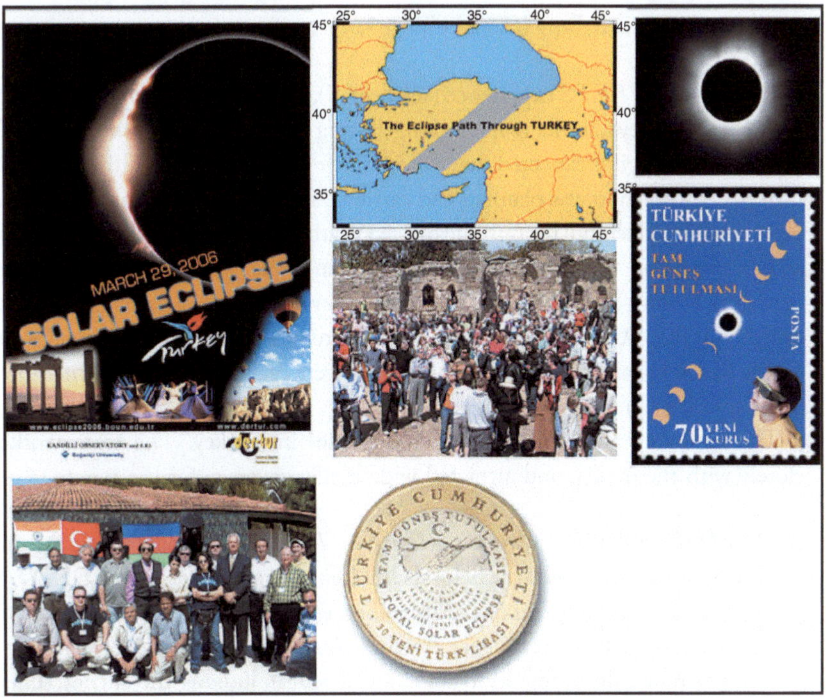

Fig. 54. *An IHY Viewing Station for the 2006 Solar Eclipse: more than one million pairs of sunglasses were distributed along the totality path.*

The IHY-Turkey activities were initiated with the meeting of "Second Annual Meeting of the Balkan, Black Sea and Caspian Sea Regional Network on Space Weather Studies" which was held at Sural Saray Hotel in Manavgat, Antalya, Turkey, between March 30 and April 1, 2006. The meeting was organized by Bogazici University, Kandilli Observatory and Earthquake Research Institute, Istanbul, Turkey right after the Total Solar Eclipse on March 29, 2006. The time and the place of the meeting were chosen at the occasion of the Total Solar Eclipse of March 29, 2006, whose line of totality was passing through the meeting place. Most of the participants witnessed this exceptional, natural event for the first time in their lifetime. Some others performed some experiments related with the eclipse. Eighty-one scientists including leading experts in various research areas of Space Weather from many countries, including Armenia, Austria, Azerbaijan, Belgium, Bulgaria, Croatia, France, Germany, Georgia, India, Japan, Poland, Romania, Russia, Slovakia, Turkey, Ukraine, and USA attended this meeting and presented 59 exciting presentations (Figure 54).

7.4.9.1. Space Weather and Europe – an Educational Tool with the Sun (SWEETS) in Turkey

Based on the SWEETS experience (see Sect. 5), there has been a special activity, initiated in Turkey in order to have a long-lasting positive impact of the SWEETS program. In turn, it is expected that this initiative would create a positive impact in Europe. A workshop was planned and carried out as the first step for a horizontal development in Istanbul to continue Space Weather activities in science, education and outreach on 6–7 December 2007. During this workshop, some invited participants from several universities in Turkey have agreed to set up a group called UHUEG (**U**zay **H**avasi **U**lusal **E**ylem **G**rubu – the National Action Group on the Space Weather). The working groups were established and information – communication net was set up.

The following are some SWEETS activities in Turkey. The team worked closely with the media, and attracted wide media coverage for the events.

(1) *"I LOVE MY SUN" outreach activity in Turkey*: The Space Weather and the Sun as conceived by the School Children of age 7–11, as a *SWEETS/IHY/COST 724* Case Sub-project. Altogether, there were 51 students involved in the exercise. The "Art Gallery Armoni" (*http://www.armoniartgallery.com*) in Ankara kindly accepted to act as an advisor in judging the entries. There was a shortlist of four children paintings of the Sun and the Sun–Earth relation.

(2) *SWEETS-related art:* Inspired by the SWEETS ballet "Sonnensturm," Y. Tulunay has taken every opportunity to attract the attention of the artists to view space weather in relation to the SWEETS activity. The artist Mr. Yalçın Gökçebağ was chosen for a program training an artist on the space weather.

There was also the Art Istanbul Fair: "Art Days" at the "Istanbul Museum of Modern Art" – 5th Antrepo on 10–18 November 2007. A painting specifically devoted to the FP6 SWEETS and the poster "Space Weather" were among the other pieces of paintings and sculptures exhibited by the Art Gallery Armoni.

The comic Books: "What is the Geomagnetic Field?" and "What is the Solar Wind?" were translated into Turkish and were made available to students.

(3) *World-wide web quiz on space weather and the IHY – in Turkey*: IHY-Turkey team members completed the translation and competition of a World-wide Web Quiz on Space Weather and IHY – *Web Quiz all over Europe.*

(4) *Public events and fairs*
  - The 7th Middle East Technical University "Promotion Fair" has been devoted to the high school graduates in Turkey who are at the stage of deciding about their future career, after having fulfilled positively the requirements of an extremely competitive university entrance examination. As part of such activities, the candidate children are introduced to the academic programs offered at the METU.
  - High school graduates visited the Cultural and Conference Center at the METU, which had a special exhibit on IHY and FP6 SWEETS (26–28 July 2007) (Figures 55 and 56).

Fig. 55. *An artist at work: Discussing the artistic interpretation of space weather with faculty at METU.*

Fig. 56. *The SWEETS Space Weather art exhibit at Art Gallery Armoni in Ankara.*

- The International Information Fair in Istanbul, Turkey (CeBIT Eurasia), has been held regularly every year. In 2007, there was a special conference sponsored by the TÜRKSAT Company, dedicated to the World Space Week. There were also World Space Week celebrations at the University.
- There was an "Open Day" at the METU Department of Aerospace Engineering (AEE) to promote SWEETS on 24 October 2007.

7.5. Western Europe

Participating Nations: *Austria, Belgium, Czech Republic, Denmark, Finland, France, Germany, Hungary, Ireland, Israel, Italy, Norway, Poland, Portugal, Slovakia, Spain, Sweden, Switzerland and United Kingdom.*

http://calys.obspm.fr/IHY/

The IHY-Western Europe initiative is comprised by a number of extremely active subgroups, many of which are also involved in the international planning of the IHY. IHY-Europe team members are active in all four of the IHY Programmatic Thrusts, including serving as Principle Investigators for the UNBSS Observatory Development Program. Several new missions to be launched, plus a rich network of ground-based observatories provide the basis for many of the IHY science campaigns. IHY planning discussions and special scientific sessions have occurred at several scientific meetings with the first IHY European General Assembly to be held in Paris, France January 10–13, 2006. Please refer to the IHY-Europe home page for news, updates, and an interactive map of IHY national coordination activities in Europe.

One of the biggest events of the IHY was the June 10, 2007 "Open Doors Day" event, which invited the public to visit IHY observatories and institutes. There are many references to this event in the following national reports. Additionally, several nations hosted activities that invited children to create artwork representing the Sun, heliosphere, and geospace.

7.5.1. Austria

Submitted by Helmut O. Rucker, IHY-Austria National Coordinator, Space Research Institute, Austrian Academy of Sciences, Graz

http://ihy-austria.oeaw.ac.at

IHY-Austria was active in both the science and outreach activities throughout 2007–2008. Among the many scientific activities, the official IHY Opening Ceremony was held at the United Nations in Vienna on February 19, 2007, followed by the IHY Science Workshop on February 20, 2007 hosted by the Austrian Academy of Sciences. Organizers representing all IHY regions, along with many IHY partner organizations, attended the day-long workshop, which covered science and outreach plans for the upcoming 2 years.

In the context of the "International Heliophysical Year 2007" several institutions in Austria perform scientific research in the field of solar and heliospheric physics:

- IWF Graz: Institut für Weltraumforschung/Space Research Institute of the Austrian Academy of Sciences located at Graz
- KF Uni Graz: Institute of Physics, Division of Geophysics, Astrophysics and Meteorology (IGAM) of the Karl Franzens University Graz
- Uni Vienna: Institute for Astronomy of the University of Vienna
- Innsbruck: Institute for Astro- and Particle Physics, University of Innsbruck.

In the area of outreach, IHY activities were held from 2007 to 2009. Fourteen institutes and observatories in Vienna hosted "Open Doors" events (see Sect. 5), from June 6 to 11, 2007. The event attracted considerable media attention, and articles on this event appeared in several newspapers.

Fig. 57. *Brochure advertising the 2009 "Within the Fire of our Sun" exhibition in Vienna.*

As IHY-Austria hosted the Opening Ceremony for IHY, it will also serve as the host for the Closing Ceremony in February 2009. In conjunction with this event, from January 26 to April 24, 2009, the Space Research Institute of the Austrian Academy of Sciences is organizing ESA's touring-exhibition entitled "*Im Feuer der Sonne*", or "Within the Fire of our Sun" (see Figure 57). It reviews the ongoing European solar system research and shows a set of spacecraft models, such as Ulysses and SOHO, and corresponding experiments. An opening event for the media, for teachers and the public was given on January 23, 2009, with an inaugural address of H.O. Rucker. Further lectures are given on April 3, 2009, on "The Giant Planets". From May 1 to July 27, 2009, the exhibition will be stationed at the planetarium of the city of Judenburg, Styria, finally in the planetarium of the city of Klagenfurt, Carinthia, until the end of September 2009.

7.5.2. Belgium

Submitted by Stefaan Poedts, IHY-Belgium National Coordinator, Katholieke Universiteit, Leuven, and Norma B. Crosby, IHY-Belgium Outreach Coordinator, Belgian Institute for Space Aeronomy, Brussels

http://gauss.oma.be/ihy2007/

7.5.2.1. International Heliophysical Year at Belgium

Belgian IHY Steering Committee

An "IHY Steering Committee" was formed for the organization of the IHY activities in Belgium. This committee consisted of members of the major Belgian institutes involved in 'heliophysical' research. The Belgian IHY Steering Committee was chaired by Prof. Stefaan Poedts (Centre for Plasma-Astrophysics of the K.U. Leuven) and collaborating institutions (members) were:

- Belgian Institute for Space Aeronomy (Johan De Keyser, Vivianne Pierrard, Norma Crosby, Stephanie Fratta, Tim Somers)
- Royal Observatory of Belgium (David Berghmans, Frédéric Clette, Jan Cuypers, Sophie Raynal, Petra Vanlommel)
- Royal Meteorological Institute (Steven Dewitte, François Brouyaux)
- Belgian National Committee for Geodesy and Geophysics (Véronique Dehant)
- Planetarium of the Royal Observatory of Belgium (Anne-Lyze Kochuyt, Rodrigo Alvarez).

IHY web pages

The Belgian IHY Steering Committee set up the tri-lingual Belgian IHY website with general information (in Dutch, French and English) on the goals of IHY and with a list of Belgian IHY activities and links to other (international) initiatives. Belgian activities under the umbrella of the IHY were both of national and international character. They are described in the following and have ranged from specialized IHY events to general public outreach activities.

Open Doors event

The Belgian IHY steering committee organized and coordinated an "Open Doors" event (see Sect. 5) during the weekend (6–7 October 2007) at the Space Pole (Plateau of Ukkel, Ringlaan 3, 1180 Brussels). The event was very successful and attracted more than 10,000 visitors. It was aimed towards both the French-speaking community "Portes ouverts 2007 6 & 7 Octobre" and the Flemish-speaking community "Opendeurdagen 2007 6 & 7 Oktober". In the framework of the IHY, the "Open Doors" theme was "The Sun and the Interaction of the Sun with the Earth".

The buildings of the Belgian Institute for Space Aeronomy (BISA), the Royal Observatory of Belgium (ROB) and the Royal Meteorological Institute (RMI) were accessible to the public and contained exhibitions and demonstration setups. In addition, tents were set up with activities for children (e.g. quiz sites, poster exhibitions). Specifically, six dedicated PCs allowed participants to take a scientific quiz that had been developed in three languages (Dutch, French and English). Also, a special issue of "SPACE Connection", the space annex of "SCIENCE Connection" which is the information magazine of the Belgian Space Policy (BELSPO), was made for this occasion and was entirely devoted to heliophysical research in Belgium. This issue of "SPACE Connection" was printed in 16,000 copies in French and Dutch and was a great success at the "Open Doors" event.

One of the highlights of the "Open Doors" was the participation of the "Space Weather and Europe – an Education Tool with the Sun (SWEETS) Space Weather Mobile Bus" (see Sect. 5 and Figure 58). The SWEETS expo bus made the tour of Europe and visited Brussels in the period 6– 8 October 2007. It was a great success at the Belgian IHY "Open Doors" event. The "Open Doors" was covered extensively by the media in news reports in Belgian newspapers and on the five main TV-channels in Belgium.

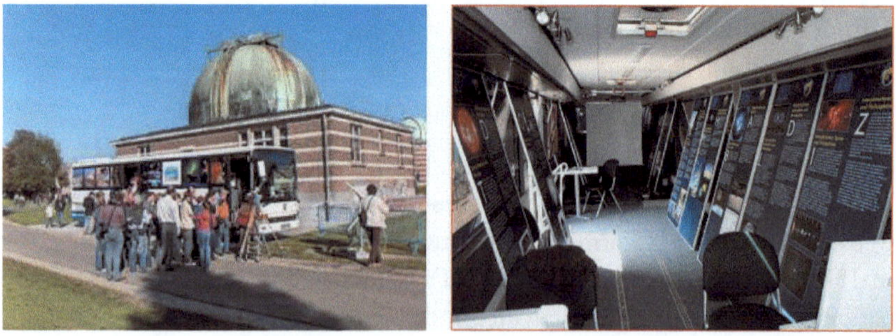

Fig. 58. The "*SWEETS Space Weather Mobile Bus*" *makes a special appearance during the* "*Open Doors Days*".

Conferences and workshops

Various workshops and conferences were held in Belgium under the umbrella of IHY2007 and they are listed here:

- The workshop "Comparing Earth and Venus" was given by Ms. Rosalyn Pertzborn (Director for the Office of Space Science Education) and Dr. Sanjay Limaye (senior scientist), both from the University of Wisconsin-Madison, at the Planetarium of the Royal Observatory of Belgium in Brussels, 25 April 2007. The event was organized under the umbrella of IHY, by Anne-Lize Kochuyt from the Planetarium in collaboration with Norma Crosby from the Belgian Institute for Space Aeronomy.
- The General Scientific Meetingof the Belgian Physical Society (BPS) was held on 30 May 2007 at the University of Antwerpen (Aula Maior buildings (Q and R) of Campus Drie Eiken). The parallel session "Astrophysics, geophysics and plasma physics" chaired by Viviane Pierrard, Belgian Institute for Space Aeronomy, was especially dedicated to IHY, with an invited talk introducing IHY and heliophysics by David Berghmans from the Royal Observatory of Belgium, and several oral and poster contributions.

IHY Gold Club medals

In connection with the welcome reception on the first day of the "Fourth European Space Weather Week" meeting in Brussels, a ceremony in the framework of the IHY was held. Three Belgian scientists had been proposed for the IHY Gold Club:

- Rene Dogniaux (worked at the Royal Meteorological Institute of Belgium, was the head of their radiometry division, and was an European expert in solar radiation)
- R. Gonze (worked at the Royal Observatory of Belgium in solar physics)
- Andre Koeckelenbergh (worked at the Royal Observatory of Belgium in solar physics and was the founder of the SIDC)

They had all fulfilled the criteria to become IHY Gold Club members and became recipients of the gold "IGY 50th anniversary" pin at this dedicated ceremony.

Exhibitions

A two-part exhibition was produced at the Planetarium of the Royal Observatory of Belgium (Boeckoutlaan 10, 1020 Brussels), with the first part focusing on the Sun as a star and global climate change. The second part focused on space weather and heliophysical research and was associated with the 50th anniversary of the Sputnik launch as well as linked to IHY. The exhibition included poster panels, a power point presentation running on an LCD screen and a 3D model of the Sun. This exhibition was visited by hundreds of school children and their teachers that visited the Planetarium and, of course, also by the many other people that visit the Planetarium every week. The exhibition ran until August 2008 and part of it was also used at "Science fairs", Space exploration days, "Open Doors").

The IHY Belgian Steering Committee also contributed to the contents and layout of the poster panels on the above-mentioned SWEETS bus and provided translations of the panels in Dutch and French.

TV documentary

Unfortunately, the advanced plans for a TV documentary on the Sun and solar and space environmental research in Belgium had to be cancelled due to a major reorganization of the Belgian state TV. In spite of already made agreements with the EU Descartes-prize winning producer, Jos Van Hemelrijck, and also with the director, Philippe Bijvoet, the plans had to be abandoned due to a lack of budget after the major restructuring. A small chance exists that a BBC documentary, already purchased for this occasion, will still be translated and broadcast on Belgian TV, but a special Belgian documentary is no longer feasible.

IHY and space weather DVD

The IHY Belgian Steering Committee also contributed to the contents and provided images, movies (visualizations of numerical simulations) and text for the IHY and space weather DVD that was produced as part of the SWEETS project in eight languages. The entire DVD was translated into Dutch by the CPA/K.U. Leuven team.

Web quiz

The IHY Belgian Steering Committee contributed to the setup of the SWEETS web quiz – formulating the questions – and translated all the questions to Dutch. The quiz was put on-line on the Belgian SWEETS web page and was made accessible both in Dutch and in French. The prize to win was a full day VIP visit to the Royal Observatory in Brussels including the Belgian space weather facilities and prediction center at the Solar Influences Data Analysis Center (SIDC). However, due to technical reasons, it was informed afterwards that there was no Belgian winner of the quiz.

Space Weather studies in Eastern Europe and Western Asia

An article by Norma B. Crosby, Belgian Institute for Space Aeronomy, Brussels, Belgium and Elchin S. Babayev, Shamakhy Astrophysical Observatory, Baku, Azerbaijan appeared in the AGU Newspaper Eos, Vol. 89, No. 15, 8 April 2008. It described the IHY Balkan, Black Sea and Caspian Sea Regional Network and the international refereed scientific journal, "Sun and Geosphere" journal that have been established to promote space weather studies in southeastern and east central Europe and in the Caucasus region under the umbrella of IHY. The main impetus behind establishing the Balkan, Black Sea, and Caspian Sea Regional Network on Space Weather Studies is to strengthen activities in the region in the field.

7.5.3. Czech Republic

Submitted by Frantisek Farnik, IHY-Czech Republic National Organizer, Astronomical Institute, Ondrejo
http://ihy2007.astro.cz

7.5.3.1. IHY in the Czech Republic: concluding report

At the end of the International Heliophysical Year we would like to briefly summarize achievements and results of Czech activities. As already declared at the very beginning of the IHY period we stressed aspects of education and popularization in our program, i.e. we preferred popularization to scientific collaboration which is already at a high level in the Czech Republic. Therefore, as the first step, we opened IHY web pages in the Czech language.

Then, in chronological order, we organized the following actions:

January 2007

- IHY-Czech Republic Kickoff Meeting, hosted by the Academy of Sciences of the Czech Republic (Figure 59).
- Press conference on IHY and its purpose.
- Lectures on astrophysical topics for secondary school students.

Fig. 59. *A photo from the IHY kickoff meeting taken at the headquarters of the Academy of Sciences of the Czech Republic. Shown from left to right are Dr. Frantisek Farnik, IHY-Czech Republic National Coordinator, Dr. Eva Markova, Director of the Upice Observatory, and Dr. Petr Heinzel, Dr. Sc., Director of the Astronomical Institute of the Academy of Sciences.*

February 2007

- A member of our Coordination Committee represented our country at the IHY 2007 opening ceremony in Vienna.
- A newly prepared exhibition "The Sun – a star of life and death" was officially opened and accompanied with lessons for different schools.

March 2007

- Three different contests were announced for children and students:
  (e) Children's art "Universe as seen through children's eyes".
  (f) Observations of the Sun, solar spots and application of solar observations.
  (g) Web pages on solar energy.

April 2007

- Following the recommendation of the Czech IHY Committee, a special edition of science and technology magazine *PORT* (broadcast by the Czech TV) was produced. The magazine was dedicated to IHY activities, science, and outreach events.

May 2007

- Evaluation of the Children's art contest. We received several hundreds of children's drawings.

June 2007

- "Open Doors Day" (see Sect. 5) at the Astronomical Observatories of Ondrejov and Upice.

September 2007

- A seminar on the Sun, taking place in Brno and aimed at professionals working for small observatories and planetaria.

October 2007

- An International solar seminar on the Sun–Earth relations at the Valasske Mezirici Observatory.
- Evaluation of the (b) and (c) contests.

November 2007

- An IHY meeting in the headquarters of the Academy of Sciences of the Czech Republic in Prague. During the meeting the winners of all the three contests received awards. A lesson on the Sun followed. Children then visited the Public observatory in Prague and students had an opportunity to attend an excursion entitled Astronomical Prague taking them to places connected with astronomy.

January 2008

- A set of lessons on the Sun given by Dr. Rusin from Slovakia.

February 2008

- A press conference winding up the IHY in the Czech Republic was held in the headquarters of the Academy of Sciences in Prague.

During the whole year several articles on IHY were published in our newspapers and radio as well as our TV broadcasted different information about IHY activities. The list of those "media activities" is available on the Czech web pages, unfortunately in Czech language only. The following links contain the media activities:

http://ihy2007.astro.cz/aktivity/tisk/, *http://ihy2007.astro.cz/aktivity/radio/*, *http://ihy2007.astro.cz/aktivity/televize/*

All the Czech activities were organized by the Steering Committee which had 13 members, which are given on the Czech web pages.

Eight partners joined in the project:

- Astronomical Institute of the Academy of Sciences of the Czech Republic
- Czech Astronomical Society
- Observatory of Upice
- Observatory of Prostejov
- Institute of Physics of the Academy of Sciences of the Czech Republic
- Institute for Atmospheric Physics of the Academy of Sciences of the Czech Republic
- Institute of Geophysics of the Academy of Sciences of the Czech Republic
- CEZ (Czech Electricity Provider Group)

IHY-Czech Republic also had the following Media Partners:

- Czech TV – Port
- Lidove noviny (Czech newspaper)
- Czech Radio Leonardo
- Ucitelske noviny (newspaper for teachers)
- Astropis (astronomical journal)
- 21.stoleti (technical journal)

7.5.4. Finland

Edited from contributions of the Finnish IHY Team by Rami Vainio (IHY-Finland National Coordinator), Department of Physics, University of Helsinki and Kirsti Kauristie (Co-Chair, ICESTAR), Finnish Meteorological Institute, Helsinki

http://www.ihy2007.fi

7.5.4.1. International Heliophysical Year: a report on Finnish activities

Heliophysical research in Finland is conducted in the *University of Helsinki (UH)*, the *University of Turku (UT)*, the *University of Oulu (UO)*, and the *Finnish Meterological Institute (FMI)*. The IHY activities in Finland were started by a Kick-off meeting in Helsinki in February 2006. All research teams were invited to and represented in the meeting. The meeting elected Dr. Rami Vainio (UH) as the National IHY coordinator and Dr. Kirsti Kauristie (FMI) as the person responsible for the national website. After the meeting, the website was created and a dedicated mailing list (*ihy-2007@helsinki.fi*) established as a means of communication of the groups.

The groups and their heliophysical research activities

The two space physics groups located in Helsinki – the UH and FMI groups – are located at the same Campus. They belong to a single umbrella organization, the *Kumpula Space Centre*, and together cover most of the topics of relevance to IHY. The group at UH concentrates on modeling and data analysis of solar eruptions and their manifestations in the interplanetary medium, and on the geoefficiency and other space weather effects of the eruptions. The group at FMI designs and manufactures spaceborne instruments, has long traditions in auroral data acquisition, and makes versatile basic research. Examples of ESA and NASA missions with FMI/UH participation are *SOHO*, *Cassini/Huygens*, *Mars Express*, *Venus Express*, and *BebiColombo*. FMI's endeavours in theoretical modeling include the only European global magnetohydrodynamic simulation code for the Earth's magnetosphere and hybrid simulation codes for the Mercury, Venus, and Mars plasma and neutral gas environments.

FMI started the systematic auroral recordings during the International Geophysical Year in 1957–1958. The present successor of those IGY activities is the *MIRACLE* (Magnetometers-Ionospheric Radars-All sky Cameras Large Experiment) network (see Figure 60) of versatile ground-based auroral instrumentation ranging from Estonia to Svalbard.

Heliophysics in the *University of Turku* is conducted in two laboratories. The *Space Research Laboratory* is a part of the Department of Physics/UT. Founded in the end of 1980s, the laboratory has its roots in the preceding 30-year activities of the Cosmic Ray Research Group of the university. The objectives of the laboratory are to explore space physics, particularly solar and heliospheric physics related to energetic particles, and to develop instrumentation for the research. The laboratory participates in the *SOHO* mission as the PI institute of the energetic particle

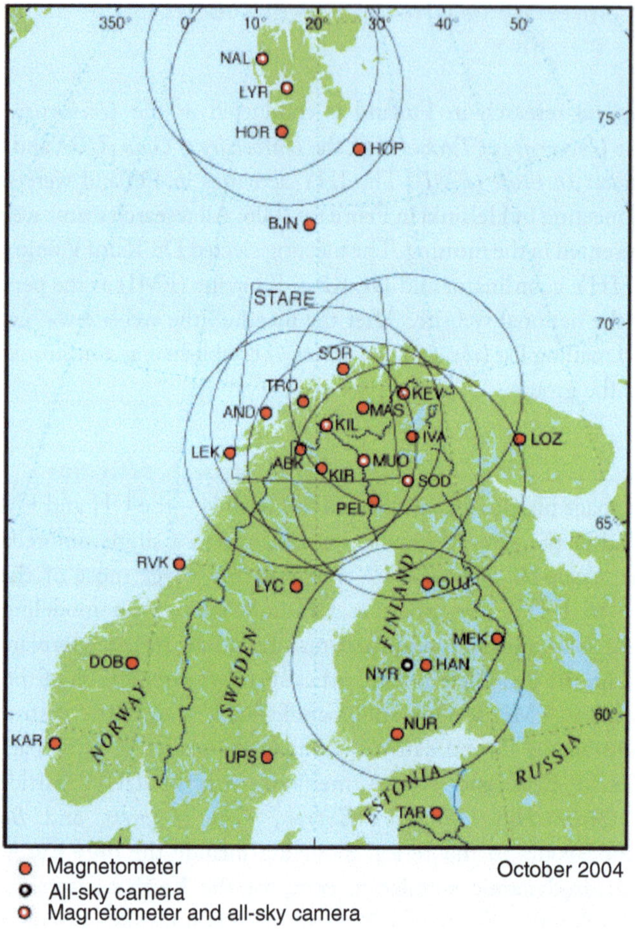

Fig. 60. *Locations of the MIRACLE network of ground-based auroral instrumentation.*

instrument *ERNE* and in the *AMS* experiment to be installed on-board the *International Space Station* in 2009.

The other group in UT performing heliophysical studies, located in *Tuorla Observatory*, concentrates on studies of solar eruptions using a variety of space-borne and ground-based measurements. The observatory also conducts measurements of the Sun at radio wavelengths. Figure 61 depicts the multi-frequency solar radio polarimeter system, earlier operated by IAP at the University of Bern in Switzerland, installed in Tuorla in 2005.

The space research group of the *University of Oulu*, Finland, consists of the *Space Physics Division* of the *Department of Physical Sciences* and the *Sodankylä* Geophysical Observatory (*SGO*). The group has an active research program in

Fig. 61. *The Tuorla multi-frequency solar radio polarimeter observatory.*

ionospheric, magnetospheric and heliospheric physics including cosmic ray modulation in the heliosphere, terrestrial effects of cosmic rays and long-term variations of the solar–terrestrial relations. This is based mainly on experimental data both from ground and from satellites.

A very important heliophysical observatory maintained by the group is the *Oulu Cosmic Ray Station*, which is a part of the SGO. Cosmic ray measurements in Oulu started in 1964 in a wooden barrack with a standard 9-NM-64 neutron monitor consisting of three units each of three counters. The monitor was moved to its present site (65.05° N, 25.47° E) in September 1973. Because of its northern latitude, the building has a pyramid shaped steep roof to prevent snow from accumulating on it. The Oulu station is a part of the World Network of Neutron Monitors. The counts of the neutron monitor are accumulated over 1-min and 1-h intervals. A separated data collecting system (built by the group in UT) with 10-s recording time is also available.

Public and educational outreach and organization of scientific meetings during IHY

Two main public outreach projects were organized during the IHY. In June 2007, the Finnish IHY organization took part in the European IHY "Open Doors" Day on June 10 (see Sect. 5). Together with the *Ursa Astronomical Association* (*http://www.ursa.fi*), the day was announced as the National Sun Day, and Ursa observatories in the whole country organized open-doors events with

Fig. 62. *Ursa Observatory in Helsinki hosted hundreds of visitors as part many "Open Doors" events.*

possibilities for the public to see the Sun safely through telescopes and ask questions from experts of solar observations and/or solar physics. The events were organized in the following sites: Helsinki, Kirkkonummi, Järvenpää, Lahti, Turku and Seinäjoki. In Helsinki, the Ursa observatory (see Figure 62) is located in the beautiful park Kaivopuisto, which hosted a rock concert free to the public at the same time. As a result, the Sun Day event attracted hundreds of people during the afternoon, most of them in their teens or twenties.

The second major public outreach event in Finland has been organized during the last year of IHY. We have organized, as a collaborative action of the three universities, FMI and Ursa, a series of six public lectures on the Sun, the heliosphere and the planetary environments with the Finnish title *Monimuotoinen Aurinko*, which could be translated as *The Versatile Sun.* The lecture series was held in Helsinki (Fall 2008), Turku (Fall 2008), and the final event will be held in Oulu (Winter 2009), and speakers are experts from all research teams participating the IHY. Figure 63 is an excerpt from the poster advertising the lecture series in Turku. The titles of the lectures translated to English are *The Moody Sun: Solar Eruptions and Space Weather*; *The Permeative Sun: The Space Dominated by the Sun*; *The Protective Sun: The Sun and the cosmic radiation*; *The Unusual Sun: Long-Term Solar Activity and Climate Effects*; *The Electrifying Sun: Solar Impact on Planets and Moons;* and *The Entertaining Sun: The Sun and the Aurora.* We expect hundreds of people from all cities to attend the lectures and obtain increased awareness of the solar impact on its surroundings and beyond.

Fig. 63. *"Monimuotoinen Aurinko" lectures given by IHY researchers cover all of the scientific aspects of IHY.*

As an international education initiative directed to university students, the FMI organized together with the University of Bergen, Norway, a winter school on Auroral substorms in Iceland in November 2007. In addition to daytime lectures and laboratory exercises the students got the opportunity to spot auroras with their own eyes during the night time.

In March 2009, the University of Oulu will organize a Space Climate School (*http://spaceweb.oulu.fi/spaceclimate/*) in Saariselkä, Finnish Lapland. The school covers a wide range of topics ranging from solar dynamo models to cosmogenic isotopes and geomagnetic activity. Right after the school follows the *3rd Space Climate Symposium* at the same location. The Symposium aims to cover a wide range of topics under the overall concept of Space Climate, i.e., the long-term change in the Sun, and its effects in the heliosphere and upon the Earth, including the atmosphere and climate.

IGY Gold Club activities

Professor Christian Sucksdorff (born 1928) has obtained the membership of the IGY Gold Club. Professor Sucksdorff retired in 1991 as the head of the Department of Geophysics of the FMI. He was a leading figure in geomagnetism in Finland since the time of IGY. He was also one of the key persons during the 1980s when the era of modern space research began in Finland.

Connections with other multinational programs related to heliophysics

The Finnish research groups have also been very active in other multinational research programs that have shared objectives with the IHY.

FMI, together with the Rutherford Appleton Laboratory (STFC, UK), coordinates the International Polar Year (IPY) Project 63 ("Heliosphere Impact on Geospace"), which forms the official link between IHY and IPY-4 activities. Cluster 63 conducts multinational research on solar-generated events which affect the composition and dynamics of the atmosphere in the terrestrial polar areas. This cluster is a collective effort of 29 international consortia which run versatile bi-polar space research instrumentation to support the IPY campaign. Besides IHY, the ICESTAR program endorsed by SCAR is involved in the Cluster 63 activities. The two programs have complementary roles in the work: IHY coordinates the use of spacecraft missions with ground-based observatory instruments to study the Sun's influence on the heliosphere, including effects at the Earth while ICESTAR coordinates research on magnetospheric and upper atmospheric responses to solar inputs, with emphasis on inter-hemispheric relationships. The science program of Cluster 63 has three main themes: (1) coupling processes between the different atmospheric layers and their connection with the solar activity, (2) energy and mass exchange between the ionosphere and the magnetosphere, and (3) inter-hemispheric similarities and asymmetries in geospace phenomena. Besides the management duties FMI has also been active in all the three science themes of Cluster 63 (Figure 64).

All the Finnish research teams have also been in an active role in the actions of the *European Co-operation in the field of Scientific and Technical Research*

Fig. 64. *ICESTAR is a program central to IHY, providing a vital link to the International Polar Year.*

(COST). One of the largest COST actions, participated by the Helsinki, Turku and Oulu groups, was COST Action 724, *Developing the Scientific Basis for Monitoring, Modelling and Predicting Space Weather.* The objectives of the Action were to develop the science underpinning space weather applications, as well as exploring methods for providing a comprehensive range of space weather services to a variety of users, based on modeling and monitoring of the Sun–Earth system. Twenty-eight countries and six institutions participated in this action, which extended from November 2003 till November 2007. The Finnish groups were especially active in working groups focusing on *The radiation environment of the Earth* and the *Interaction of solar wind disturbances with the Earth.* The COST Action 296, *Mitigation of Ionospheric Effects on Radio Systems* (MIERS), has been actively participated by the Helsinki and Sodankylä groups. The main objective of this action is to develop an increased knowledge of the effects imposed by the ionosphere on practical radio systems, and for the development and implementation of techniques to mitigate the deleterious effects of the ionosphere on such systems. The action runs from February 2005 till February 2009.

7.5.5. France

Submitted by Brigitte Schmieder, IHY-France National Coordinator, Observatoire de Paris Meudon, Paris
http://www.astro.uio.no/~schmiede/index_IHY.html

7.5.5.1. IHY France 2007–2008

The main activities in France for IHY concern outreach and capacity building. Many exhibitions, conferences were performed in different towns in France organized or sponsored by the Centre National de la Recherche Scientifique (CNRS), the Centre National d'Etudes Spatiales (CNES), individual observatories, or by different private organizations.

First of all, France and mainly the Observatoire de Paris were members of the SWEETS consortium (see Sect. 5) supported by European Union in 2007. SWEETS (Space Weather and Europe – an Educational Tool with the Sun) is a public outreach activity and encompasses some of the activities of the International Heliophysical Year (IHY). SWEETS has different activities; France participates in some of them: the SWEETS exhibition, a quiz, a tour bus, and a DVD.

Space Weather on the SWEETS tour-mobile bus

The SWEETS (Space Weather and Europe – an Educational Tool with the Sun) exhibition on space weather effects was presented from 18 to 24 June, 2007 at the Observatoire de Paris at Meudon. The event coincided with a celebration, with a sound and light show celebrating the 100th anniversary of astronomer Jules Janssen was projected over the Meudon castle façade. It explained how this historical castle became an Observatory with Jules Janssen initially a solar astronomer, then the first director in 1875. On Saturday 23 June, the bus was open until 10 pm, and all the people (several thousand) who attended the show could not avoid to look at the external pictures of the bus. Some of them stopped to get more information on the exhibition. The visitors were leaving the bus with IHY posters, the quiz and advertisements on the Observatoire of Paris (Figure 65).

During the previous week, the scientific Space Weather mobile bus tour was visited by *more than 1000 people* including school chidren. For 2 days the bus visited secondary and technical schools and the pupils could get explanations on the sources and effects of the Space Weather. During these 7 days, a team of seven scientists of the Observatoire guided the visitors. The Sun was shining nearly all the week and the visitors could observe the Sun with two telescopes and a spectrograph. Each day the bus tour was guided by 3–4 high-level solar physicists and outreach-educated European space weather scientists.

On the first day, an opening ceremony took place in presence of Dr. F. Jansen and two German journalists, the director of the solar department of the Observatoire and the president of the national program on the Sun and Earth funded by the CNRS. We also had a visit from the local director (Paris-South region) of CNRS.

Fig. 65. *The SWEETS Space Weather Bus at the Observatoire de Paris (Meudon).*

The quiz was advertised on the SWEETS web pages and France had five winners. The first winner was invited at the Observatoire de Paris at Meudon for an afternoon and at the Observatoire du Pic du Midi for an astronomical night. The other four winners were invited to the Observatoire de Paris (in Paris) for visiting the two exhibitions SWEETS and "Du Soleil à la Terre".

The French team of SWEETS participated in the production of the SWEETS DVD and provided new inputs on the Sun (such as new movies from TRACE and Hinode), and updated the text of the "Fundamentals" chapter. The French translation of the DVD was shown at the Observatoire de Paris with a team of seven astronomers.

Space Weather/Sun/Heliospheric public science days

In France the SWEETS exhibition, with panels in both French and English, had been duplicated by the Observatoire de Paris and were presented during different public events occurring either in Meudon or downtown in Paris.

IHY exhibition "Du Soleil la Terre" or "From the Earth to the Sun"

This exhibition was been developed by the Observatoire de Paris and its French and Swiss partners (see Sect. 5). The exhibition was produced by a Swiss agency Hartmann in three languages (French, German, and Italian) and is travelling through large shopping centers and museums in Swiss, Germany, France and other countries. The exhibition consists of 15 panels of 2 m high, a planetarium, several modules in 3D of the Sun, the planets, a ¼-size scale model of the satellite SOHO. It is an interactive exhibition with experiments and games for pupils above 8 years old. An authentic meteorite, more than 360,000 years old is presented (Figure 66).

The exhibition "Du Soleil à la Terre" was displayed from November 8 to December 15, 2007 at the Observatoire de Paris in Paris, jointly with the SWEETS exhibition and the SWEETS-CD show (6 h per day and 5 days per week). On November 8, occurred the opening of this exhibition with 200 invited participants (scientists, journalists, members of the government, offices and agencies like CNRS and CNES). More than 50 schools visited the exhibition. The number of visitors was estimated to be around 5000.

During the opening of the exhibition, IGY Gold certificates were received by 10 scientists who have worked for IGY 1957: Drs Y. Avignon, A. Boischot, J. F. Denisse, Lacroix, J. Legrand, M. J. Martres, M. Pick, P. Simon, J. L. Steinberg. The certificates were given by the IHY European Regional

Fig. 66. *The "Du Soleil à la Terre" exhibition in the Cassini Room at the Observatoire de Paris.*

Coordinator Jean-Louis Bougeret in the presence of the President of the Observatoire de Paris and the IHY French National Coordinator Brigitte Schmieder.

The exhibition "Du Soleil à la Terre" was presented also 6 weeks in the Muséeum d'histoire Naturelle à Orléans with the participation of Researchers of Orléans University and of the Observatoire de Nançay.

7.5.5.2. European IHY initiative Open Doors Day

Three French institutes participated in the European initiative of the Open Doors Day of June 10, 2007: Institut d'Astrophysique Spatial (Orsay), Nancay Radio observatory, and Paris Observatory (on its Meudon site). About 2000 persons have visited the three sites.

Conferences, exhibitions (Sun, the space weather, space instrumentation – with presentation of real instruments – solar wind and relationships between the solar wind and the different object of the solar system), activities for kids and observations of the Sun for to the public.

Researcher Night

(*La nuit des chercheurs*) took place in Paris on September 28, 2007. Brigitte Schmieder, and Milan Maksimovic gave presentations on Space Weather and on the STEREO mission.

The Science Fair

(La fête de la Science) took place on October 15–16, 2007, in Meudon.

The Day of the Amphitheater

(Envie d'amphis) took place at the Observatoire de Paris in Paris on November 24, 2007. Four presentations were given on the interior of the Sun, on the solar activity, on the climate, and on the auroras.

Fiftieth anniversary of SPUTNIK

An exhibition organized by the CNES presented amazing large-format pictures retracing the Sputnik venture, for two weeks on the grid of one of the largest gardens in Paris (Luxembourg garden, where the Senate stands). These exhibitions have been seen by many thousands of persons due to popular location (both for Parisians and foreigners). To inaugurate this exhibition and commemorate the Sputnik venture, a conference was organized at the French Senate. Historical talks from Russian colleagues, as well as prospective discussions on the space development and meeting with astronauts were some of the activities that transpired.

Several other events have been performed in the frame of IHY. The first one is a new planetarium show devoted to the sun and space weather. This show, opened in November 2007, was based on a movie developed in the USA but adapted to include new space weather features. The making of this show involved an IHY researcher (J. Lilensten), a professional writer (C. H. Buffard) and four technician staff. It is performed weekly at the Vaulx en Velin planetarium (in the Lyon vicinity). This planetarium, the third largest in France, is fully digitized, so this show was the first digitalized one devoted to space weather in Europe.

An experiment called "Planeterrella" was developed during IHY by Lilensten et al. (2008). It makes it possible to simulate, in a naive yet useful way, the formation of polar lights. It involves shooting electrons at a magnetized sphere placed in a vacuum chamber. This experiment was inspired by the Terrella of K. Birkeland, built at the turn of the 19th century. However, many very new features extend dramatically its capabilities, and allow the visualization of very many geophysical and astrophysical phenomena: stellar winds, coronal holes,

Fig. 67. *The "Planeterrella" demonstrates geomagnetic processes by which space weather phenomena cause auroral lights.*

interaction between magnetized bodies, exoplanets falling on a star. This experiment has been very successful and has been demonstrated in numerous public events. The main ones have been the "Week of the Sun" in *Vaulx en Velin*, along with the start of the show mentioned above (12–20 November, 2007). Another important event where the Planeterrella was exhibited was during an event which took place from May 4 to June 17, 2007 in Grenoble. The title of the event is "la Bastille in the Stars." The Sun and the solar environment played a dominant role. The weekend of 9–10 June 2007 was devoted specifically to publicize IHY Open Doors Days. We welcomed a very large audience of the public and school children. There was an inflatable planetarium installed permanently, and observations of the Sun where performed with a Coronado. The exhibition included also interactive workshops, discovery of the sky and of astronomical instruments, and an exhibition in the form of posters titled "The Discovery of Sky" and "Billions of Planets (Figure 67)."

7.5.6. Germany

Submitted by Bernd Heber, IHY-Germany National Coordinator (Christian-Albrechts-Universitat, Kiel) and Alexander WARMUTH, IHY-Germany Outreach Coordinator (Astrophysikalisches Institut Potsdam)

http://www.ihy2007.de

7.5.6.1. IHY in Germany

The IHY Germany activities were organized by a small committee formed in early 2006. The committee focused on public outreach and educational topics as well as science planning. The IHY German opening ceremony took place during the German Physical Society assembly in Regensburg on 25 April 2007. Prof. Dr. Carl Rawer was honored with the IGY gold medal for his substantial contributions to the IGY in 1957. Activities and plans for activities were presented, and several talks and interviews were given in many places all over Germany.

Science activities resulted in an international conference "The Sun, the Heliosphere, and the Earth" held in Bad Honnef, Germany, from May 14 to 18, 2007. Seventy-four scientists from 21 nations and all continents participated. Each session of the conference started with a series of reviews, followed by contributed talks, and was accomplished by a poster session. Each of the five days had a main topic: (1) Structures, (2) Magnetic Fields, (3) Solar Activity, (4) Helio- & Astrospheres and (5) Influence on Earth. Because of the numerous experts from different fields, an almost complete survey of these topics was achieved, and also an interdisciplinary discussion between solar-,

Fig. 68. *Talks and discussions at the IHY conference "The Sun, the Heliosphere, and the Earth" held in Bad Honnef, Germany, May 14–18, 2007.*

helio-, and geo-physics was accomplished. To improve the sustainability of the conference the contributions were not published in special proceedings, but in the scientific online journal ASTRA (*http://www.astrophys-space-sci-trans.net*) (Figure 68).

Coordinated Investigation Programs (CIPs)

Combined with the international conference, mentioned above, there was a heliospheric and cosmic ray CIP meeting. German researchers have contributed to several CIPs.

Public Outreach

The scientific activities have been supported by a series of talks given to the public at different places. Several institutions, like the Max-Planck-Institut für Sonnensystem Forschung in collaboration with the University of Göttingen's Institute for Astrophysics, and the Max-Planck-Institut für extraterrestrische Physik, the Astrophysikalisches Institut Potsdam, the Planetarium Nürnberg, the Planetarium Hamburg, the Ruhr Universität Bochum and the Christian-Albrechts-Universität zu Kiel were part of different international days. On April 23 the Planetarium Hamburg and the Institute for Astrophysics of the University of Göttingen organized in a joint collaboration with NASA and the Naval Research Laboratory, Washington, D.C., a press conference during which the 3D observations of the Sun from STEREO were presented for the first time to the public. The response of the media was overwhelming and included releases of the material in all major German newspapers, TV and radio stations and various science TV series, such as Odysso and Nano.

Exhibitions

In collaboration with the Arbeitsgemeinschaft Deutscher Planetarien the exhibition "*Das Reich der Sonne – Heimat der Menschheit*" ("The Empire of the Sun – Home of Mankind") was organized. Instruments built by different German institutions flown on satellites like Azur in 1969, up to recent instrumentation from SOHO, are shown together for the first time. Thanks to ESA, generous support by the Hereaeus Stiftung and the EGU these items together with spacecraft models and a set of explanatory posters are presently displayed in different German cities. The exhibition was first shown at the computer fair CeBIT in Hannover, before it began its official public tour in Bochum on 03.04.2007. Highlights were Girls and School days when more than 300 pupils of local schools visiting the exhibition. Since then it was shown in Mannheim, Berlin, Kiel, Nürnberg, and Halle. For the rest of 2008 it will travel to Cottbus and Berlin (Figures 69 and 70).

Fig. 69. *Potential future scientists study a model of the trajectory of the Ulysses probe in the Carl Zeiss Planetarium in Bochum.*

Fig. 70. *The exhibition "Das Reich der Sonne – Heimat der Menschheit" as presented in the Planetarium of the Wilhelm-Foerster Public Observatory in Berlin.*

Together with the Astrophysical Institute of the Georg-August-University Göttingen the Max-Planck-Institut for Solar System Research in Katlenburg-Lindau organized the exhibition "Our Sun – Fire of Life, History of Investigation and Current Research". The exhibition co-sponsored by the German Committee of the IHY, took place in the famous Paulinerkirche church building in Göttingen and attracted people at all ages.

Public lectures

The Göttingen-based IHY activity was accompanied by a well attended series of public lectures from April to June 2007 in the same Paulinerkirche church building in Göttingen covering topics like "The Enigma of Solar Neutrinos", "Helioseismology: A View into the Sun", "The Sun – Source of Life", "Sun–Earth-Space Weather", "The mysterious magnetosphere of the planet Mercury", "Aurora in the Solar System", "Does the Sun control the Earth's Climate?", "The Heliosphere in Space and Time", "The Search for a Second Earth" and "Recycling in The Universe". A "*Ringvorlesung*" (lecture series) was organized at the Christian-Albrechts-Universität Kiel covering topics which are of interest for the public like the Sun, Cosmic Rays, the Magnetosphere, Solar-terrestrial relations, climate changes as well as local activities like instrumentation made in Kiel. Most of the presentations and additional information can be found on the IHY-Germany home page.

Evening lectures on topics such as "Solar storms and space weather", "The Sun – our nearest star" or "Sun, Moon and Earth" have been given by members of the Astrophysikalisches Institut Potsdam at several planetaria in and around Berlin. During the "Long night of science" on July 14, 2008, over 1200 people visited the Einstein tower in Potsdam, where they were informed about the Sun, heliosphere and solar–terrestrial relations by staff members of the Astrophysikalisches Institut Potsdam.

7.5.6.2. IHY-Germany in the media

TV

The German TV station RBB produced a 10-min documentary on the IHY and solar–terrestrial relations featuring members of the Astrophysikalisches Institut Potsdam for its science series "OZON". This film has since been broadcast by other TV stations and has recently been awarded with the film prize of the science series "Nano" of the German/Austrian/Swiss TV station 3sat. Only five contributions out of 1200 have won such a prize.

Publications

In Germany, the IHY has created significant interest in solar–terrestrial relations in popular media. In addition to the special edition of the popular amateur journal "Sterne und Weltraum" (see Appendix III) articles based on interviews with members of the German IHY committee have appeared in the magazines "natur + kosmos" and "Bild der Wissenschaft" (both popular science magazines), as well as in the lifestyle magazine "GQ". But not only the general public was informed about the IHY – we also reported about various scientific aspects of the IHY in the "Physik Journal" of the German Physical Society (Deutsche Physikalische Gesellschaft, DPG). With the corresponding five articles on Heliophysics the Sun and its activity, the heliosphere as a plasma laboratory and the significance of both for the terrestrial climate were presented to the large community of physicists in Germany.

Web

The German web site became operational in late 2006. The page has not only been used to inform the public about the different activities but also to provide educational material like talks and a movie about our star the Sun as well as all posters used in the exhibition "Das Reich der Sonne – Heimat der Menschheit". This material has been used, e.g. by the museum am Schölerberg in Osnabrück as input for their own activities. In April 2007, the Online service of the national German TV station ARD published an article on solar activity based on an interview with members of the Astrophysikalisches Institut Potsdam.

Open Doors Days

On June 6, 2007 the International Open Doors Day was celebrated at several places in Germany. Among them the Planetarium Nürnberg, the Max-Planck-Institut for Solar System Research in Katlenburg-Lindau and the University of Applied Science together with the Christian-Albrechts-Universität organized this well recognized event for the public. Additionally, on April 12, 2007 a Yuri's Night World Space Party (see Sect. 5) and a family day took place at the Christian-Albrechts-Universität.

Educational activities

Germany takes also part in the Space Weather Monitor (Sudden Ionospheric Disturbances) program (see Sect. 5), which is an educational project to build and distribute ionospheric monitors to students around the world, led by the University of Stanford. The monitors detect effects of solar storms on telecommunication signals as well as local Ionospheric disturbances, e.g. from thunderstorms. The program is led by the Institute for Astrophysics of the University Göttingen in collaboration with the German Space Agency, DLR, with financial support from Astrium/Friedrichshafen and public outreach support by the Planetarium Hamburg. The monitors have been installed at six high schools in Northern Germany and continuous measurements are collected and scientifically analyzed.

7.5.7. Hungary

Submitted by K. Kecskemety, IHY-Hungary National Coordinator, KFKI (Central Research Institute for Physics), Budapest
http://www.rmki.kfki.hu/kffo/IHY/

7.5.7.1. IHY Activities in Hungary

The activities in the frame of the International Heliophysical Year are centered around 8 scientific institutes and observatories. These take place in Budapest, Debrecen, Sopron, and Tihany, all carried out in worldwide international collaboration. Ground-based observations of the Sun are being performed continuously at the Debrecen Heliophysical Observatory (photoheliograph program, Hα measurements), whereas geomagnetic observations are made in Sopron and Tihany. The Nagycenk Geomagnetic Observatory near Sopron was established in 1956–1957 for the International Geophysical Year. Since the advent of space era Hungarian institutions have been involved in the following fields of heliophysics (Figure 71):

- Activity of the Sun (Debrecen), solar interior (Eötvös University, Astronomy Department, Budapest), coronal mass ejections (Konkoly Observatory, Budapest)

Fig. 71. *The Nagycenk Geophysical Observatory, established for the IGY, celebrated its 50th anniversary in 2007.*

- Solar energetic charged particles, interplanetary medium (KFKI RMKI),
- Interaction of solar wind with planets and comets (KFKI RMKI and KFKI Atomic Energy Research Institute, Budapest), and
- Magnetosphere and ionosphere of the Earth (Eötvös University Space Research Group Budapest and Geodetic and Geophysical Institute Sopron, KFKI RMKI), Mars, Venus, Jupiter, Saturn, and Titan (KFKI RMKI).

Hungarian institutes have been involved in various space programs analyzing magnetometer, plasma detector, and energetic particle spectrometer data in missions in:

- Russian collaboration (Intercosmos, Venera, VEGA, Phobos, Compass-2, ISS Obstanovka, etc.),
- ESA (Ulysses, Cluster, Mars Express, Rosetta, Venus Express, SOHO, Cassini),
- French (Demeter), and
- NASA (SOHO, International Space Station, Cassini) programs.

The most recent one is the STEREO mission o study the 3-dimensional evolution of coronal mass ejections (IMPACT, low energy charged particles). Future missions include BepiColombo to planet Mercury (PICAM ion mass spectrometer), and the Solar Orbiter (magnetic field).

Hungarian scientists have been involved in theoretical studies of the solar atmosphere and interior, magnetic coupling of solar atmosphere and its effects on interplanetary space and Earth, numerical simulations of the solar wind-nonmagnetic body interactions, propagations of solar energetic particles and modulation of galactic cosmic rays.

Symposia and meetings

On 21 September 2007 the Symposium on Earth Electromagnetism: Geophysical Symposium to celebrate the 50th anniversary of the Nagycenk Observatory took place in Sopron. Hungary will also host the 11th Scientific Assembly of the International Association of Geophysics and Aeronomy in Sopron, 23–30 August 2009.

Public outreach

The institutions participating in the public outreach include the Budapest Planetarium, the Urania Observatory, Polaris Observatory, and the House of Future (all in Budapest).

March 2007 was the month of Heliophysics at the Polaris Observatory with four part series of public talks about topics including the operation of the Sun (E. Dajka), sunspots and solar activity (B. Kálmán), solar wind and space weather (K. Kecskeméty), solar influence on the Earth's magnetosphere and atmosphere (E. Illés).

The June 10, 2007 "Open Doors" Day (see Sect. 5) at IHY Observatories and Museums was held at various institutes in Hungary. The most successful among them was the program at the Geodetic and Geophysical Research Institute in Sopron on June 8 with about 60 participants. The presentations included a short review about the Sun and phenomena of solar activity; effects of solar activity on the terrestrial environment (with a movie and commentaries); the 11-year solar activity cycle and Schumann resonances, amplitude and frequency of the Earth-ionosphere cavity resonator indicating indirectly solar activity changes by the ionizing galactic cosmic rays.

A popular science movie entitled "The Invisible Solar System" was filmed with the participation of scientists of KFKI RMKI. The movie (directed by Z. Hever) provides introductory presentations on solar activity, space weather, interaction of solar wind with the Earth and other planets, and cosmic rays. It was publicly broadcasted by the television channel Pax Tv in April 2008 and competed at the 39th Hungarian Film Review.

7.5.8. Ireland

Submitted by Peter T. Gallagher, IHY-Ireland National Coordinator (School of Physics, Trinity College Dublin) and Miruna POPESCU, IHY-Ireland Outreach Coordinator (Armagh Observatory, Armagh)

http://www.ihy2007.ie

7.5.8.1. Ireland's eye on IHY

Summary

International Heliophysical Year (IHY) was a great success in Ireland. During 2007 and 2008, we visited nearly 30 schools, reaching over 2000 school children at both primary and secondary level; developed a traveling exhibit that was viewed by about 14,000 members of the public; launched a large-scale exhibit entitled "Heliosphere" at Ireland's newest form for public understanding of science, the Science Gallery, which was viewed by nearly 17,000 visitors, and; installed a number of space weather monitors for students to make their own measurements of the ionosphere over Ireland. In total, our outreach activities engaged around 40,000 members of the public, from primary school students to senior citizens, in exploring and understanding our Heliosphere. The IHY events organized in Ireland are listed on the IHY-Ireland Home Page.

Introduction

One of International Heliophysical Year's goals was to demonstrate the beauty, relevance and significance of Space and Earth Science to the World. During the IHY activities in Ireland we aimed at inspiring the next generation of scientists and creative thinkers with the wonder of the Sun and its effects on the planets. IHY presented us with an excellent opportunity to showcase this fascinating area during what was a high profile global event. The project was also a great opportunity to raise awareness within schools and among the general public of the world-leading research carried out by Irish scientists, and to develop links between schools and university researchers.

We operated the Irish IHY on a series of complimentary themes: school visits; initiating a schools conference; the development of two interactive exhibits; the deployment of space weather monitors; and involving ourselves also in smaller-scale outreach events (open doors, receiving visitors, giving invited lectures). These exciting activities are described in more detail below.

School visits

Over the course of 2007–2008, we visited nearly 30 schools throughout Ireland, reaching more than 2000 students from primary through secondary levels. "Ireland's Eye on IHY" lectures were developed by providing inspirational role models in third level education, scientific research and creative technological innovation and enterprise. School visits used an exciting audio-visual slide show including huge projected images of the Sun and the Earth's atmosphere live from NASA and ESA spacecraft. The aim of the lectures was to introduce key science concepts related to the Sun's influence on the Earth and the Sun's relevance to our daily lives. The specific resources developed are available from the IHY Ireland website.

Themed events

We have reached several thousand people through small-scale outreach events we have organized throughout Ireland in 2007–2008. Although most of these events did not involve a lot of people at once, the quality of the "message" given was high. The events we have organized included:

Cross-border schools conference: On March 8 and 9, 2007, Armagh Observatory organized an IHY Cross-Border School Conference "Discover the Stars at Armagh". The conference brought together about 260 students (130 on each day) aged 13–14 (Key Stage 3) from 16 schools on both sides of the Border, who learned about astronomy and mathematics using the facilities of the Armagh Observatory and the Armagh Planetarium.

IHY Open Doors Days: As part of the IHY Open Doors Day on 10 June 2007 (see Sect. 5), about 60 people from Counties Antrim, Armagh, Monaghan and Tyrone attended an event at Armagh Observatory. The tour included seeing the Observatory's historic building, the Astropark, the Human Orrery, the Hill of Infinity, and the telescopes. It continued with a live view of the Sun using an H-alpha telescope and a lecture by solar physicist Dr. Miruna Popescu, entitled "Surviving in our Sun's Explosive Atmosphere".

IHY summer evening tours: About 130 people attended guided tours of the Armagh Observatory Grounds, Astropark and Human Orrery in July and August. The visit concluded with an illustrated talk on the Sun given by Dr. Miruna Popescu, and a question-and-answer session in the Observatory library.

Transition year students visits: About 80 students who took part in two TYPE weeks organized at Trinity College Dublin in 2007 and 2008 visited the Armagh Observatory and Planetarium and worked in small teams to complete the questionnaires about the Human Orrery and the Astropark, and so learning about our Solar System in an interactive and fun way.

Work experience students: The activities of (about 30) work experience students that were on 2–5 days placements at Armagh Observatory were on the direct supervision of Dr. Miruna Popescu and were concentrated on projects about the Heliosphere ("The Influence of the Sun on Climate", "What are the Auroras", "Types of Solar Cycles", etc.).

Invited lectures: We increased the number of lectures we normally give to local amateur astronomical societies. The lectures highlighted the newest findings within our Heliosphere, and displayed to members of the public the front-line research carried on in Irish Universities and Research Institutes. Astronomers from Trinity College Dublin and Armagh Observatory were "star" speakers to all the national professional and amateur astronomy events organized during 2007–2008.

Traveling Exhibit: "Living with a Star"

A major display put together for the IHY was a traveling exhibit that was viewed by about 14,000 people on six locations in Ireland: Newgrange, Armagh, Belfast, Birr, Ballyclare and Dublin. The interactive "Living with a Star" Exhibit was first shown at the Royal Society in London, in July, and was put together by Dr. Lucie Green from University College London in collaboration with other research institutes and universities across the UK. Dr. Miruna Popescu adapted this exhibit for use as a traveling exhibit in Ireland. She also used this occasion to conduct a survey for assessing people's knowledge about the Sun and our Heliosphere, to which she had about 300 responses (Figure 72).

Fig. 72. *A montage of photos showing outreach events at Newgrange, Armagh, Belfast and Birr.*

Fig. 73. *Members of the public viewing the Sun in 3D at the Science Gallery (www.sciencegallery.com) in Dublin.*

The traveling exhibition contained: 3D Sun movies, an interactive computer display with up-to-date movies taken with the SoHO, STEREO and Hinode satellites, magnetic and UV light hands-on displays, and various other posters highlighting the beauty of the Sun, as well as freebies (cards, posters, fridge magnets, information sheets) and questionnaires. In August and September Dr. Popescu also displayed the STEREO and Hinode satellite models (Figure 73).

More than 10,600 visitors saw the "Living with a Star" exhibit that was set up in Newgrange, at the Bru na Boinne visitor center, as part of Ireland's Heritage week, from 25 August to 2 September. The exhibit, organized together with the Irish Astronomical Association (IAA), was highly acclaimed by its many visitors.

About 350 people (including organized school visits) saw the solar exhibit and attended the lectures given by Dr. Miruna Popescu at W5 Belfast Odyssey from 24 to 30 September. In addition, Dr. Simon Jeffrey, from Armagh Observatory, gave a lecture entitled "Inside the Stars".

The exhibit then travelled to the biggest annual amateur event in Ireland: the Birr Whirpool Star Party, from 5 to 7 October. The exhibit, which was spread in the lecture theatre as well as through the Dolley's Hotel, was seen by about 200 people. In connection to this, Dr. Peter Gallagher gave a "Living with a Star" lecture to over 100 people.

On 5 November, the exhibit was part of a visit to Ballyclare High School. About 300 students attended the lectures given by Miruna Popescu during the day and explored the interactive exhibit. On 13–14 November, the exhibit was featured during Ireland's Science Week at Birr Castle Demesne. More than 1100 students (organized school visits) attended this event and explored the interactive solar exhibit. The final location of the exhibition was in Dublin over the weekend of 25 November. The event, called Astro-Expo 2007 and organized by Astronomy Ireland, attracted several hundred members of the public. Dr. Popescu also gave a lecture about the Sun to an audience of about 50 people.

Interactive Exhibit: "Heliosphere"

The second large-scale exhibit put together for the occasion of the International Heliophysical Year, entitled "Heliosphere", was displayed at Ireland's newest form for public understanding of science, the Science Gallery. This interactive exhibition was viewed by nearly 17,000 visitors. Launched in 2008, the Martin Naughton Institute at TCD houses a completely new kind of center for science – the Science Gallery. The Science Gallery is a public venue inviting everyone to engage with science and technology and its impacts on our everyday lives. Space artist Anna Hill of Space Synapse Ltd. and space scientist Dr. Peter Gallagher of TCD developed an interactive exhibition to highlight the beauty and relevance of the Sun to our every-day lives in the Science Gallery.

The exhibit included:

- A huge display of 2D and 3D movies of the Sun's atmosphere projected onto the roof of the gallery. This was visible from the street and road below and was therefore viewed by many thousands of commuters in Dublin's city center.
- Interactive suit – feedback from the ionosphere's of Earth and Saturn and from solar wind measurement made at Earth.
- Auroral synapse: a BAFTA-nominated film and soundscape installation that used a spatialised soundtrack of low radio frequency recordings and the human voice to interact with the senses connecting the audience's breath with audio and visual experience. Anna Hill led the development of a multi-sensory exhibit.

The exhibit was launched to much acclaim as part of the Science Gallery's Lightwave exhibit in January 2008. Dr. Gallagher gave a lecture entitled "Solar Voyage" during the Lightwave festival at the Science Gallery.

Space Weather monitors

During IHY we installed an AWESOME space weather monitor (see Sect. 4) at the School of Physics in Trinity College Dublin. The monitor was set up by an undergraduate student (Sophie Murray) and will be used as a teaching tool for undergraduates for many years to come. During the project, Sophie succeeded in detecting lighting strikes from as far away as northern Africa and most spectacularly, identified disturbances in the ionosphere above Ireland which we were resulted from a series of flares on the Sun. Data from AWESOME is now transmitted in near-real time from Trinity to the central archive at Stanford University's Solar Center.

We have also made a number of Sudden Ionospheric Monitors (SIDs) available to primary and secondary schools throughout Ireland. These low-cost monitors enable secondary school students to build their own antennas and then make measurements of the Earth's ionosphere from their school.

Acknowledgements

The IHY events in Ireland were arranged and run by teams from Trinity College Dublin, Armagh Observatory and Space Synapse Ltd. This would not have been possible without the support of the Irish Government's Discover Science & Engineering (DSE) program (*www.science.ie*). Special thanks go to Claire Raftery, Paul Conlon, Jason Byrne, Larisza Krista, James McAteer, Shaun Bloomfield and Chia-Hsien Lin for giving talks at schools and setting up the Heliosphere installation at the Science Gallery, and to the Irish Astronomical Association for the help given to display the traveling exhibition in Newgrange.

7.5.9. Italy

Submitted by Francesca Zuccarello, IHY-Italy Outreach Coordinator, NAF-Dipto Fisica-Astron Sez Astrofis, Osservatorio Astrofisico di Catania, Catania
http://ihy.oata.inaf.it

7.5.9.1. Report on IHY activities in Italy

IHY-Italy hosted the 2nd European IHY General Assembly "European Implications to the Large Infrastructures of the Future" in Turin Italy, June 18–22, 2007. There were 89 participants coming from 16 different countries and the Proceedings are going to be published in a Special issue of the "Surveys in Geophysics" Journal.

Institutes from all over Italy were involved in IHY, including the National Institute for Astrophysics (INAF) sites at Turin, Trieste, Florence, Rome (2), Naples, Palermo and Catania, the National Institute of

Geophysics and Volcanology, and universities in Turin, Florence, Urbino, Perugia, L'Aquila, Rome, Naples, Cosenza, Palermo and Catania.

7.5.9.2. Open Doors Day

There were many successful events held as part of "Open Doors Day" June 10, 2007 (see Sect. 5):

- The SVIRCO observatory for cosmic rays (Rome) was opened to the public and has hosted special presentations on the heliophysical environment: more than 200 participants among which high-school students and teachers, undergraduate and Ph.D. students and several families with children (Department of Physics – Roma Tre University and the Istituto di Fisica dello Spazio Interplanetario – INAF).
- The University of L'Aquila and Consorzio Area di Ricerca in Astrogeofisica organized a public conference on the solar–terrestrial relationships. In this occasion, the students of the Istituto Superiore "D. Cotugno" discussed the results of their analysis of several events of CMEs and consequent geomagnetic storms.
- The University of Rome "Tor Vergata" organized observations and talks for the public.
- The Catania Astrophysical Observatory was opened to the public for solar observations and special conferences (~110 participants). During this event, there was the award ceremony for six students of the primary school winners of the competition "Observe the Sky and Draw Your Emotions" (see Figure 74).

Fig. 74. *Winning entries from the "Observe the Sky and Draw Your Emotions" competition.*

The competition was very successful and a total of 256 drawings (all shown on the IHY-Italy web site) were presented. Winners were chosen by a committee formed by astronomers and teachers and by members of the Catania Astrophysical Observatory mailing list for outreach activities.

- In Catania, on June 13 the second part of the Open Doors Day took place in the Theatre "Scenario Pubblico", where the students of secondary schools, winners of the Competition "A solar presentation" were awarded of the prizes by Authorities of the Catania University (~100 participants). During this ceremony, the ballet "Sun Mit", inspired by the Sun, was performed (Figure 75).
- The Torino Astronomical Observatory was opened to the public for solar observations and ad-hoc presentations: an unexpected large number of people (250–300) participated in the event with much interest and most of them were families with children.

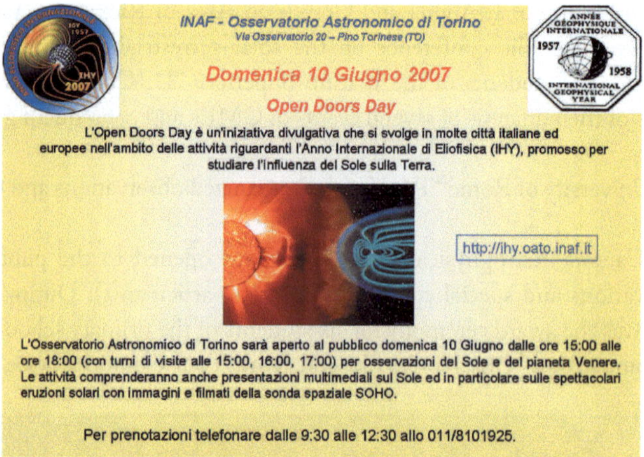

Fig. 75. *The ballet "Sun Mit" won the competition for IHY Open Doors Day in Catania.*

- The Astronomical Observatory of Capodimonte-Naples was opened to the public with visits, observations and special conferences devoted to "the Sun and its role in the Universe".
- The University of Calabria organized public conferences and solar spots observations open to the public. Solar observations have been carried out with several telescopes with Mylar filters on the Bridge P. Bucci of the University of Calabria, and a large number of students (about 200) have had the possibility to observe solar spots. Also, several high school teachers have participated with their students and have illustrated their teaching activities regarding the building of sundials and other solar observations. A public conference on the aim of the IHY has been organized in the afternoon, which has attracted a large audience.
- The Astronomical Observatory of Bologna, in collaboration with other institutes, organized a day devoted to the Sun with experiments, conferences and activities also for children.
- Arcetri Observatory organized conferences about the Sun.

7.5.9.3. Initiatives for schools

- Educational Program "Solar activity and geomagnetic storms": students of the Istituto Superiore D. Cotugno of L'Aquila (other schools will be involved in the next future) have analyzed, with the support of researchers, solar activity, interplanetary structures and geomagnetic activity on the basis of space and ground-based heliophysical data during the last two years. The first results were reported during the Open Doors Day on 10 June 2007.
- Catania Astrophysical Observatory: Competition "Observe the Sky and draw your emotions": children till 10 years were invited to send drawings regarding the Sun, the solar system and other astronomical objects. Competition for the Secondary Schools: "A solar presentation" (see results above).
- School calendar COST296/IHY (2007–2008): European competition of drawings for children of the elementary schools, coordinated by the National Institute of Geophysics and Vulcanology (INGV).
- Project "Students around the world through IHY" coordinated by the Astronomical Observatory of Turin (2008–2010): 30 high-school of Piemonte and Liguria regions have already started, or will start in the next year, solar observations through three instruments (solarscope, PST-Coronado telescope and SID monitor) in order to develop the interest of the students to the scientific method by comparing different data of the solar activity.

7.5.9.4. Comic book translation

Scientific Committee on Solar–Terrestrial Physics (SCOSTEP) has translated in Italian some comic books with the main purpose to improve the knowledge of general public on scientific topics concerning the Sun–Earth interaction.

7.5.9.5. Public conferences

- "International Heliophysical Year: A look to the Sun" Prof. Umberto Villante. January 24, 2007 – L' Aquila; November 30, 2007, Pescara.
- "The Sun: living with a star", Prof. Ester Antonucci; May 10, 2007, Turin.
- "Living with our star" Dr. Lucia Abbo; April 7–11, 2008, Turin.
- "The Sun", Prof. Fabio Reale, IX National Week of Astronomy: conference and solar observations with telescopes; May 11, 2008 – Palermo.

7.5.9.6. Temporary exhibition: "In the fire of the Sun"

"In the fire of the Sun" is a temporary exhibition (see Sect. 5) conceived and carried out by European scientists in the frame of the outreach activities of IHY with the aim to point out the advances on the heliophysics knowledge. It has been hosted two times: September 25–October 2007 in Turin and April 10–29, 2008 in L'Aquila.

7.5.9.7. Advanced training

The "SERSES: Series of Events on Relations in the Sun–Earth System and Space Weather" (L'Aquila) program is carried out from 2006 to 2009 and is organized by the International School of Space Science and the Consortium Area di Ricerca in Astrogeofisica. It includes six high-level courses (*http://www.cifs-isss.org*):

- N. 1. Spring 2006: The Physics of the Sun (i.e. The Active Sun on your Active Desktop) 27 March–1 April 2006
- N. 2. Fall 2006: Solar Terrestrial Physics 10–16 September 2006
- N. 3. Spring 2007: Magnetospheric dynamics 9–15 April 2007
- N. 4. Fall 2007: Turbulence and Waves in Space Plasmas 9–14 September 2007
- N. 5. Spring 2008: Geomagnetism and Ionosphere 7–12 April 2008
- N. 6. Fall 2008: Solar–terrestrial relations in Antarctica.

7.5.9.8. Scientific activities

Nine of the CIPs (see Sect. 3) are led by Italian research institutions. Additionally, many Italian institutions have joined the Whole Heliosphere Interval (see Sect. 3), the observations concerned the period 20 March–16 April 2008 (Carrington Rotation 2068). These include the team of L'Aquila, the network SEGMA, the Observatory of Arcetri and the Observatory of Turin.

7.5.10. Norway

Submitted by Nikolai Ostgaard, IHY-Norway National Coordinator, Department of Physics and Technology, University of Bergen
http://web.ift.uib.no/Romfysikk/RESEARCH/PROJECTS/IPY_ICESTAR/

7.5.10.1. Norwegian IHY activity

The main IHY activity in Norway is through the program IPY-ICESTAR, which is under the IHY-ICESTAR Project ID #63 (see Appendix V) and was funded by the Norwegian Research Council.

The three main objectives of the IPY-ICESTAR projects have been (1) public outreach and education, (2) continuous run of the EISCAT Svalbard incoherent scatter radar throughout one year, and (3) investigation of how various constituents of the magnetosphere-ionosphere system respond to different heliospheric conditions, by utilizing conjugate measurements in both hemispheres from space and ground.

One of the public outreach initiatives was a conference that took place in September 2001, where scientists and journalists came together to exchange experiences. Following this conference a number of articles about new findings in space research have appeared in Norwegian newspapers. Another public outreach initiative is the Space Suitcase, a suitcase with various instruments used by researchers, such as magnetometer, all-sky camera, Geiger counter, and a solar telescope, which is being lent to Norwegian high school classes. The idea for the Space Suitcase is to let high school students experience first hand how research in a complex field such as space science may transpire. Many schools have expressed an interest in lending the Space Suitcase, and it has been booked for several months ahead.

The EISCAT Svalbard radar long run took place from March 2007 until the campaign ended one year later, in February 2008. During this time, the radar ran almost continuously, and produced unprecedented data set, which allows scientists to follow long-term trends in the upper atmosphere. This intitiative also lead to the effort of operating continously but not so frequent as Svalbard, other radar facilities like PFISER, Poker Flat and Sondrestrom and Millstone. The International Space Science Institute has provided funding and premises for a group of scientists to meet and investigate this data set further.

The main data set used during IPY-ICESTAR to investigate the interaction between the magnetosphere-ionosphere system and the heliosphere is auroral images from the IMAGE and Polar spacecraft. On some occasions, these two satellites were in a position to monitor the aurora in both hemispheres simultaneously. These data can be used to study the global system in a way that has not previously been possible. The IPY-ICESTAR program is funding one Ph.D. student for this research. Three masters degree students are involved in the program.

7.5.11. Poland

http://ihy,cbk.waw.pl

Heliophysical studies are mainly developed in Warsaw, Wroclaw, Torun, and Cracow. In particular, during nearly thirty years the Space Research Centre of the Polish Academy of Sciences (in Warsaw, with Solar Physics Division in Wroclaw) in broad and international cooperation carries out pure and applied studies in space physics, in particular:

- solar physics,
- heliosphere and its interaction with the interstellar medium,
- physical research of planets and environment,
- comets and small planetary bodies, and
- magnetosphere and ionosphere of the Earth and planets.

Ph.D. studies in various heliophysical topics are carried out in the Centre. The Centre has incorporated into the NASA projects, the cornerstone ESA missions, e.g., Cassini-Huygens, Mars Express, Rosetta, Venus Express, and Cluster projects, as well as has participated in many European Union Framework grants and program. One should also note research in solar physics and space physics, including physics of sunspots and hydrodynamic simulations, at Astronomical Observatory of the Warsaw University, Nicolaus Copernicus Astronomical Centre

(Warsaw, Torun), and Astronomical Institute of the Wroclaw University. Investigations at Astronomical Observatory of the Jagiellonian University in Cracow and at the Centre for Astronomy of the Nicolaus Copernicus University in Torun concentrate on solar radio observations and radiation in plasmas in the Solar System. Space Research Centre of the Polish Academy of Sciences also coordinates IHY activities of the Polish COST (Cooperation in Scientific and Technology) 296 community.

7.5.12. Spain 

Submitted by Javier RODRIGUEZ-PACHECO, IHY-Spain National Coordinator, and José Antonio BONET, IHY-Spain deputy National Coordinator, Instituto de Astrofsica de Canarias, Tenerife

http://www.iac.es/IHY

7.5.12.1. IHY Spanish report

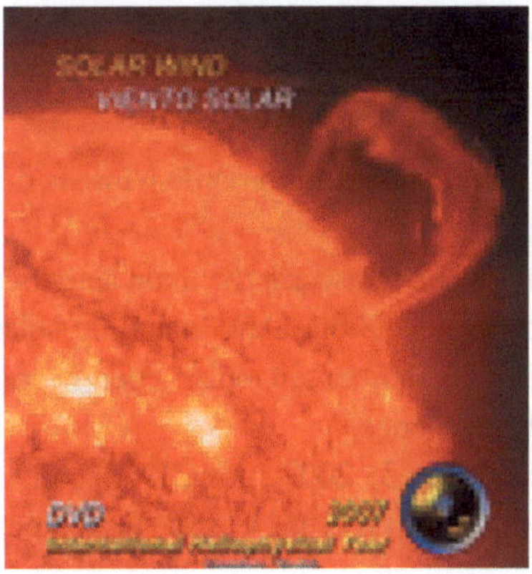

The Spanish community participated in many research and outreach activities throughout IHY. A special documentary on IHY, titled "Solar Wind/Vento Solar" is available from the IHY-Spain website. The Instituto de Astrofisica de Canarias (IAC) produced the 6-min long film, and was widely circulated around the English- and Spanish-speaking world. The documentary is available on the IHY-Spain website.

IHY-Spain hosted many successful events for the "Open Doors Day" (see Sect. 5). A summary of many events follows:

(1) Open Doors Day at Observatorio del Teide, Tenerife, Canary Islands (Figure 76):

Roughly 1000 people visited the observatory. There were guided tours to the main solar telescopes: Vacuum Tower Telescope (VTT), THEMIS and Solar Lab and

Fig. 76. *Live Solar Observations at Observatorio del Teide (left) and educational activities (right) were some of the highlights of Open Doors Day.*

introductory talks about the Solar Dynamics and to the Solar Telescopes, and live observations.

(2) Open Doors Day at Observatorio del Roque de los Muchachos, La Palma, Canary Islands

Roughly 300 people visited the observatory. There were guided tours to the main solar telescopes: Dutch Open Telescope (DOT) and Swedish Solar Telescope (SST), introductory talks about the Solar Dynamics and to the Solar Telescopes, and live solar observations with the Swedish telescope.

(3) Museo de la Ciencia y el Cosmos del Organismo Autónomo de Museos del Cabildo de Tenerife

June 2007 was named "El mes del Sol" at the Museum, the main activities during this month were associated with Open Doors Day, including solar observations at the Museum Terrace, experiences on solar energy and Erastotenes experience performed by the audience.

Ciclo de cine AluCINE con el Sol: June 3, 2007, AluCINE con el futuro: "Supernova (1999)," "Viaje al fondo del mar (1961)," and "Frequency (2000)."

Conferences at the Museum:

- Wednesday June 6, 2007: "El Sol, nuestra estrella", Dr. Jorge Sánchez Almeida (IAC)
- Thursday June 7, 2007: "El Sol, el clima, y el cambio climático", Dr. Enric Pallé Bagó (IAC)
- Friday June 8, 2007: Children's talk: "El Sol, la estrella de tu vida", Dra. Inés Rodríguez Hidalgo (MCC, ULL, IAC), and "Auroras boreales observadas en Tenerife", Dr. Manuel Vázquez Abeledo (IAC)

- Friday June 22, 2007: "El Sol de los Incas. El secreto de las LLamas Blancas", Cipriano Carrillo González, Gotzon Cañada y Rubén Naveros Naveiras
- Friday June 29, 2007: "Tormentas solares: la ira del Sol", Dr. Javier Rodríguez-Pacheco, SRG/UAH (Space Research Group, Dpto. Física, Universidad de Alcalá de Henares)

(4) Activities at the L'Hemispheric, Museo de la Ciencia Principe Felipe, Ciutat de les Arts i les Ciènces, Valencia
"La Ira del Sol", Dr. Javier Rodríguez-Pacheco, SRG/UAH (Space Research Group, Dpto. Física, Universidad de Alcalá de Henares), Tuesday June 5 2007.

(5) Conference at the Agrupación Astronómica de Sabadell:
"Avances en Física Solar", Ricard Casas, January 2007. This club made leve solar observations throughout the year.

List of Planetariums and Museums that collaborated with the IHY

- Planetario de Santander. Activities: Collaboration with the local Astronomy Club "Cantabria": Solar Research, Conferences, Expostions
- Casa de las Ciencias del Ayto. de Logroño: Activities: Open-Doors Day 10 June, Practical applications of the Solar Energy, Live Solar Observations (solar image projected on a big screen), Solar/Heliospheric Conference, Exposition on Solar/Heliospheric Physics
- Planetario de Pamplona: Activities: Open-Doors Day 10 June, Live Solar Observations coordinated with local Astronomic Associations, including screen projections, spectrograph, and workshops on the street, Exposition Conferences and Courses on Solar/Heliospheric Physics
- Agrupación Astronómica de Sabadell
- Museu de la Ciencia I de la Técnica, Terrassa: Activities: Open-Doors Day 10 June, Practical applications of the Solar Energy
- L'Hemispheric/Valencia
- Museo de la Ciencia y el Cosmos del Organismo Autónomo de Museos del Cabildo de Tenerife
- Planetario de Madrid
- Cosmocaixa
- Museo de las Ciencias de Valladolid: Activities: Open-Doors Day 10 June, Live Solar Observations, Exposition on Solar and planetary Astronomy (2 March–27 June), Solar/Heliospheric Conference.

List of University Departments and Institutes that collaborated in the IHY

- DMG (Data Mining Group, Univ. Málaga)
- GACE (Grupo Astrofísica y Ciencias del Espacio, Univ. Valencia)
- IAA Instituto de Astrofísica de Andalucía
- IAC (Instituto de Astrofísica de Canarias)
- OE (Observatori de l'Ebre)
- SRG/UAH (Space Research Group, Univ. Alcalá)
- STPSW (Solar–Terrestrial Physiscs and Space Weather Group, Univ. Barcelona)
- UE (Universidad de Extremadura)
- UIB (Universitat de les Illes Balears).

7.5.13. Switzerland

Submitted for the IHY-Switzerland National Committee by A. O. Benz, C. Monstein, Svetlana Berdyugina, Jan O. Stenflo (Institute of Astronomy, ETH Zurich), M. Bianda, R. Ramelli (Istituto Ricerche Solari Locarno), and W. Schmutz (World Radiation Center, PMOD/WRC, Davos)

7.5.13.1. International Heliophysical year in Switzerland – a review

The International Heliophysical Year (IHY) announced by the United Nations Organization (UNO) for 2007 initiated several activities in Switzerland. They included an international project in the frame of Observatory Development of the United Nations Basic Space Science (UNBSS) Initiative, a public exposition "In the Fire of the Sun" traveling through continental Europe, Open Doors days at observatories, conferences, and reports in the media in Switzerland. The activities have enhanced public awareness of Sun–Earth interactions and heliophysics, as well as communicated the fascination of the Sun. Some of the initiated activities will continue in the coming years.

Introduction

Major inputs to the IHY in Switzerland to the IHY have come from academic institutions as well as from groups of amateur astronomers and public observatories. The impact varied much, depending on the local effort. Here we summarize the activities of the professional astronomers at the four major solar centers in Switzerland: The Institute of Astronomy at ETH Zurich, a university at academic level, Istituto Ricerche Solari (Institute of Solar Researches, IRSOL) in Locarno, the Physikalisch-Meteorologische Observatorium Davos and World Radiation Center (PMOD/WRC) in Davos, and the International Space Science Institute (ISSI) in Bern three research laboratories at the federal and international level.

Observatory Development

Radio spectrometers of the CALLISTO type (see Sect. 4), designed to observe solar flares, have been distributed to nine locations around the globe in the frame of an UNBSS instrument array. The instruments observe automatically, their data is collected every day via internet and stored in a central data base. A public web-interface exists through which data can be browsed and retrieved. The nine instruments form a network called e-CALLISTO. It is still growing in the number of stations, as redundancy is desirable for full 24 h coverage of the solar radio emission in the meter and low decimeter band. The e-CALLISTO system has already proven to be a valuable new tool for monitoring solar activity and for space weather research. It will be operating in the coming years, hopefully up to the solar maximum in 2011. A detailed description of the program can be found in Sect.4.

Exposition "In the Fire of the Sun"

The exposition "In the Fire of the Sun" (see Sect. 5) was realized by a commercial company (Hartman Events, GmbH, Au, Switzerland), explicitly in view of the IHY. The scientific collaborators were mainly from ETH Zurich and the Observatory of Paris (Meudon). The exposition appears in three languages (German, French, and Italian). It is aimed at the very large public in shopping malls. The exposition consists of objects, a movie theatre, posters, hands-on experiments, and a children's corner. It covers an area of up to 200 m^2. The exposition is built up in the main hall of shopping centers and moves every two weeks from one place to the next. Switzerland, Germany, Austria, Italy, and France are on the program. The tour will continue at least into 2009, the International Year of Astronomy. On the opening day at a given location, it is often inaugurated by a press conference. It was installed also in a mall of a large shopping center at Tenero, 11–22 September. IRSOL and Specola Solare Ticinese collaborated with the mall and Centro Stefano Franscini in Ascona in order to optimize the outreach also under the denomination "IHY event". It is estimated that the exposition will be seen by several million people.

Open Doors Days

The Physikalisch-Meteorologische Observatorium Davos (PMOD/WRC) celebrated its 100th anniversary in 2007. It was combined with the IHY activities. An exposition on the 100 years of PMOD was displayed at the local historic museum all year long. Days of open house were on March 21 and June 23. Public lectures were held at PMOD/WRC on June 10, the day of International Heliophysical Year, and on June 22. Lecture days for local schools were from June 18 until June 21, 2007. A star party event for amateur astronomers took place on September 15, and the event "Culture meets the Sun" on December 22, 2007.

On 10 June, the Istituto Ricerche Solare Locarno (IRSOL) participated to the Open Doors day organized by IHY-Europe (see Sect. 5). The event was announced in the most popular regional newspapers. The weather was good; the number of participants (around 200) was ideal, taking into account the size of the institute. It was possible to present good quality information, and people appreciated the effort.

The Institute of Astronomy at ETH Zurich met the people at the most popular promenade at the shore of Lake Zurich. Thousands of people strolled on the sunny afternoon and more than a thousand looked at the posters, videos, and live telescopes, informing about the interior, the surface activity and the corona of the Sun.

Doors were open also at the Specola Solare Ticinese in Locarno. Today the observatory located above town is serving the public at large. It was opened 1957 as a scientific station by Max Waldmeier (ETH Zurich) as a Swiss contribution to the IGY. The Open Door Day thus was also celebrated as the 50th anniversary.

Conferences

ETH Zurich and IRSOL organized the workshop SPW5 in Ascona (17–21 September 2007). During the workshop a public lecture in Italian was given by Prof. Egidio Landi degl'Innocenti (Florence) with the title "Influenze astronomiche sul clima terrestre" (Astronomical influences on the Earth climate). There were about 70 participants, and Landi degl'Innocenti was interviewed by the most listened local radio, and more articles were written on local newspapers.

An international CAWSES meeting was organized in Davos, 18–20 September 2007, on the topic "One century of UV Radiation Research" by the PMOD/WRC.

The International Space Science Institute (ISSI) in Bern was addressing heliophysical themes with its Workshop program and publications. Two volumes resulting from earlier Workshops have already appeared in the Space Sciences Series of ISSI (SSSI): "Solar Dynamics and its Effects on the Heliosphere and Earth" (Vol. 22) and "Solar Variability and Planetary Climates" (Vol. 23).

A Workshop entitled "From the Outer Heliosphere to the Local Bubble: Comparisons of New Observations with Theory" on 15–19 October 2007. One particular highlight of the Workshop was the report of E. C. Stone (former director of JPL) about the two Voyager spacecraft having crossed the heliospheric termination shock. Another Workshop is currently in progress on the theme of "Solar Magnetism" it is the first in a trilogy on Magnetism in the Solar System. The highlight of this Workshop is a keynote lecture by E. N. Parker (University of Chicago) on the centenary of the discovery of solar magnetism by George Ellery Hale. Both workshops will result in book publications later in 2008. Finally, ISSI will be hosting the second "The Sun, The Heliosphere and the Earth" IHY conference (after Bad Honnef in May 2007) on 6–10 November 2008.

Heliophysics is also featured in the SPATIUM series published by the Pro-ISSI association, which aims to convey space science themes to the general public: No. 17 deals with "The Heliosphere: Empire of the Sun" by A. Balogh (Imperial College, London, and ISSI), and a forthcoming issue will deal with "Solar Magnetism" based on Parker's lecture mentioned above.

News related to heliophysics on the newspapers, radios and TV

Reports on the open doors day, the SPW5 public conference, and the exposition were publicized in newspapers and by local radios. Most noted were two interviews on national radio on e-CALLISTO at 26 May and IHY on 10 June 2007. It was thus possible to inform a wider public about the IHY and provide outreach news.

7.5.14. United Kingdom

Submitted by Andrew R. Breen, IHY-UK National Coordinator (University of Aberystwyth, Wales), Richard A. Harrison, IHY-UK National Coordinator (Rutherford Appleton Laboratory, England) and Lucie M. Green, IHY-UK Outreach Coordinator (Mullard Space Science Laboratory, Surrey, England)

http://ihy2007.org.uk; http://www.heliophysical.or.uk

7.5.14.1. International heliophysical year activities summary

In summary, IHY scientists performed research involving a wide range of scientific topics, and outreach activities were run to raise awareness of IHY and IHY science to scientists, schools, teachers and the general public across the three nations of Britain,

Northern Ireland and the Republic of Ireland. Communicating the term 'heliophysical' presented some difficulties so IHY was discussed mostly in terms of the UK's contribution to Solar System science. In summary the numbers of people who had direct contact with an IHY scientist or activity are thought to exceed:

- 45,000 members of the public
- 21,000 school students.

7.5.14.2. UK scientific exploitation and the IHY

The UK has been an active member of the IHY activity since the beginning and was a key player in defining the Coordinated Investigation Programme (CIP) structure which was adopted as a mechanism for defining and running campaigns within the IHY framework. IHY-UK hosted the CIP program, and UK research Richard Stamper served as the lead coordinator for the CIPs. Thus, the key scientific elements of the UK's approach to the IHY activity have been intimately related to a number of CIPs. These have already resulted in two dozen papers in refereed journals, encompassing specific areas of research which we discuss below, namely:

- CME onsets and propagation in the heliosphere (utilising STEREO, SOHO, Hinode);
- IPS investigations of the solar wind and heliosphere;
- Radio tomographic imaging; and
- Participating in the extensive ionospheric monitoring period with EISCAT.

It should be noted that there are a range of smaller-scale activities, some of which relate to collaborative efforts in a range of non-UK-led CIPs. However, the 'headline' topical areas listed above relate mainly to leadership of the IHY activities through the Rutherford Appleton Laboratory and Aberystwyth University, but also involving a number of other groups such as the Mullard Space Science Laboratory, and Southhampton University.

The topical area of *Coronal Mass Ejection onset and propagation studies* relates to CIP 1. In many ways, CIP 1 was drafted as the forerunner for outlining the IHY CIP campaign approach, but it has also been an active research effort led from the UK, specifically addressing the CME onset and propagation of CMEs in the heliosphere. It was loosely built around the fact that through 2007 and 2008 we anticipated the operation of the newly launched NASA STEREO and Japanese / Hinode spacecraft (Solar-B). The CIP 1 Website at *http://ihy2007.org.uk/CIP01.htm* outlines the basic aims and opportunities. However, with the UK's unique instrument involvements both in STEREO and Hinode, and including the ESA/NASA SOHO spacecraft, the aim was do perform very specific, strategic studies to advance two areas of study during IHY.

Concerning the CME onset, which is shrouded in so much mystery, the extreme-UV (EUV) spectroscopic observations from SOHO have been used to extend considerably on past observations of what we call coronal dimming. An emphasis on the depleted plasma resulting from these emissions has shed light into model observations and mass estimates of the eruptions.

The other side of the investigation, the wide-angle heliospheric imaging observations from STEREO are being used now to detect and track CMEs in the heliosphere. This has become routine since the start of scientific operation of the UK-led HI instruments on the two STEREO spacecraft. The first success was early in the IHY period with the announcement that the first interplanetary CME had been detected in this way and tracked to Earth-like distances from the Sun. Other ongoing studies on CME origin, propagation and structure are also part of the investigation.

A second major area of study relates to the *Aberystwyth radio observations.* Solar wind observations carried out by Aberystwyth University as part of IHY contributed to four CIPs (CIPs 1, 11, 12 and 43). The emphasis was on two main science areas, namely, the acceleration profile of the solar wind inside 15 solar radii (CIP 12) and the large-scale structure of the solar wind inside 0.5 AU, including the interaction between the solar wind and CMEs (CIPs 1, 11, 43). The Aberystwyth contribution to these studies used a combination of radio scintillation (IPS) observations from the EISCAT and MERLIN facilities, white-light observations from the STEREO heliospheric imagers and in-situ measurements from the plasma instruments on Venus Express (VEX). Data analysis and interpretation to date has concentrated on IPS/STEREO/VEX observations during April and May 2007 (forming part of CIP 1). The ability of these extremely long-baseline observations to resolve small deviations from radial in solar wind direction is proving to be of great value in understanding stream interaction regions and the interaction of CMEs with the background solar wind.

Another area in which the UK has participated is the *EISCAT extensive ionospheric monitoring* during the IHY/IPY period. Specifically, the UK has been an active component of the CIP 25 campaign whose aim was to provide extensive ionospheric monitoring during the IHY/IPY period, from March 2007. This activity will be reported elsewhere but, in particular, for the UK this has centerd on the UK's role, through the Rutherford Appleton Laboratory group, in the EISCAT operation for which the Svalbard radar has been operated in near continuous mode. It is too early to judge the scientific impact of such a comprehensive period of coverage but the campaign to perform such a scheme during IHY/IPY is absolutely appropriate to the goals of IHY.

Finally, we report on the activities relating to *radio tomography imaging*. The aim of CIP 41, which is concerned with radiotomographic imaging of the arctic ionised atmosphere by the International Ionospheric Tomography Community (IITC), was to investigate the impact of space weather processes on the terrestrial ionised atmosphere and identify the roles of geophysical processes that influence the large-scale plasma distribution. Radio tomography is a relatively new technique for imaging the large-scale ionospheric plasma on horizontal scales of tens to hundreds of km. The technique gives wide latitudinal coverage from a limited number of ground stations and so is ideal for remote inaccessible polar regions.

For the UK, we have reported on a few areas of major activity. We must stress that the nature of IHY and the CIP activities, which involve large, multinational collaborators, means that this report for the UK cannot be complete. However, it gives a flavour for the IHY-related science activities in the UK over the IHY period, a period of great scientific advances.

7.5.14.3. IHY-UK outreach

The travelling exhibit developed for the Royal Society Summer Exhibition was one particular highlight of the year. This exhibit was extremely well received by both the public and the science community. The UK events were coordinated by Dr. Lucie Green of UCL's Mullard Space Science Laboratory. Other events include:

- Teacher INSET at National Maritime Museum, London on 8 November 2006: teachers attended to learn the most recent solar system science and how to communicate it in the classroom.
- Association for Science Education annual conference on 6 January 2007: A talk on IHY and activity ideas were given out.
- British Rocket Oral History Programme on 12 April 2007: The opening talk for the conference was on IHY.
- National Astronomy Meeting at University of Central Lancashire in April 2007: A special session on IHY science, and presentation on IHY outreach activities.
- For the IHY Open Doors Day (also see Sect. 5) four events attracted nearly 1000 participants:
    - Armagh Observatory opened up their doors and ran tours of the facilities, talks and activities.

Fig. 77. *Hartland Observatory "Open Doors Day" press story in the Hartland Times.*

- The Hartland Centre ran 3 days of activities for schools and the general public. The event was organized by the British Geological Survey, which is based in Edinburgh and who run the Hartland Magnetic Observatory, along with Plymouth University. An exhibition, observatory tours and talks were held (see Figure 77).
- The Kindle Centre ran 4 days of activities and worked with school students, the general public and the Royal National College for the Blind.
- The Roseland Observatory had their 20″ heliostat feeding a 10″ Newtonian for public solar viewing. The magnetometer was running and they had a local firm showing sundials and an armillary sphere.

- Association for Science Education teacher training day on 13 June 2007: Another successful teacher training event.
- Royal Society Summer Exhibition: The well-attended event attracted many to the posters, models and interactive exhibits organized and staffed by a consortium of UCL, RAL, Imperial College, Aberystwyth University, University of Central Lancashire, Cambridge University and Armagh Observatory.
- IHY had a unique opportunity at the Royal Horticultural Society's annual flower show at Tatton Park, which attracted 150,000 visitors. Nigel Marshall worked on a garden celebrating 50 years of space science as an extension of his 'Spaced-

out UK' project. Adjacent to the garden was a tent containing information about the UK's solar system science research.

- Lucie Green attended over 2 days talking to visitors about IHY. IHY "goody bags" were handed out containing posters, information sheets and fridge magnets. The Spaced-out UK garden won a silver award and attracted press coverage in light of this.
- Kids in Space Tour: A PPARC large award was given to Space Fund, a company who takes solar system science shows into primary schools. The shows build on lesson content to present the science in an exciting way. The Kids in Space Tour included IHY science in the show and also saw the development of a special IHY activity pack which was given to every school on the tour. This enabled pre and post-visit activities to be run by the teachers across four science themes. The tour has reached 150 schools and 20,000 students and teachers across the most deprived areas of England including those in Lancashire and on the Sussex and Kent coast.
- The BA festival in York was held on 12 and 13 September 2007 and IHY workshops were run during the week-long festival. The workshops introduced IHY and allowed students to investigate UV light from the Sun through the use of colour change beads. Teachers' packs and promotional materials were given out. The workshops were run as part of the Young People's Programme.
- The IHY exhibit at Jodrell Bank for Space50, held 1–5 October 2007.
- IHY Scottish Schools Lecture Tour: A series of lectures promoting IHY science was organized by Clare Parnell, David Pontin, and Lyndsay Fletcher. Talks were given by researchers, and handout materials and teacher packs were also provided for every school on the invite list. Three events in Dundee, St Andrews and Glasgow were held. Teaching packs were sent to all schools invited.
- Secrets of the Sun: Wynyard Planetarium ran a series of schools and public events which included solar observing sessions, talks, classroom activities and lessons in the virtual learning environment. A new website was also developed to promote IHY information and resources, see *http://www.heliophysical.org.uk/*
- IHY exhibit at climate change meeting: The RS Summer Exhibition stand returned to the Royal Society for one day. 100 people attended the stand to learn about the Sun and its relation to climate change.

Along with the above main activities the many popular articles were written, in media such as Astronomy and Geophysics Magazine and Science Technology Magazine.

7.6. Latin America and the Caribbean

Participating nations: Argentina, Brasil, México, Puerto Rico and Perú.
http://www.alage.org/ihyla.html

The International Heliophysical Year in Latin America

Submitted by Cristina Mandrini, IHY-Latin America Regional Coordinator, Universidad de Buenos Aires, Buenos Aires, Argentina

Latin America took part in different types of IHY activities. Starting in 2005, with the First Regional Planning Meeting held at the Centro de Rádio Astronomía e Astrofísica MacKenzie (CRAAM) of the Universidade Presbiteriana MacKenzie in São Paulo, Brasil, 15 institutions including national observatories, national research institutes, and university laboratories, groups or departments decided to join the IHY efforts. These institutions are devoted to different IHY scientific disciplines via basic research and/or development of instrumentation, such as: heliosphere and cosmic rays, solar physics, planetary ionospheres, thermospheres, mesospheres and climate studies, and Geomagnetism.

After this first regional meeting others were held: the second one in Rio de Janeiro, Brasil, during March 2006 (in conjunction with the 11th Quadrennial Solar Terrestrial Physics Symposium of SCOSTEP), and the third one in Puerto Vallarta, México, during November 2006 (in conjunction with the meeting The Physics of Solar-Wind/Magnetosphere Coupling). During these meetings the Latin American IHY Web page, hosted by the Latin American Association of

Fig. 78. *IHY Logos in Portuguese, English and Spanish.*

Space Geophysics (Asociación Latino Americana de Geofísica Espacial, ALAGE), was organized, decisions were taken to promote joint Coordinated Investigation Programs and the organization of the Latin American IHY School was discussed. The last meeting took place during July 2007 in Mérida, México (see below).

Latin American IHY activities are mainly concentrated in Argentina, Brasil, México and Perú (Figure 78).

7.6.1. Argentina

Argentinean researchers participated in the United Nations Basic Space Science (UNBSS) Initiative, led by Brasil, called South American Very Low Frequency Network (SAVNET, see Sect. 4). Antennae belonging to the network were installed at Complejo Astronómico El Leoncito (CASLEO), San Juan, Argentina. This project is described in the next Section (Brasil).

During September 2006, a workshop and school were organized within the frame of IHY educational activities. The school and workshop were held at Instituto de Astronomía y Física del Espacio (IAFE), Buenos Aires. A total of 40 participants (see picture above left) including Latin American researchers, Ph.D. and post-doc students from Argentina, Brasil and Chile, attended lectures on Solar Radio Physics, Physical Processes in the Interplanetary Medium and Planetary Magnetospheres. Participants had the chance to present posters and discuss their scientific results during the workshop (see Figure 79).

IHY Kickoff activities in 2007 included conferences for the general public during the 33rd Feria Internacional del Libro within the "A Space for Youth" under the motto "Honoring the Sun", in April and May. Furthermore, during

Fig. 79. *Participants in the September 2006 IHY School and Workshop.*

November four conferences to the general public were delivered at IAFE under the title "Full IHY Month".

7.6.2. Brasil 

CRAAM (Centro de Rádio Astronomía e Astrofísica MacKenzie São Paulo, Brasil) leads an international joint program within the frame of UNBSS, together with Argentina and Perú. The objective of the project, called SAVNET, is to monitor long-term solar activity and to study the South Atlantic Magnetic Anomaly region using very low frequency detectors. The network, described in Sect. 4 of this document, comprises seven detectors located at Piura and Punta

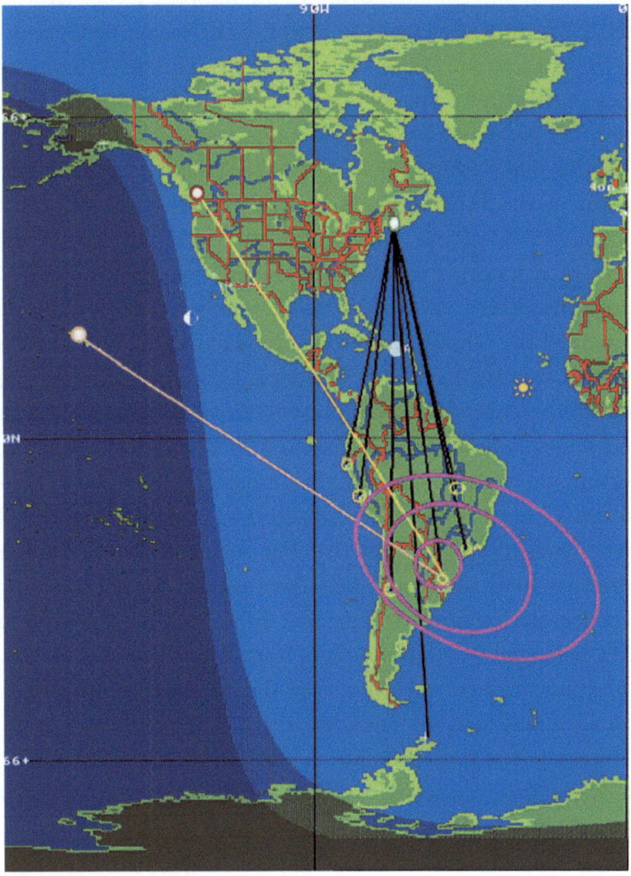

Fig. 80. *Sites of the SAVNET network.*

Lobos (Perú), CASLEO (Argentina), Palmas, São Paulo and Santa Maria (Brasil), and the Antartic Brasilian station Comandante Ferraz (EACF), as shown in the Figure 80. The institutions participating in SAVNET are: Centro de Rádio Astronomía e Astrofísica Mackenzie (CRAAM) of the Universidade Presbiteriana Mackenzie (Brasil), Universidade do Vale do Paraiba (UNIVAP, Brasil), Universidade de Taubaté (UNITAU, Brasil), Observátorio Espacial do Sul (OES, Brasil), Universidade Federal de Santa Maria (UFSM, Brasil), Centro Regional Sul de Pesquisas Espaciais (CRSPE/INPE, Brasil), Complejo Astronómico El Leoncito (CASLEO, Argentina), Radio Observatorio de Jicamarca (JRO, Perú), Universidad Nacional Mayor de San Marcos (UNMSM, Perú), Comisión Nacional de Investigación y Desarrollo Aeroespacial (CONIDA, Perú), and Universidad de Otago (New Zealand).

Educational activities in Brasil started with the Simpósio Brasileiro de Geofísica Espacial e Aeronomia (Brasilian Symposium on Space Geophysics and Aeronomy) at INPE (Instituto Nacional de Pesquisas Espaciais, São José dos Campos) from 23 to 26 October, 2006. During the symposium 110 students and 61 researchers participated. The symposium included tutorials, oral and poster presentations, together with visits to laboratories.

One of the main educational IHY activities in Latin America was the organization of the IHY Latin American School. The school was held at Universidade Presbiteriana MacKenzie from 14 to 20 February 2008. Eighty-five participants, including researchers, post-doc and Ph.D. students from Argentina, Brasil, Chile, Cuba, Germany, México, Perú, U.S.A. and Russia attended the school. Lectures were given by 17 specialists in different IHY topics. The school activities included 17 lectures, 26 oral presentations by advanced students, 36 poster presentations, 2 data analysis laboratories (solar and ionospheric observations were processed) and 2 visits to scientific and observational facilities (Instituto Nacional de Pesquisas Espaciais – INPE, São José dos Campos, and Rádio Observatório do Itapetinga, Atibaia). The school was sponsored by several Brasilian agencies and educational institutions. Students from different Latin American countries were supported by their local agencies, while lecturers were either supported by the institutes to which they belonged or received grants from Brasil. The school was extremely successful. Students were able to discuss their research projects with first level international researchers, to exchange their experiences and start collaborations with researchers from different countries in the region. At the end of the school all participants received a DVD containing lectures, oral and poster presentations (Figure 81).

As part of the outreach activities in Brasil, researchers from CRAAM also contributed short articles to children's magazines.

Fig. 81. *Poster from the 2008 IHY Latin American School.*

7.6.3. Mexico 

As part of UNBSS, one instrument from the Compound Astronomical Low-cost Low-frequency Instrument for Spectroscopy and Transportable Observatory (CALLISTO), provided by the Institute of Astronomy, ETH-Zentrum, in Zurich, was installed at Instituto de Geofísica (IGf) of the Universidad Autónoma de México (UNAM). The instrument (described in Sect. 4 of this document) is devoted to continuous coverage of solar radio spectra in the range 51–858 MHz. Its observations are used to study fine structures during solar flares and the radio counterpart of propagating solar disturbances.

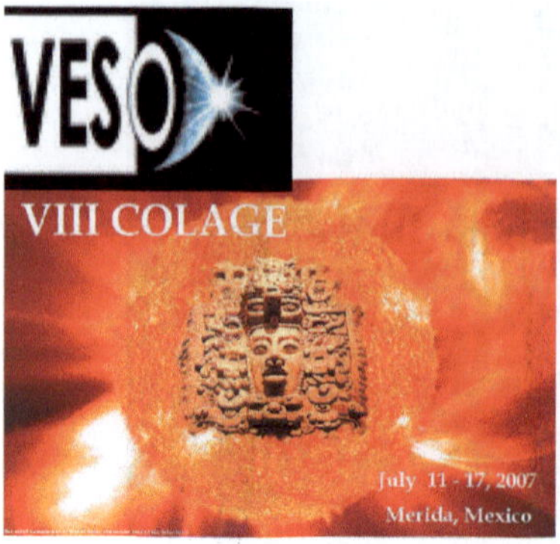

The IHY kickoff in México was led by IGf from UNAM. Several press conferences were delivered on the same date (February 19, 2007) of the IHY official opening in Vienna, Austria. Press releases were published to raise awareness in the general public of the relevance of heliophysical sciences. Other outreach activities took place along 2007 and still continue.

A contribution of IGf to the international community is the implementation of the Observatorio Virtual Tierra-Sol (VESO). A web page (*http://www.veso.unam.mx*) where real-time data from instruments covering data from the Sun to the Earth are available. The instruments are: Radio Interferómetro Solar (Solar Radio Interferometer, RIS), Observatorio de Centelleo Interplanetario (Mexican Interplanetary Scintillation Array, MEXART), Observatorio de Rayos Cósmicos (Cosmic Ray Observatory) and Observatorio Magnético from Teoluyacan (Teoluyacan Magnetic Observatory).

From July 11 to 17, 2007, the VIII Conferencia Latino Americana de Geofísica Espacial (VIII Latin American Conference on Space Geophysics) of ALAGE was held in the city of Mérida. During the conference a full session was devoted to IHY, several talks highlighted the participation in IHY of different Latin American countries (via instrumentation, educational and/or outreach programs). The session finished with an open discussion of IHY activities and possible collaborations in the region. The second Escuela Latino Americana de Geofísica Espacial (ELAGE) was organized in simultaneity with the VIII COLAGE (Latin-American Conference on Space Geophysics). The school was part of IHY educational activities. Ph.D. and postdoctoral students from all participant countries attended lectures on: Solar Physics, Auroral Physics, Space Weather, Ionospheric Physics, Planetary Magnetospheres.

7.6.4. Peru

Two programs have been developed in Perú as IHY activities. The first one is an educational project through the Office of Outreach and Education (Oficina de Divulgación y Educación). A number of high schools are involved in this project. The main objective of this program is to raise awareness in the educational community at high school level of the relevance of space science, as well as to provide a methodology and the necessary elements to do some basic scientific experiments. This program started with conferences including topics related to IHY main objectives, mainly our Sun, its activity and its impact on human technologies. At the end of these conferences material was provided to the students in the form of posters (Figure 82).

Fig. 82. *High school students attend a lecture explaining how to perform some basic scientific experiments.*

Fig. 83. *IHY scientists set up a new SAVNET station at Punta Lobos.*

Two basic experiments were proposed, a written guide was provided and teachers were trained. These experiments were: (1) the construction of a radiometer and the measurement of solar irradiance using a glass bottle, a cylinder with 200 ml of water, a thermometer and black ink to assure the efficient absorption of radiation, and (2) the counting of sunspots using two Russian 4″ telescopes from Comisión Nacional de Investigación y Desarrollo Aeroespacial (CONIDA).

The second program is the SAVNET project. The installation of two SAVNET stations was completed at Punta Lobos (see Figure 83) and Piura. These stations, as those in Argentina and Brasil, are constantly monitoring solar activity. The Punta Lobos station has proven to have a great sensitivity to very low intensity solar flares: it has detected 31 B class flares higher than B3.5, 21 C class flares and 2 M class flares, as compared to the 93 solar flares detected by the GOES spacecraft during the same period of time.

7.7. North America

Participating nations: *Canada and the United States*

Submitted by Donald M. HASSLER, IHY-North America Coordinator, Southwest Research Institute, Boulder, Colorado

The IHY North American activities are coordinated by a planning team for the entire region working to synthesize the activities of the region's two constituent nations, Canada and the United States of America. For most efforts, Canada and the USA are cooperating directly instead of having separate programs. There have been joint Canada-USA activities for each of the four major main components of the IHY (Science, Observatory Development, History and Outreach), and the primary emphasis of the coordination has been the science. The IHY North American Planning Workshop held February 16–18, 2005 finalized the science priorities for IHY, and synthesized the inputs from the international teams into a clear science plan. The IHY North American planning team, through numerous planning teleconferences and several community workshops and town hall meetings, were able to establish comprehensive research and instrumentation programs that have been successfully implemented.

In addition to the development of scientific priorities, launching scientific initiatives, and the convening of numerous special sessions on IHY at scientific meetings, the IHY North American team was able to recruit the participation of hundreds of individual institutions and observatories. These individual groups anchored many of the science campaigns and CIPs, as well as Education and Public Outreach activities throughout North America.

A significant amount of the effort of the IHY North American teams has been focused on cooperation with other IHY regions. Principle Investigators and team leaders from the U.S. and Canada are playing major roles in the Observatory Development Program (Sect. 4). The launch of the IGY Gold History Program (Sect. 6) in North America coincided with the launch of the international IHY History program. Additionally, our scientists have placed a high priority on organizing and supporting international meetings and programs that emphasize collaboration with scientists in developing nations. A brief summary of IHY activities in Canada and the USA follows, though it is important to mention that other contributions of IHY North America can be found throughout this report.

7.7.1. Canada

Submitted by Ian MANN, IHY-Canada National Coordinator, University of Alberta, Alberta

7.7.1.1. Executive summary

Canada has pursued an active program in support of the objectives of the International Heliophysical Year (IHY) during 2007–2009. As the world's third nation in space, with the flight of the Alouette I satellite launched on September 29, 1962, Canada has a long history as a space pioneer. Consistent with the continued development of excellence and expertise in space science in Canada, significant progress was obtained toward all three of the IHY objectives. Scientific activities have included the development of significant new ground- and space-based infrastructure for heliophysical science. Highlights from infrastructure expansions include:

(1) the extensive deployment and operation of new ground-based instrumentation which exploits Canada's position as the world's most accessible auroral landmass,
(2) new partnerships,
(3) instrument deployment, and
(4) scientific collaboration with new missions.

Canada also contributed to the development of the IHY Observatory network (see Sect. 4) and to the IHY goal of deploying "Low-Cost Instrumentation in Developing Nations" by leading the CARISMA program and participating in the AMBER (African Meridian B-Field Education and Research) program lead by

UCLA. In relation to IGY, many Canadian researchers who were active in 1957–1958 were identified, and a celebration of their many achievements as part of IHY Gold has transpired. A review of "Canada's Fifty Years in Space" was also published, and a number of outreach activities brought space science to the public and into classrooms, including the AuroraWatch service.

7.7.1.2. Introduction

Canada views its participation in space science as an investment that produces many tangible benefits on Earth. In addition to scientific discovery, they enhance the scientific and technological skills of Canadians, as well as their ability to compete in an increasingly knowledge-based economy. They also generate spin-off technological and economic benefits, and inspire young people to pursue advanced education and careers in science and technology. These objectives have been pursued actively by Canadians during the IHY, and have been summarised in two Canadian reports on IHY activities presented at the United Nations.

7.7.1.3. Highlights from Canadian IHY activities

Canadian Scientific Research and Instrument Deployment during IHY

Canadian GeoSpace Monitoring (CGSM) Program: The Canadian Space Agency recently issued nearly $10 million in new contracts for continuing operations of the Canadian Geospace Monitoring (CGSM) Network, an extensive series of ground-based stations used to observe disturbances in geomagnetic space. Many of the CSGM arrays have been significantly expanded during the IHY, including the expansion of the Canadian Array for Real-time Investigations of Magnetic Activity (CARISMA) magnetometer array (see Sect. 4). This includes the addition of 15 new fluxgate magnetometers to the previous network of 13, as well as the deployment of a new network of eight induction coils magnetometers, the expansion of the NORSTAR array comprising the addition of five all-sky-imagers (ASIs) to a pre-existing network of five ASIs and the operation of four meridian scanning photometers (MSPs). A new network of 13 riometers is also being added to the magnetometer instrumentation operated by Natural Resources Canada (NRCan). Similarly, the Canadian High Arctic Ionospheric Network (CHAIN), operated and deployed by the University of New Brunswick, is a distributed array of ground-based radio instruments comprising ten high data-

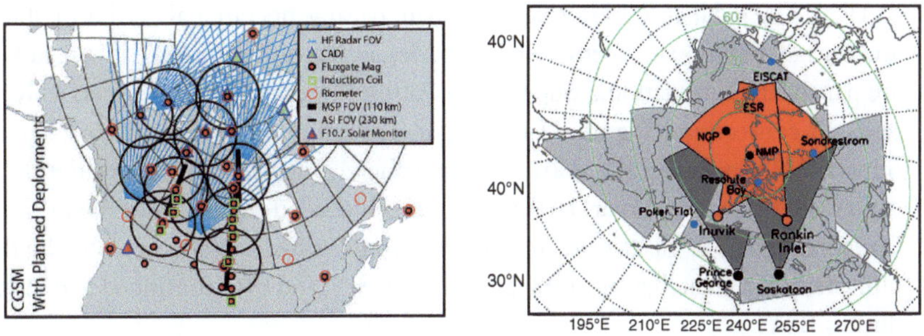

Fig. 84. *Left panel: Canadian Geospace Monitoring Instrumentation. Please note the most recent deployments in the CHAIN array are not shown (see below). Right panel: Schematic of the northern hemisphere PolarDARN (red) and Canadian SuperDARN (dark grey) radar fields of view.*

rate GPS ionospheric scintillation and total electron content monitors and six Canadian Advanced Digital Ionosondes were deployed in the high-latitude arctic during IHY. Also included in the instrument set are solar monitors that measure F10.7 radio waves emitted from the sun, and digital ionosondes and superDARN radars (see below) that send out bursts of radio energy into the atmosphere and measure the reflections, providing information on the state of the ionosphere and the auroras. Finally, CGSM includes a modelling program in the Facility for Data Assimilation and Modelling (FDAM) which draws these multi-instrument datasets together through modelling, simulation, and data-assimilation efforts. Collectively these arrays represent a world-unique continent-scale network, and Figure 84 (left panel) shows the CGSM instrument coverage.

Observing how the Sun and Earth interact through CGSM will help refine the forecasts of highly charged space weather effects, which may adversely affect satellite-based communications, GPS navigation, electric power grids, and the health of humans living and working in space. The data collected through the CGSM is also playing an important role in forecasting space weather, disturbances in the space environment above the Earth, including contributing to Canadian Space Weather Forecast Centre in Ottawa, operated by Natural Resources Canada. A large and diversified science team will collect and analyze CGSM data, including more than 80 scientists led by principal investigators from the universities of Alberta, Calgary, Saskatchewan, Waterloo and New Brunswick. In addition to the Canadian Space Agency, Natural Resources Canada and the National Research Council of Canada are also partners in this program.

SuperDARN and PolarDARN Radars: During IHY, the University of Saskatchewan has greatly increased the emphasis on polar cap research with the installation of two new over-the-horizon radars in the Canadian High Arctic, called Polar-

DARN. The radar at Rankin Inlet, Nunavut, became operational in May of 2006, and the radar at Inuvik, Northwest Territories, in November of 2007. These radars are part of the international Super Dual Auroral Radar Network (SuperDARN) research consortium. Figure 84 (right panel) illustrates the current coverage of SuperDARN radars in the northern hemisphere. The viewing areas of the polar cap radars (known as PolarDARN) are the red shaded regions, and the original Canadian radars at Saskatoon and Prince George, which are also operated by the University of Saskatchewan, are shaded darker grey, and have continued to operate during IHY.

The PolarDARN radars at Rankin Inlet and Inuvik are designed to measure plasma convection in the polar cap, which is the region where the Earth's magnetic field lines are directly connected to the interplanetary magnetic field and where the Sun's influence on the Earth's magnetosphere is most direct. The PolarDARN radars respond practically instantaneously to changes in the solar wind and interplanetary magnetic field. The polar cap is largely unexplored, in terms of ionospheric plasma convection. The potential for scientific output from a pair of PolarDARN radars is immense. Recent research has focussed on convection response to solar wind driving, studies of ULF waves and field line resonance phenomena, as well as a new technique examining temporal gradients in the ionospheric convection flow as a diagnostic for the study of magnetospheric boundaries and their dynamics.

Resolute Bay Incoherent Scatter Radar (RISR): The Resolute Bay Incoherent Scatter Radar (RISR) is a USA-Canada initiative to develop a phased array Incoherent Scatter Radar in Canada's arctic. The University of Calgary leads the Canadian RISR consortium which includes researchers from Athabasca University, University of Saskatchewan, York University, University of New Brunswick, and University of Western Ontario. RISR has strong synergies with CSA's ePOP (see below) and Atmospheric Chemistry Experiment (ACE) and ESA's Swarm missions, and will be complimented by the more synoptic observations afforded by Canadian GeoSpace Monitoring. The project will lead to improved GNSS accuracy, and advances in our understanding of how space weather affects climate. The primary USA partner is SRI International, with involvement from MIT, Cornell, and University of Alaska. Funding for the first phase came from the National Science Foundation of the USA. A Canada Foundation for Innovation (CFI) application has been submitted to secure funding for the Canadian part of RISR.

Enhanced Polar Outflow Probe (ePOP): In June 2009, Canada plans to launch a small satellite called CASSIOPE into polar orbit around the earth. It will carry the enhanced polar outflow probe (ePOP), a collection of eight scientific instruments that will investigate the role of the "polar wind" in causing space weather. This "polar wind" or "polar outflow," which consists of electrically charged and neutral particles flowing up from the Earth's ionosphere into the magnetosphere, has not been well-studied so far; in fact, scientists didn't even know of its existence until the last few decades. ePOP measurements from the ground-based CGSM network will help Canada improve its ability to forecast space weather.

Time History of Events and Macroscale Interactions During Substorms (THEMIS): Canada is an active participant in the NASA MIDEX mission THEMIS, which studies the mysterious auroral substorms caused by an explosive release of stored energy from the solar wind into the Earth's magnetosphere. In February 2007, NASA launched five small THEMIS satellites to observe the aurora from above. Meanwhile, Canadian researchers are using a network of ground stations to take images of the aurora every three seconds and provide magnetic monitoring on the ground of these space weather disturbances on an unprecedented scale. Canada is uniquely positioned to participate in this study because it has the most accessible landmass underneath the auroral zone anywhere in the world, and has a long history of auroral research. Sixteen of the twenty THEMIS ground stations are located in Canada, with the remaining four in Alaska. These data are being coordinated with the measurements taken by the satellites to provide a comprehensive look at the dynamic events that are responsible for driving the most powerful aurorae during magnetic substorms.

JOULE-II Sounding Rocket: The ionosphere is a region of transition between Earth's neutral atmosphere and space. The crossover altitude, where ionospheric ions are influenced equally by electric fields and neutral winds, takes place at an altitude of $\sim$120 km, which is too high for balloons and too low for satellites. On January 19, 2007, the NASA JOULE-II sounding rocket mission was launched into this transition region to make the first-ever simultaneous measurements of the key parameters whose interplay defines this region: electric and magnetic fields, ionospheric flow, and neutral winds. The University of Calgary provided an imaging charged particle detector that measured ion flow to an accuracy of 10 m/s, and ion temperature to a resolution of 0.01 eV.

Virtual Observatories (VxOs): In 2008, CANARIE Inc., Canada's advanced network organization, announced funding for the CANARIE *Canadian Space Sciences Data Portal* (CSSDP) as a Virtual Observatory (VxO). The CANARIE CSSDP will enhance and extend a data portal project by integrating data from heterogeneous sources into a single analytic platform, developing a virtual observatory environment for constructing analytic and visualization tools by and for space scientists, and establishing standards for data access and science work-flows. The data registry allows for resources to dynamically move into and out of the federated system without programmer or administrative intervention, with multiple ways to access and obtain information from resources.

Global Auroral Imaging Access (GAIA) is a University of Calgary-led international virtual observatory program which enables rapid searching of summary data from instruments that remote-sense auroral precipitation. At present GAIA has nodes in Canada, the UK, and Sweden, and provides access to data from dozens of instruments operated by researchers in North America, Europe, and Asia, and is rapidly growing. Ongoing initiatives include, for example, the establishment of additional nodes in Finland (November 2008) and Norway (December 2008), the development of data access capabilities, and enabling connectivity to other VxOs around the world.

Radiation Belt Science: IHY scientists at Memorial University of Newfoundland are seeking to extend our knowledge of the dynamics of the Earth's radiation belts, and in particular to understand the generation of high-energy (relativistic) electrons during magnetic storms. These electrons (colloquially called "killer electrons") can cause serious damage to spacecraft, therefore forecasting relativistic electron events is a key priority of space weather science. The measured wave modes are thought to be instrumental in governing the dynamics of Earth-orbiting energetic radiation belt electrons, and play a fundamental role in the energization and scattering loss of killer electrons.

Space Weather in Solar System (SWISS): The work on this project, corresponding to IHY CIP #20, is proceeding at Natural Resources Canada (NRCan), including funding for a Visiting Fellow at NRCan in May 2007. Currently, none of the existing space weather models are capable of providing the near-real time determination of the propagation of the solar disturbances (Coronal Mass Ejections). In order to fulfill the need for the near real-time modeling of the CME propagation for applications in space weather forecasting, the Visiting Fellow is currently working on the development of the proto-operational MHD code. The University of Waterloo is also developing an unstructured grid MHD model for applications of the propagation of the disturbances from the Sun into the Heliosphere.

Athabasca University Geophysical Observatory (AUGO): Athabasca University (AU) has continued to benefit from its location in the southern auroral zone through the internationally connected activities of its Geophysical Observatory (AUGO). Collaboration with STELAB of Nagoya University, Japan, has made AUGO the most comprehensively instrumented auroral observatory in Canada, and this combination of magnetic and optical instruments has given new insight into the proton aurora and its connection to EMIC waves in the inner magnetosphere. AU also operates the AUTUMN magnetometer network, a network of subauroral and auroral magnetometers, which is a collaborative effort with UCLA, and has supported the continued operation of some University of Tokyo STEP network magnetometers. A major thrust in innovative instrument development has resulted in unique, low-cost, high-precision magnetometers suitable both for research and for deployment in schools.

IHY-Canada Education and Outreach

Canadian Space Agency Educator Workshop, 5–7 August 2008: A three-day conference was organized by the Canadian Space Agency where educator participants and CSA engineers and scientists explore space-related, curriculum-relevant topics at the primary and secondary levels. CSA led a highly successful session on the upcoming solar maximum that directly addresses the IHY's third objective.

Canadian Solar Workshops: Following a significant growth in the number of Canadian solar scientists and space weather applications researchers, especially from the University of Montreal, Queens University, Dominion Radio Astronomy Observatory, and NRCan Space Weather forecast centre, Canada organized a number of special annual Canadian Solar workshops (Figure 85).

High School Classroom Teaching Resources: The Centres for Research in Youth, Science Teaching and Learning (CRYSTAL) is a program of Science and Engineering Research Canada (NSERC), headquartered at the University of Alberta Faculty of Education. In collaboration with the Department of Physics,

Fig. 85. *Participants at the Canadian Solar Workshop 2007.*

CRYSTAL developed classroom learning resources related to IHY including a unit of work focusing on the solar coronal heating problem. This problem was used to illustrate and demonstrate that real-life controversy exists in science, and that scientific theories are established on the basis of debate and logical argument based on the available scientific evidence.

Alberta Physics Teachers Network, Cosmic Rays, and ALTA: During the IHY, networking and outreach to the Alberta Physics Teachers network promoted the development of classroom built magnetometers for teaching high school science. A collaboration between ALTA ("search in **A***lberta* over a **L***arge area* for cosmic ray shower **T***ime coincidences* using an **A***rray* of detectors; *http://csr.phys.ualberta.ca/alta/*) at the University of Alberta is also being established. This would enable those High Schools already involved in ALTA and hosting cosmic ray shower detectors to also monitor the geomagnetic field, and hence students to examine links between heliophysical disturbances and the incidence of cosmic ray showers.

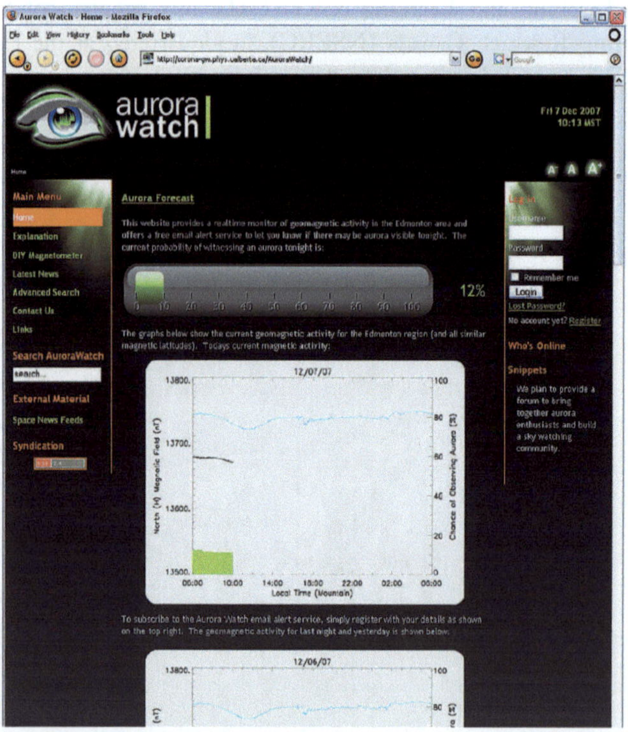

Fig. 86. *The AuroraWatch service provides now- and fore-casting of auroral viewing opportunities using real-time magnetometer data from the CARISMA array.*

AuroraWatch Canada: AuroraWatch is a public website which provides a local forecast for observing the aurora in the night sky which is operated by the University of Alberta (see Figure 86). By monitoring near-Earth space weather events, specifically large scale perturbations of the Earth's magnetic field using CARISMA magnetometers, the AuroraWatch website is able to provide a now- and fore-cast of the likelihood of observing bright aurora for the current night. The website algorithm used statistics from more than a decade of co-located magnetic and auroral monitoring using data from the CGSM infrastructure from the CARISMA and NORSTAR arrays. The website, located at www.AuroraWatch.ca, also provides a free automated email alert notification to subscribers during periods when the potential for observing bright auroral displays is significantly increased.

7.7.2. United States

Submitted by Roger Smith, IHY-USA National Coordinator, Geophysical Institute, Fairbanks, Alaska and Deborah Scherrer, Chair of the IHY-USA Education Advisory Committee, Stanford University, Stanford, California

The United States IHY Program would not have been a success without the efforts of the hundreds of scientists and education professionals who planned and launched the IHY activities, and the many thousands of people who participated in them. The United States provided much of the support for the IHY Secretariat, and was involved in IHY planning activities from the beginning. The National Aeronautics and Space Administration (NASA) provided support for IHY-USA and IHY's international programs, including substantial funding through competed proposals. These resulted in the support for many projects, including several of the observatory activities mentioned in Sect. 4.

7.7.2.1. National Planning efforts

IHY planning began in 2001, with a small committee of scientists who organized a special session on IHY at the 2002 World Space Congress in Houston, TX. Planning began in earnest in 2004, beginning with the IHY Planning Workshop held April 20–22, 2004 in Sacramento Peak, NM. This was an extremely well-attended event, with balanced participation of scientists from all areas of heliophysics. The concept of "Universal Processes" was born, and an ambitious plan was set in motion. Later that year, the IHY/UNBSS Science Organizing

Committee Planning Meeting laid forth the principles of IHY's observatory development program, and the recruitment of scientists for the UNBSS program became a top priority. The science investigations resulting from this effort (described in Sect. 4) became the topic of the UNBSS workshops, starting in 2005.

The United States hosted the IHY North American Planning Workshop February 16–18, 2005 in Boulder CO, where more detailed science plans were established. A more formal USA organizing committee was established, and this committee began launch a series of special sessions and workshops in preparation for IHY.

With the launch of IHY, the USA initiated a wide range of activities, and supported hundreds more. International partnerships, particularly with developing nations, have been a major emphasis of the USA IHY program. A brief summary of these activities follows.

7.7.2.2. Science activities

As mentioned previously, USA scientists made major contributions to the establishment of the IHY science program through its planning workshops. The execution of the science programs proceeded through a number of workshops, meetings, and investigation programs. USA scientists led 22 Coordinated Investigation Programmes (CIPs), including the Whole Heliosphere Interval held March 20–April 16, 2008. WHI (see Sect. 3), which involved over 200 scientists from around the world, coordinated observers and modelers focusing on a full solar rotation and its resulting influence on the Earth. The WHI Data and Modeling Assessment Workshop, held August 25–29 2008, was held in Boulder, CO and focused on early scientific results and the coordination of scientific efforts for deeper analysis. A follow-up science workshop, with a broader scope, is planned for 2009.

The two most significant IHY science meetings held in the USA were the January 23–29, 2008 "Polar Gateways" conference in Barrow, AK, and the Chapman Conference on Universal Heliophysical Processes held November 10–14, 2008 in Savannah, GA. The Polar Gateways conference held in conjunction with the International Polar Year, focused on interdisciplinary studies throughout the solar system. It was held at the time of the "Arctic Circle Sunrise" and included virtual participants around the world and also included a strong educational component supported by the Passport to Knowledge program. The Chapman Conference on Universal Heliophysical Processes was unique in that all oral presentations were on topics of Universal Processes, with extended discussion periods to allow the participants to integrate their own research topics with the information presented by the speakers.

There were several IHY workshops hosted in conjunction with CEDAR (Coupling, Energetics and Dynamics of Atmospheric Regions) meetings, allowing atmospheric and ionospheric scientists to focus on a particular IHY topic. These included a 2005 session on the international science years, and the 2008 IHY/CEDAR Workshop titled "Geospace Response to Solar Minimum Drivers." A summary of the 2008 workshop results, dealing with one aspect of the WHI campaign (see Sect. 3) is given below.

As of December 2008, there have nearly 20 special sessions on IHY at American Geophysical Union (AGU) meetings, as well as special sessions at other conferences, report on IHY activities and inviting scientists to participate in the discourse of the new discipline of heliophysics.

Workshop Report for "Geospace Response to Solar Minimum Drivers – All Solar Minima Are Not the Same"

Organizers: Janet U. Kozyra (Univ. of Mich) and Larry J. Paxton (JHU APL)

During the recent IHY Whole Heliosphere Interval (WHI), which took place 20 March–16 April 2008, isolated, trans-equatorial coronal holes were present on the solar surface. The high-speed solar wind that arrived at Earth originated from deep within the coronal holes and thus was exceptionally fast and long-lived. In contrast, last solar minimum with focus on Whole Sun Month (10 August–8 September 1996), the Earth experienced wind from the equatorward extension of a polar coronal hole called the "Elephant's Trunk". The resulting high-speed streams, coming mostly from the edges of the narrow coronal hole extension, were weak and disorganized. Since the speed and duration of coronal hole winds have been shown to control the nature and severity of geospace and atmospheric disturbances, we expect interesting differences in the geospace response comparing solar minima in solar cycles 22 and 23.

The purpose of this workshop was to: (1) take a first look at the upper atmosphere and geospace response to solar wind driving during the current solar minimum with particular focus on WHI; (2) compare to conditions during the last solar minimum in 1996 with focus on WSM; and (3) identify interesting ITM (ionosphere/thermosphere/mesosphere)-focused science questions for continuing collaborations with the WHI post-event analysis.

Below are the observation topics and physical processes around which the first ITM questions are being formulated for the WHI campaign analysis:

(1) Periodicities in ITM parameters in this solar minimum compared to last.
(2) Neutral atmosphere and ionosphere in this solar minimum compared to last.
(3) Coupling to the lower atmosphere
(4) Coupling to geospace

- Ion and Neutral Outflows
- Subauroral and Equatorial Electrodynamics

Another major science effort within the USA was the Observatory Development Program. Seven of the participating instrument programs are headquartered in the USA (see Sect. 6), as well as the tremendously successful AWESOME educational space weather monitors (see Sect. 5). NASA sponsored a special "Announcement of Opportunity" whereby scientists and outreach professionals were invited to submit proposals and compete for IHY funding. The result of the NASA proposals was a significant boost to the instrumentation activities, and expanded outreach programs, including the "Passport to Knowledge" program (see Sect. 5).

7.7.2.3. Outreach activities

The Space Physics and Aeronomy Education and Public Outreach Committee of the American Geophysical Union (in spite of the name, a world-wide organization) has served as the USA Advisory Committee for Education and Public Outreach for the International Heliophysical Year (IHY). The committee, chaired by Deborah Scherrer of Stanford University, includes Cristina Rabello-Soares, the IHY International Coordinator for Education, plus several other individuals also directly involved in IHY activities. The committee hosted several events to highlight IHY science, including teacher workshops in Ethiopia and Mexico, Sun-Earth Day events, and family science events in Mexico and Florida, USA. There were also lectures and public events held around the country, inviting the public to participate in IHY.

Most of the efforts of this committee had an international scope. The events and activities reached millions of people, and few of these efforts were confined to the

United States. Educational activities that occurred on an international level were described in Sect. 5; a result, this section seems deceptively small. IHY-USA outreach team members organized the teacher workshops, Sun–Earth Days, eclipse activities, several IHY Summer Schools, AWESOME space weather monitors, and World Space Week activities described in Sect. 5, and the reader is directed to that section to learn more about the impact of the IHY-USA outreach program. An example of a local USA event follows.

Family Science Event "Exploration Station" – Ft. Lauderdale, FL, USA

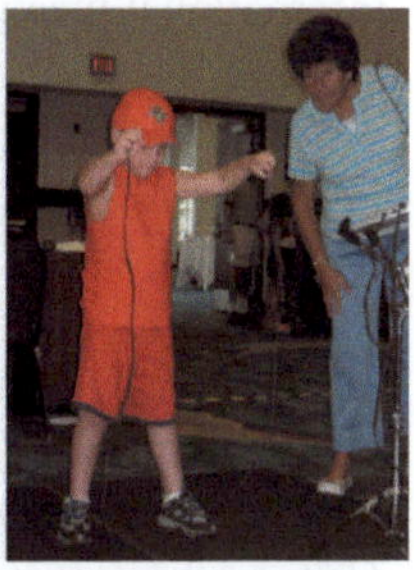

On 31 May 2008, the USA IHY EPO Advisory Committee, the American Astronomical Society's Solar Physics Division (SPD), and Rochester Institute of Technology Center for Imaging Science Insight Lab, experimented with a new outreach program held in conjunction with the American Geophysical Union's Meeting of the Americas 2008 Joint Assembly. Our "Exploration Station" was a 4-h open house for the local community to learn about the cool science that meeting attendees are currently involved in. Participants had a chance to meet scientists, do hands-on Earth and space science activities, and take home science related resources collected during their visit. This program was aimed at changing perceptions of what science is, who scientists are, and what they do. There were 13 exhibitors present at Exploration Station, most of them representing the Solar Physics community, including: the AGU, Florida International University, Stanford Solar Center, SPD, NASA, South Florida Water Management District, the National Solar Observatory, Rice University, and the Design Rhythmics Sonification Research Lab. Activities presented included build-your-own spectroscopes, experimentation with UV beads, a magnet discovery station, STEREO and solar data sonification exhibit, and two inflatable planetariums. Although turnout was low, participants stayed anywhere between 1 and 3.5 h when accompanied by children, a large amount of time compared to exhibits at a standard science museum.

7.7.2.4. Support of international IHY activities

As mentioned previously, a significant amount of the IHY-USA effort prioritized the support of international activities and programs supporting developing nations. Many of the members of the IHY Secretariat, including the Executive Director and Outreach Coordinator, are USA scientists. In addition to the Science and Outreach international activities described above, USA scientists led the IAU Working Group on IHY, assisted in the coordination of the ICESTAR program, served as the regional coordinator for North America, organized the IHY-Africa workshops, led three IHY summer schools, and hosted special sessions at many international meetings.

7.8. Western Asia

Participating nations: *Bahrain, Iraq, Jordan, Kuwait, Lebanon, Oman, Palestine, Qatar, Saudi Arabia, United Arab Emirates, Yemen*

Submitted by Hamid M. K. Al-Naimiy, IHY-Western Asia Regional Coordinator, College of Arts and Sciences, Sharjah University, Sharjah, United Arab Emirates

The IHY-Western Asia initiative has played a fundamental role in each of the four Programmatic Thrusts of IHY. The first UNBSS/ESA Workshop on IHY was held in Abu Dhabi and Al-Ain, United Arab Emirates 20–23 November 2005. Representatives from 44 U.N. member states were present, representing a large portion of the Western Asian regions as well as Northern Africa. This allowed extensive discussions of the state of Heliophysical research in these regions, and facilitated planning for IHY national and regional regions.

IHY has coordinators in place for many of the nations in the Western Asian region, with the goal of establishing a coordination team in each nation in early 2006. In addition to hosting the first UNBSS/ESA Workshop on IHY, IHY-Western Asia team members are active in coordinating scientific activities which will lead to many scientific campaigns. The region's extensive activities for the World Year of Physics 2005 (*http://www.physics2005.org*) have resulted in experience and a foundation for the continued development of educational programs, summer schools, and historical events.

During the First UN/ESA/NASA workshop hosted by UAE during the period November 20–23, 2005, a regional (West Asia) committee formed for

IHY Western-Asia regional committee.

Name	Location
Jammal Maymoni	Centre de Recherché en Astronomie Astrophysique et Geophysique
Samir Nait Amor	CRAAG Algiers, *Algeria*
Musalam Shaltot	The National Research Institute of Astronomy and Geophysics
A. A. Galal	Helwan, Cairo, *Egypt*
Rashid Al-Naimi	Department of Atmospheric Sciences, College of Science
	Al Mustansiriyah University of Baghdad, *Iraq*
Saleh Al-Shedhani	Physics Dept./College of Science, Sultan Qaboos University, Al-Khoud, *Oman*
Hanna Sabat	Institute of Astronomy and Space Science, Al Al Bayte University, Mafraq, *Jordan*
Rojer Hajjar	Physics and Astronomy, Notre-Dame University, Zouk Mikael, *Lebanon*
Emad Barghouthi	Professor of Physics, Department of Physics, Faculty of Science, Al-Quds University, Jerusalem, *Palestine*
Abdul Haq Sultan	Head of Physics Department, Faculty of Science, Sanaa University, *Yemen*
Nour-Eddine Najid	Université Hassan II Ain Chock, Faculté des Sciences Ain Chock, Casablanca, *Morocco*
Fodil Mammar	CRAAG/Observatoire d'Alger, Algiers, *Algeria*
Hassan Basurah	Astronomy Department, King Abdulaziz University, Jeddah, *Saudi Arabia*
Ismail Sabbah	Department of Physics, Faculty of Science, Kuwait University, *Kuwait*
Abdul Qader Abseim	Libyan Remote Sensing and Space Center, *Libya*
Shawqi Al-Dalal Mohammed Al Othman	Physics Department, Bahrain University, *Bahrain*

IHY. Since then the committee has met many times, first meeting was on November 23, 2005 and the second meeting on August 14, 2006 in Amman, Jordan. The third was on March 2008 in Sharjah/UAE. The meetings were chaired by Prof. Hamid M. K. Al-Naimiy, IHY-West Asia National Coordinator, and around 50 participants attended representing 14 countries from Iraq, Palestine, Bahrain, Egypt, Jordan, Kuwait, Libya, Oman, Saudi Arabia, Syria, Qatar, UAE and Yemen.

7.8.1. IHY & BSS cooperation between Arab countries

(1) Enhancing IHY & Basic Space Science facilities and research centres in Arab countries requires cooperation with international institutions in general and among many Arab universities & organizations in particular. Establishing good IHY & BSS activities in any region cannot be accomplished without strong cooperation between different communities. In Arab countries, everything for IHY & BSS development is available (such as budget, personal, sites, environments, etc.), the only thing missing is cooperation and scientific support from international organizations and scientists.

(2) Building good, modern observatories and IHY stations in the region jointly by Arab solar physicists, astronomers and scientists is essential and will be an excellent step toward developing IHY, astronomy and astrophysics, particularly when the stations & the observatories contain modern-sized solar and optical telescopes with their auxiliaries, including a millimetre radio telescope to be part of any international VLBI (Very Long Baseline Interferometry) system. There are very good sites for observatories and IHY stations in different parts of Arab countries, particularly in those places were observations can be made of both southern and northern parts of the sky. It is most useful if the observatories are built in collaboration with the Arab Union for Astronomy and Space Sciences (AUASS) and the IHY West Asia Committee.

7.8.2. Conclusions and future scope

In the Western Asian region, it is very important to establish two main Education and Research Centres in fields of IHY & BSS. These centres could currently be associated with this initiative: the Arab Space City (ASC), and the Arab Astronomical and Solar Facility (AASF). The AASF would be oriented towards supporting higher education and university level astronomical and space science research and development, hence its centre-piece would be the solar, optical and radio observatories, located on a good astronomical site, with a major support centre comprising assembly and lecture halls appropriate to hosting sizable conferences as well as laboratory and classroom facilities, which would be built in association with or near to the already established infrastructure of a University.

The ASC would be established in the vicinity of the capital city of any Arab country, and would focus on providing the larger general public with inspiration and awareness of the importance of science and technology on their own lives and especially on the lives of their children, and on inspiring younger students towards a career in science through the message of astronomy and space science. These themes would be developed via the introduction of the important history of Arabic and ancient Middle-Eastern discoveries and contributions to the science of astronomy up to and including the present epoch, via the Solar physics, Astronomy and Space Museum, the Interactive Astronomical Image Gallery, the deeply moving Planetarium and continued solar activities. The show experience, using a range of topics, changing according to monthly themes, will bring many viewers back time and again to be refreshed, recharged with enthusiasm, and (re)inspired to seek their own way in the quest of knowledge and discovery. With a 500–600 seat planetarium, for example, there would be every reason to encourage classroom groups of school children from virtually every nation of the Arab World to come to ASC to have this experience. It would be an experience to see the wonders of the universe, and participate in personal hands-on space and astronomy activities, space-camps, flight simulators, observe real-time solar activity with a dedicated solar imaging telescope, and see special exhibits depicting current space activities such as comet sightings, asteroid missions, lunar and Mars exploration activities, space satellites, and activities on the International Space Station. There would always a connection to the activities at the suggested AASF. This aspect of the ASC would both stimulate and educate the public at large, and give rise to new generations of students to populate the science curriculum of the Arab Universities and indeed other international colleges and universities as well. Such a rising awareness and affinity with sciences, taken on balance with the great moral and spiritual strengths imbued through Islam, give tremendous promise to a most fulfilling future for all the peoples of region, the entire Arab World, a vision of harmony combined with maximum personal achievements for everyone: a vision which we all dream of together, deep within our hearts of prayer, God so Willing. In a country encouraging the development of science and technology, it is natural to build a large and important IHY and SSA centre.

One of the main goals of the IHY and space initiative is to promote the Arab astronomy heritage and to provide a core scientific research focus not only for Arab countries, but for all the Middle East and perhaps for Asia. It will also provide life information for the space sciences to the public community. It will show how important the IHY, space science and technology is for humans. The project will concentrate on scientific research, education, learning, knowledge and scientific entertainment. This will be through the exploration of the universe and the discovery of the cosmos and celestial objects, which gives great achievements to the human community.

7.8.3. National summaries

The participants discussed the current research projects and the future cooperation for IHY, and gave overviews for their activities in their countries. Below are summaries of IHY and Basic Space Science (BSS) activities in some countries of this region.

7.8.4. Iraq

7.8.4.1. Science activities

Professor Rashid Al-Naimi is the national coordinator for IHY. His group has initiated studies on Sunspots, Ionosphere, Climate change and develop devices for measuring Solar Radiation. Currently they are conducting the following research:

1. The effects of sunspots on the ionosphere
2. Measurements and modeling of the spectral ultraviolet solar radiation
3. Developments of simple and low cost devices for the solar radiation measurement.
4. The role of solar effects in the climate change of the earth.
5. Development of computer aided learning modules for teaching meteorology and astronomy
6. Modeling the atmospheric effects on radio signals.
7. Accurate determination of the sun position relative to Earth.

7.8.4.2. United nations basic space science observatory development

Iraqi astronomers and space scientists started with good steps in 1980, by developing and establishing "The Iraqi National Astronomical Observatory (INAO)", in Kurdistan, the Northern part of Iraq. The INAO project was built as a West German joint venture, in a "Turn-Key" contract, which includes civil work and a residential complex. Work started in 1981 and the last section of the project as finished in 1986. The observatory comprises the following telescopes:

(1) 30 m Millimetre Radio Telescope with a receiver system plus its auxiliaries
(2) 3.5 m Optical Telescope and 1.25 m Optical Telescope with other astronomical observing equipment is also included such as IR and UBVIR

photometers and different types of spectrographs such as Echelle, Code, and Nasmyth with other needed auxiliaries.

The observatory, which cost U.S. $150 million (in 1980), was built on top of a Korek mountain of 2200 m above sea level, with a very good observational site conditions. Unfortunately, this observatory was damaged during the two wars (1980 and 1991). Recently we had some correspondence with the local scientists and the Government of Kurdistan for rebuilding the observatory at Korek Mountain. A solar telescope and 5–6 m optical telescope of modern design has been suggested which probably fit in the 3.5-m dome.

The current situation of Basic Space Science in Iraq is as follows:

- Space Research Centre: It is part of the Ministry of Science & Technology in Baghdad. The main research programmes are remote sensing, wave propagation, communications, astronomy and astrophysics.
- Physics departments in most Iraqi universities offer courses in astronomy and astrophysics. The College of Science at the University of Baghdad has a good astronomy department, established in 1998 for undergraduate and postgraduate studies in space science and astronomy (SSA). The College also has a small observatory, "Al-Battani Observatory", which contains two telescopes: a 40 cm reflector and a 20 cm refractor, purchased from Goto Company of Japan. The observatory is located in Tarmia, about 50 km north of Baghdad.
- Secondary schools include small general topics about astronomy in the physics courses that they offer, covering such topics as day and night, moon–Earth system, solar system, and stars.

7.8.5. Jordan

The Second IHY Western Asia Regional Meeting was held on August 14, 2006 in Amman, Jordan. The following organizations are concerned with Basic Space Science in Jordan:

(1) *Universities*: Astronomy and astrophysical courses are offered to undergraduate students in more than five universities, but the main Space Science and Astronomy (SSA) activities are concentrated in the Al-Bayt University in Mafraq. In 1994, this university established the Institute of Astronomy and Space Science, which offer a M.Sc. in SSA, beside it has a small optical observatory with the name Maragha Astronomical Observatory.

(2) *Arab Union for Astronomy and Space Sciences (AUASS) (http://www.jas.org.jo/union.html)*: This union established in Amman in 1998. More than 150

astronomers and scientists from 19 Arab countries are members in the Union. The aim of the union is to develop SSA in Arab countries through conferences, meetings, publications, joint research projects, etc. with the cooperation of international SSA institutions. Recently the AUASS council decided to involve with the IHY activities through many members, their specialty related to IHY topics.

(3) *Jordanian Astronomical Society* (*http://www.jas.jo*): An organization that promotes amateur and sometimes professional SSA.

(4) *Teaching BSS in Jordanian Schools*: General astronomy courses are taught in secondary schools. The courses contain general information about the Earth–Moon system, day and night, seasons, solar system, stars, clusters, the Milky Way, galaxies and the universe.

7.8.6. Kuwait 

Dr. I. Sabbah installed the first university robotic Air Pollution Remote Sensing station in the Middle East in collaboration with NASA. In addition, he has completed the Global Muon Telescope Network (4 Telescopes worldwide) by installing a muon telescope at the Kuwait University in collaboration with the University of Delaware... He also recorded the first solar storm ever detected from the Middle East, on 14 December, 2006. The work resulted in the first published paper about remote sensing of desert dust over Saudi Arabia and the first paper published by the Muon Telescope Network team.

In April 2007, the Kuwaiti Yuri's Night World Space Party was hosted by the Space Science center in the Kuwait Science Club (*http://astronomy.ksclub.org/*) (Figure 87).

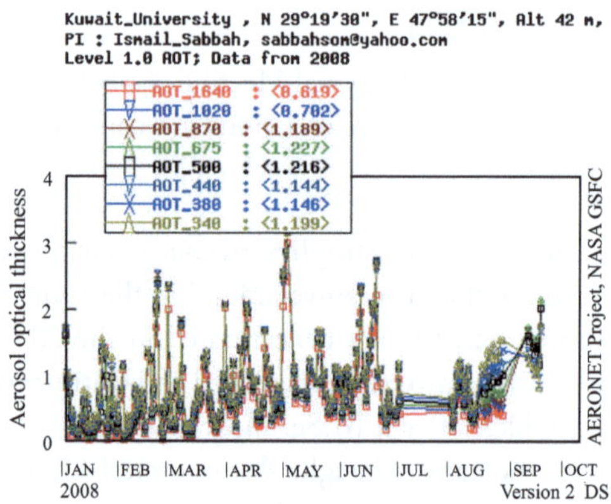

Fig. 87. *Aerosol optical thickness data collected at Kuwait University.*

7.8.7. Oman

Dr. Saleh Al-Sheethani installed a 14″ RCX400 telescope for site testing, along with portable telescopes: 12″, 11″ and 5″ Schmidt-Cassegrain telescopes and a 16″ Newtonian reflector. They are also in the process of building a Robotic 1.5 m telescope optical observatory on the top of Jaabel Al-Shames (2980 m).

7.8.7.1. The Gulf Observatory

An observatory with a 2-m optical telescope proposed by astronomers from the Arabian Gulf region (Bahrain, Kuwait and the Sultanate of Oman), to be built on top of the Jabal Shams (2980 m above sea level), in the Sultanate of Oman. There is a plan in the near future for a site-testing programme for this mountain.

7.8.8. Palestine

Imad Barghouthi performed research on upper atmosphere/ionosphere reactions, resulting in the following papers:

I. A. Barghouthi (2008), "A Monte Carlo study for ion outflows at high altitude and high latitude: Barghouthi model," J. Geophys. Res., 113, A08209. 1–11.
I. A. Barghouthi, N. M. Doudin, A. A. Saleh, V. Pierrard (2008), "The effect of altitude and velocity dependent wave-particle interactions on the H+ and O+ outflows in the auroral region," Journal of Atmospheric and Solar–Terrestrial physics, 70, 1159–1169.
I. A. Barghouthi, N. M. Doudin, A. A. Saleh, V. Pierrard (2007), "High-altitude and high-latitude O+ and H+ outflows: The effect of finite electromagnetic turbulence wavelength," Annales Geophysicae, 25, 2195–2202.

7.8.9. Qatar

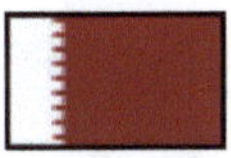

Sheikh Salman Bin Jabor Althani has been working locally with the support of the Qatar Metrology Authority on studying the effect of coronal mass ejections on the dust movements in atmospheric regions with the start of the new solar cycle. This study will be completed by mid-2009. Twenty schools will be participating in tracing and logging any sunspots that appear next year on the sun as a part of a plan to start studying the sun with the maximum number of local participants. Althani is also planning to install a solar observatory consisting of a Coronado Solarmax 40 telescope, a Coronado Solarmax 90 telescope, a Coronado Solarmax 90 CaK telescope, and an optical observatory (20″ RCX400™ on MAX Robotic German Equatorial Mount) in Jordan near Al-Betraa. He is also planning to build a 1-m telescope equipped with a special purpose monitoring instruments:

- Spectro-photometric Monitoring of Type Ia Supernovae
- Detection of Extra-Solar System Planets Using Gravitational Micro-lensing. Plus 1.2 m Near Earth Objects (NEOs).

7.8.10. Saudi Arabia

Submitted by Dr. Hassan Basurah, Astronomy Department, King Abdulaziz University, Jeddah

Dr. Hassan Basurah and his group at King Abdulaziz University (KAU) conducted daily sunspot observations in coordination with SIDC in Belgium. They recently installed a cosmic ray muon observatory and observed stellar occultation.

In addition to many science activities, the team has undertaken many outreach activities. The IHY-Saudi Arabia events are transitioning into the International Year of Astronomy, *http://astronomy2009sa.org*. The efforts focus on several

Fig. 88. *Some of Academic staff at King Abdulaziz University in Saudi Arabia teaching the use of a telescope to recognize objects in the sky.*

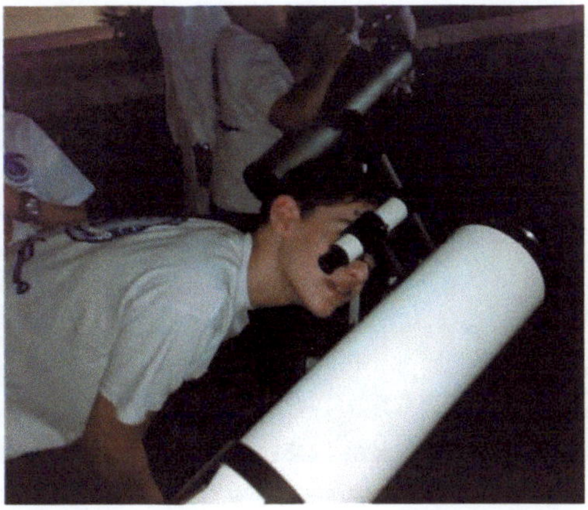

Fig. 89. *Observing the Lunar conjunction with Venus and Jupiter.*

Fig. 90. *IHY-Saudi Arabia team members set up telescopes to prepare for eclipse observations.*

aspects, centering on increasing the awareness of heliophysics and astronomy for students and also the general public. The 1 August 2008 solar eclipse and the Venus-Jupiter lunar conjunction were excellent opportunities to share the event with the public (see Figures 88–90).

The team also emphasized activities that improved the astronomy education in Saudi Arabia. Teacher training lectures and activities (Figure 91) helped develop curricula that broaden science education, and special events for secondary school students (see Figures 92 and 93) allow students to experience the university environment and learn from astronomy experts.

Fig. 91. *The KAU Astronomy Department organized a number of lectures to science teachers in Jeddah city.*

Fig. 92. *School students visiting the Astronomy Department at King Abdulaziz University.*

Fig. 93. *A general lecture at KAU for the visiting secondary school students.*

7.8.11. Syria 

7.8.11.1. Syrian Arab Republic

Plans exist for building a 2-m optical telescope. The country has active societies, the Syrian Cosmic Society and the astronomical amateur astronomy. It has an agreement with Jordan and Lebanon for building the Middle East Observatory.

7.8.12. United Arab Emirates

http://www.ihy.uaeu.ac.ae/

The UAE has been very active in the coordination of activities in IHY-Western Asia, starting with the hosting of the first IHY/United Nations workshop on Basic Space Science in 2005, and the third IHY Western Asia Regional Meeting in March 2008. Please see the summary at the beginning of this section.

7.8.13. Yemen 

Submitted by Dr. Abdul HAQ SULTAN, IHY-Yemen National Coordinator, Faculty of Science, Sana'a University, Sana'a

http://ihy2007.org/organization/yemen_ihy.shtml

The IHY-Yemen Kickoff Meeting took place on 14 May 2006 at the faculty of science Sana'a University, Sana'a Yemen. The National Organizing Committee was chaired by Dr. A. H. Sultan, Head of the Physics Department. The meeting brought together experts and stakeholders in the fields of relevance to IHY such as space science, atmospheric sciences and environmental sciences. In the meeting a national planning team was elected, which decided to launch IHY activities next month by installing a low-cost radio telescope (RADIO JOVE).

As a part of the IHY 2007 activities, the IHY Yemen National Planning Committee has organized workshops in 2006–2007. The title of each Workshop starts with "Celebrating IHY by:" and each workshop focused on one key element of IHY activities in Yemen (Figures 94 and 95).

Fig. 94. *An IHY-Yemen team member explains the use of a telescope to the president of Sana'a.*

Fig. 95. *The "What is Magnetism" experiment at the Sana'a University "Miniature Physics Mosaic" exhibition.*

7.8.13.1. January 2007: Celebrating IHY by fostering International Scientific Cooperation

Due to the visit of Prof. Oliver Wright (Department of Applied Physics/Graduate School of Engineering/Hokkaido University/Japan) to Sana'a University, the January 2007 workshop topic was "Celebrating IHY by fostering international scientific cooperation." The following public lectures were scheduled on Saturday 13 January 2007:

(1) New findings about the lunar first visibility research: Beyond Danjon's limit by A. H. Sultan
(2) Watching Ripples on Crystals by Oliver Wright
(3) IHY resources on the web by Amar Khalid
(4) Radio Jove by Sallami Mohammed

There was also an IHY poster exhibit during the afternoon, and a computer monitor displaying an animated presentation of "How Radio Jove works" and "The different steps of Sana'a University Radio Jove installation".

7.8.13.2. February 2007: celebrating IHY by using media and web

Using the Media: Through Yemen TV and Yemen newspapers, we are exerting immense efforts to enhance public consciousness to help improve and develop education and research activities in space science, to convince people that space science should be included in the science curriculum of the schools and universities, as a track of increasing public knowledge, understanding, and appreciation of space science. This activity is happening concurrent with the IHY Kickoff in Vienna.

7.8.13.3. March 2007: celebrating IHY by promoting space science among the population

At the very beginning of the second semester, we profited from the Mars Lunar Total Eclipse to continue our educational purposes through promotion of astronomy to the entire public, i.e. not only to those who have interest in astronomy, especially the students and the educators, but also to the ordinary people and more importantly to school pupils. The event featured lectures and exhibitions during the day and an observational session directed to the public mainly for observing the Lunar Total Eclipse.

7.8.13.4. April 2007: celebrating IHY by highlighting the first human in space

IHY-Yemen participated in the world-wide "Yuri's Night World Space Party". As Kuwait was the only other country celebrating the event in the region, young amateurs of astronomy in the two countries are profiting of this occasion to promote cooperation and to plan for future activities. This was accomplished with the partnership of the Kuwait Science Club.

An interview with Professor Sultan appeared in the widespread Yemeni newspaper "Athawra" on April 13, 2007, on the occasion of exactly 46 years of the first human in space and after about 22 years of the first Middle Eastern man in space.

7.8.13.5. May 2007: celebrating IHY by demonstrating that Physics is present in each moment of our lives

For the May 2007 IHY theme, the French Cultural Center in Sana'a (*http://www.ccclsanaa.com*), the Faculty of science at Sana'a University and the IHY-Yemen National Committee organized a special public event to demonstrate the role of physics in everyday life. The event was called "Miniature Physics Mosaic," and was held 29 April–9 May, hosted by the Faculty of Science of Sana'a University. The exhibit was prepared by CCSTI with the support of the French Foreign Affaires and the participation of the French Society of Physics. Visitors were invited to participate in 18 interactive experiments which were supported by 15 posters designed to highlight scientific enigmas. Many experiments and posters were supervised and demonstrated by fourth-year undergraduate students participating in IHY-Yemen activities.

Appendix I IHY Committees and Team Leaders

IHY Secretariat	
Executive Director	Joseph M. Davila, NASA Goddard Space Flight Center
International Coordinator	Nat Gopalswamy, NASA Goddard Space Flight Center
Director of Operations	Barbara Thompson, NASA Goddard Space Flight Center
UN Basic Space Science Observatory Development Programme	Hans Haubold, United Nations Office of Outer Space Affairs
Education and Public Outreach Coordinator	Cristina Maria Rabello-Soares, Stanford University
Coordinated Investigation Programs	Richard Stamper, Rutherford Appleton Laboratory
IHY Schools	David Webb, Boston College and Hanscom AFRL
Newsletter Editor	David Webb, Boston College and Hanscom AFRL
Web Designer	Emilie Drobnes, NASA Goddard Space Flight Center
Web Editor	Joseph M. Davila, NASA Goddard Space Flight Center

Regional Coordinators			
Region, subregion or area	Participating countries and areas	Regional Coordinator	Institute and location
Africa	Algeria, Cameroon, Cape Verde, Côte d'Ivoire, Egypt, Ethiopia, Kenya, Libyan Arab Jamahiriya Morocco, Nigeria, South Africa, Zambia	Marius Potgieter	North-West University, South Africa
East Asia and the Pacific	Australia, China, India, Indonesia, Japan, Nepal, Malaysia, Mongolia, Philippines, Republic of Korea, Singapore, Taiwan Province of China, Thailand, Viet Nam	Chi Wang	Laboratory for Space Weather, Chinese Academy of Sciences, China
Western Europe	Austria, Belgium, Czech Republic, Denmark, Finland, France, Germany, Hungary, Ireland, Israel, Italy, Norway, Poland, Portugal,Slovakia, Spain, Sweden, Switzerland, United Kingdom	Jean-Louis Bougeret, Co-chair; Carine Briand	Observatoire de Paris at Meudon, France
Latin America and the Caribbean	Argentina, Brazil, Mexico, Peru, Puerto Rico	Cristina Mandrini and Jean-Pierre Raulin	IAFE, Argentina; MacKenzie University, Brazil
North America	Canada, United States	Donald M. Hassler	Southwest Research Institute, United States
Eastern Europe and Central Asia	Kazakhstan, Russian Federation, Ukraine, Uzbekistan	Alexander Stepanov	Pulkovo Observatory, Russian Federation
Balkan/Black Sea area	Armenia, Azerbaijan, Bosnia and Herzegovina, Bulgaria, Croatia, Georgia, Greece, Montenegro, Romania, Russian Federation, Serbia, Turkey	Katya Georgieva	Solar-Terrestrial Influences Laboratory, Bulgarian Academy of Sciences, Bulgaria
West Asia	Bahrain, Egypt, Iraq, Jordan, Kuwait, Lebanon, Oman, Palestine, Qatar, Saudi Arabia, United Arab Emirates, Yemen	Hamid M. K. Al-Naimiy	University of Sharjah, United Arab Emirates

IHY National Coordinators		
Country	National Coordinator	Location
Algeria	Mammar Fodil, Samir Nait Amor	Centre de Recherche en Astronomie Astrophysique et Geophysique (CRAAG) and the Observatoire d'Alger
Argentina	Cristina Mandrini	IAFE, Buenos Aires
Armenia	Ashot Chilingarian	Cosmic Ray Division, Alikhanyan Physics Institute, Yerevan
Australia	Brian Fraser	University of Newcastle
Austria	Helmut Rucker	Space Research Institute, Austrian Academy of Sciences, Graz
Azerbaijan	Elchin S. Babayev	Shamakhy Astrophysical Observatory, Baku
Bahrain	Mohammed Al Othman	Physics Department, Bahrain University
Belgium	Stefaan Poedts	Katholieke Universiteit, Leuven
Bosnia and Herzegovina	Alexandra Andic	University of Banja Luka, Republika Srpska
Brazil	Walter Gonzalez, Alisson Dal Lago	Instituto Nacional de Pesquisas Espaciais (INPE), Sao Jose dos Campos
Bulgaria	Katya Georgieva	Solar-Terrestrial Influences Laboratory, Bulgarian Academy of Sciences
Cameroon	Emmanuel Guemene Dountio	Ministry of Scientific Research & Innovation, Energy Research Laboratory, Yaounde
Canada	Ian Mann	Department of Physics, University of Canada, Alberta
Cape Verde	Jose Pimenta Lima	Instituto Nacional de Metereologia e Geofisica (INMG)

(*continued*)

IHY National Coordinators		
Country	National Coordinator	Location
 China	Guangli Huang	Purple Mountain Observatory, Nanjing
 Côte d'Ivoire	Olivier K. Obrou	Laboratoire de Physique de l' Atmosphère, Université de Cocody
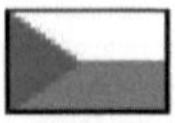 Czech Republic	Frantisek Farnik	Astronomical Institute, Ondrejo
 Croatia	V. Rusdijak, Bojan Vrsnak	Hvar Observatory, Faculty of Geodesy
 Denmark	Klaus Galsgaard	The Niels Bohr Institute, University of Copenhagen
 Egypt	A. A. Galal	National Research Institute of Astronomy and Geophysics
 Ethiopia	Baylie Damtie	Department of Physics, Bahir Dar University
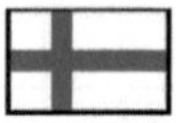 Finland	Rami Vainio	Department of Physical Sciences, University of Helsinki
 France	Brigitte Schmeider	Observatoire de Paris at Meudon
 Georgia	Marina S. Gigolashvili	Abastumani Observatory
 Germany	Bernd Heber	Christian-Albrechts-Universitat, Kiel
 Greece	Xenophon Moussas, A. Nindos	Laboratory of Astrophysics, University of Athens
 Hungary	Karoly Kecskemety	Central Research Institute for Physics (KFKI), Budapest
 India	P. K. Manoharan	Radio Astronomy Centre, Ooty
 Indonesia	Thomas Djamaluddin, Dhani Herdiwijaya	National Institute of Aeronautics and Space Department of Astronomy Institut Teknologi
 Iraq	Rashid Al-Naimi	Department of Atmospheric Sciences, College of Science, Al Mustansiriyah, University of Baghdad

(*continued*)

IHY National Coordinators		
Country	National Coordinator	Location
Ireland	Peter Gallagher	School of Physics, Trinity College, Dublin
Israel	Michael Gedalin	Department of Physics, Ben-Gurion University
Italy	Ester Antonucci	Osservatorio Astronomico di Torino
Japan	Masayoshi Kojima (STE Lab., Nagoya Univ.) Prof. Takashi Sakurai (Nat. Astron. Obs. of Japan) Prof. Kazunari Shibata (Kwasan and Hida Obs., Kyoto Univ) Prof. Shinichi Watari (Nat. Inst. of Info. and Com. Tech) Prof. Kiyohumi Yumoto (Space Env. Res. Ctr., Kyushu Univ.)	
Jordan	Hanna Sabat	Institute of Astronomy and Space Science, Al Al Bayte University, Mafraq
Kazakhstan	Nikolay Makarenko	Institute of Mathematics, Almaty
Kenya	Maurice Odondi K'orowe	Jomo Kenyatta University of Agriculture and Technology Physics Department
Kuwait	Ismail Sabbah	Department of Physics, Faculty of Science, Kuwait University
Lebanon	Rojer Haijar	Department of Physics and Astronomy, Notre-Dame University, Zouk Mikael
Libyan Arab Jamahiriya	Abdul Qader Abseim	Libyan Remote Sensing and Space Center
Malaysia	Mazlan Othman	National Space Agency of Malaysia
Mexico	Juan Américo González Esparza, Rogelio Cabellero	Instituto de Geofisica, Ciudad Universita, Coyoacan
Mongolia	Tsolmon Renchin	National University of Mongolia
Morocco	Nour-Eddine Najid	Université Hassan II Ain Chock, Faculté des Sciences Ain Chock

(*continued*)

IHY National Coordinators		
Country	National Coordinator	Location
Nepal	Jayanta Achararya	Mahendra Sanskrit University, Bakeemi Campus
Nigeria	A. Babatunde Rabiu	Department of Physics, Federal University of Technology, Akure
Norway	Bo Andersen, Paal Brekke, N. Oestgaard	Norwegian Space Center, Oslo; University of Bergen
Oman	Saleh Al-Shedhani	Physics Department, College of Science, Sultan Qaboos University
Peru	Walter Guevara Day	Universidad Nacional Mayor de San Marcos, Facultad de Ciencias Fisicas, Lima
Philippines	Daniel McNamara	Manila Observatory Ateneo de Manila Campus, Loyola Heights
Poland	Wieslaw Macek	Space Research Centre, Polish Academy of Sciences
Portugal	Dalmiro Maia	Porto University
Qatar	Sheikh Salman Bin Jabor Althani	Chairman of Astronomy Department, Qatar Science Club
Republic of Korea	Young Deuk Park	Korea Astronomy and Space Science Institute (KASI)
Romania	Cristiana Dumitrache, Georgetta Maris	Astronomical Institute of the Romanian Academy, Romanian Academy Institute of Geodynamics
Russian Federation	Geliy A. Zherebtsov: Russian National Coordinator Galina A. Kotova: Scientific Secretary Nataly F. Blagoveshenskaya: Ionospheres Anatoly A. Petrukovich: Magnetospheres Alexander V. Stepanov: Solar Phenomena Igor Veselovsky: Interplanetary Medium/Cosmic Rays Yuly I. Zetzer: Lithosphere-Atmosphere-Ionosphere Interactions	
Saudi Arabia	Hasan Basurah	Head of Astronomy, King Abdulaziz University

(*continued*)

IHY National Coordinators		
Country	National Coordinator	Location
Serbia	Istvan Vince	Belgrade Astronomical Observatory
Singapore	Phil Chan Auk Hui	National University of Singapore
Slovakia	Karel Kudela, V. Rusin	Slovak Academy of Sciences, Institute of Experimental Physics
South Africa	Marius Potgieter	School of Physics, North-West University
Spain	Javier Rodriguez Pacheco	Instituto de Astrofísica de Canarias, Tenerife
Sweden	Henrik Lundstedt	Swedish Institute of Space Physics
Switzerland	Arnold Benz	Institute for Astronomy, Federal Institute of Technology, Zurich
Thailand	Boonrucksar Soonthornthum, David Ruffolo	National Astronomical Research Institute of Thailand (NARIT)
Tunisia	Hassen Ghalila	Départment de Physique, Faculté des Sciences de Tunis, Université de Tunis
Turkey	Atila Ozguc	Kandilli Observatory and E.R.I., Bogazici University
Ukraine	Nataly Stepanyan	Crimean Astrophysical Observatory
United Arab Emirates	Hamid M. K. Al Naimiy	College of Arts and Sciences, Sharjah University
United Kingdom	Andrew Breen, Richard Harrison	University of Aberystwyth, Rutherford Appleton Laboratory
United States	Roger Smith	Geophysical Institute University of Alaska, Fairbanks
Uzbekistan	Shukhrat Egamberdiev	Ulugbek Astronomical Institute

(*continued*)

IHY National Coordinators		
Country	National Coordinator	Location
 Viet Nam	Doan Minh Chung	Deputy Director, Space Technology Institute (STI), Vietnamese Academy of Science and Technology (VAST)
 Yemen	Abdul Haq Sultan	Head of Physics Department, Faculty of Science, Sana'a University
 Zambia	Geoffrey Munyeme, Adrian Habanyama	Department of Physics, University of Zambia, Lusaka
 Palestine	Imad Barghouthi	Department of Physics, Faculty of Science, Al-Quds University, Jerusalem
 Puerto Rico	Sixto Gonzalez	NAIC/Arecibo University

National Outreach Coordinators		
Country	National Coordinator	Location
Belgium	Norma Bock Crosby	Belgian Institute for Space Aeronomy, Brussels, Belgium
Brasil	Adriana Valio Roque da Silva	MacKenzie University, Sao Paulo, Brasil
Bulgaria	Penka Stoeva	Solar-Terrestrial Influences Laboratory, Bulgarian Academy of Sciences, Bulgaria
Czech Republic	Rostislav Halas	Realne Gymnasium Prostejov, Prostejov, Czech Republic
Finland	Heikki Nevanlinna	Finnish Meteorological Institute, Space Research, Helsinki, Finland
France	Brigitte Schmieder	Observatoire de Paris Meudon, France
Germany	Alexander Warmuth	Astrophysikalisches Institut Potsdam, Germany
India	Arvind Paranjpye	Inter-University Centre for Astronomy and Astrophysics, India
Ireland	Peter Gallagher	School of Physics, Trinity College, Dublin
Italy	Francesca Zuccarello	Department of Physics and Astronomy, Catania University, Catania, Italy

(*continued*)

National Outreach Coordinators		
Country	National Coordinator	Location
Japan	Shinichi Watari	National Institute of Info. & Com. Tech., Japan
Jordan	Hanna Sabat	Institute of Astronomy & Space Sciences–Al al-Bayt University, Mafraq, Jordan
Malaysia	Azreena Ahmad	Space Science and Education Division of the National Space Agency, Putrajaya, Malaysia
Mexico	Guadalupe Cordero	Instituto de Geofisica, Ciudad Universita, Coyoacan, Mexico
Norway	Arve Aksnes	Department of Physics and Technology, University of Bergen, Bergen, Norway
Peru	Maria-Luisa Aguilar Hurtado	Facultad de Ciencias Fisica, Universidad Nacional Mayor de San Marcos, Lima, Peru
Romania	Nedelia Antonia Popescu	Astronomical Institute of Romanian Academy (AIRA), Bucharest, Romania
Slovakia	Karel Kudela	Slovak Academy of Sciences, Institute of Experimental Physics, Kosice, Slovakia
South Africa	Lee-Anne McKinnell	Department of Physics and Electronics, Rhodes University, Grahamstown, South Africa
Spain	Javier Rodriguez-Pacheco	Departamento de Fisica, Universidad de Alcala, Madrid, Spain
Sweden	Henrik Lundstedt	Swedish Institute of Space Physics, Lund, Sweden
Thailand	Busaba Kramer	National Astronomical Research Institute of Thailand (NARIT), Chiang Mai Thailand
Turkey	Yurdanur Tulunay	METU/ODIU, Department of Aerospace Engineering, Ankara, Turkey
United Kingdom	Lucie Green	Mullard Space Science Laboratory, University College London, United Kingdom
United States	M. Cristina Rabello-Soares	Stanford University, Stanford, CA, USA
Yemen	Nada Alhaddad	University of Sana'a, Yemen

IHY International Advisory Committee	
Name	Institute
Roger M. Bonnet (Chair)	International Space Science Institute, Bern, Switzerland
Richard A. Behnke	National Science Foundation, Arlington, VA, USA
Odile de la Beaujardiere	Air Force Research Laboratory, Hanscom, MA, USA
Oddbjorn Engvold	University of Oslo, Oslo, Norway
Richard R. Fisher	NASA Headquarters, Washington, DC, USA
Wing-Huen IP	National Central University, Chung Li, Taiwan, Province of China
John Leibacher	NSO, Tucson, AZ, USA
William Liu	Canadian Space Agency, Ottawa, ON, Canada
Marcos Machado	Comision Nacional de Actividades Espaciales (CONAE), Argentina
Hermann Opgenoorth	European Space Agency, Noordwijk, Netherlands
Frank McDonald	Institute for Physical Science and Technology, University of Maryland, MD, USA
G. Madhavan Nair	Indian Space Research and Organization, Bangalore, India
Katepalli R. Sreenivasan	Abdus Salam International Centre for Theoretical Physics, Trieste, Italy
S.-T. Wu	University of Alabama, Huntsville, AL, USA

IHY International Steering Committee	
Name	Institute
Joseph Davila (Chair)	NASA Goddard Space Flight Center, Greenbelt, MD, USA
Jean-Louis Bougeret	Observatoire de Paris-Meudon, Paris, France
Kumar Chitre	ANDHER, Mumbai, India
Arnab Rai Choudhuri	Department of Physics, Indian Institute of Science, Bangalore, India
Nancy Crooker	Center for Space Physics, Boston University, Boston, MA, USA
Brian Fraser	University of Newcastle, Australia
Nat Gopalswamy	NASA Goddard Space Flight Center, Greenbelt, MD, USA
Richard Harrison	Space Physics Division, Rutherford Appleton Laboratory, United Kingdom

(*continued*)

IHY International Steering Committee	
Name	Institute
Siraj Hasan	Indian Institute of Astrophysics, Bangalore, India
Hans Haubold	United Nations Office of Outer Space Affairs, Vienna International Centre, Vienna, Austria
Masayoshi Kojima	Solar-Terrestrial Environment Laboratory, Nagoya University, Nagoya, Japan
Christina Mandrini	Instituto de Astronomia y Fisica del Espacio, Buenos Aires, Argentina
Marius Potgieter	School of Physics, North-West University, Potchefstroom, South Africa
Alexander Stepanov	Central Astronomical Observatory at Pulkovo, Saint Petersburg, Russian Federation
Barbara Thompson	NASA Goddard Space Flight Center, Greenbelt, MD, USA
Ji Wu	Chinese National Space Administration, Beijing, China

IHY Discipline Coordinators	
Discipline	Coordinator
CIP Coordinator	Richard Stamper, Rutherford Appleton Laboratory, UK
Solar	Dipankar Banerjee, Indian Institute of Astrophysics, India
	Sarah Gibson, High Altitude Observatory, Boulder, USA
Heliosphere/ Cosmic Rays	José Valdes-Galicia, Instituto de la UNAM, Coyoacan, Mexico
	Bernd Heber, Christian-Albrechts-Universitat, Kiel, Germany
Magnetospheres	Kiyohumi Yumoto, Kyushu University, Japan
	Walter Gonzalez, INPE, Sao Jose dos Campos, Brazil
Ionized Atmospheres	Ian McCrea, Rutherford Appleton Laboratory, UK
	Kirsti Kauristie, Finnish Meteorological Institute, Helsinki, Finland
Neutral Atmospheres	Tim Fuller-Rowell, NOAA/University of Colorado Boulder, USA
Meteor/Meteoroids	Svetlana Kolomiyets, Kharkiv National University of Radioelectronics, Ukraine

Appendix II United Nations Basic Space Science Workshops and Projects

The United Nations, in cooperation with the European Space Agency, initiated in 1990 the organization of annual Workshops on Basic Space Science through the United Nations Office for Outer Space Affairs. These Workshops, focusing on astrophysics and space science, have been held in a variety of locations, balancing the geographical representation of the program activities. The participation in these workshops is balanced between scientists from developed nations, scientists from developing nations, and scientists from geographical areas near the site of the workshop. This appendix lists the workshops on Basic Space Science, both prior to and including the IHY workshops, as well as a brief summary of the projects initiated or pursued by the Basic Space Science initiative.

United Nations/ESA workshops on Basic Space Science (1991–2004) and United Nations/ESA/NASA/JAXA workshops on the International Heliophysical Year 2007 and Basic Space Science (2005–2009)							
Year	City	Target region	Host institution	No. of participants	No. of participating countries	Workshop title	Report
1991	Bangalore (India)	Asia and the Pacific	Indian Space Research Organization	87	19	Basic Space Science	A/AC.105/489
1992	San José and Bogotá	Latin America and the Caribbean	University of Costa Rica and University of the Andes	122	19	Basic Space Science	A/AC.105/530
1993	Lagos	Africa	University of Nigeria and Obafemi Awolowo University	54	15	Basic Space Science	A/AC.105/560/Add.1
1994	Cairo	Western Asia	National Research Institute of Astronomy and Geophysics	95	22	Basic Space Science	A/AC.105/580
1995	Colombo	Asia and the Pacific	Arthur C. Clarke Institute for Modern Technologies	74	25	Basic Space Science: from small telescopes to space missions	A/AC.105/640

(*continued*)

United Nations/ESA workshops on Basic Space Science (1991–2004) and United Nations/ESA/NASA/JAXA workshops on the International Heliophysical Year 2007 and Basic Space Science (2005–2009)							
Year	City	Target region	Host institution	No. of participants	No. of participating countries	Workshop title	Report
1996	Bonn	Europe	Max Planck Institute for Radioastronomy	120	34	Basic Space Science: ground-based and space-borne astronomy	A/AC.105/657
1997	Teguci-galpa	Latin America and the Caribbean	Universidad Nacional Autónoma de Honduras	75	28	Basic Space Science: small astronomical telescopes and satellites in education and research	A/AC.105/682
1999	Mafraq (Jordan)	Western Asia	Al al-Bayt University	95	35	Basic Space Science: scientific exploration from space	A/AC.105/723
1999	Vienna	All regions	United Nations Office at Vienna			(a) Third United Nations Conference on the Exploration and Peaceful Uses of Outer Space (UNISPACE III)	1. U.N. publication, Sales No. E.00.I.3
						(b) International Astronomical Union (IAU)/ Committee on Space Research (COSPAR)/ United Nations Special Environmental Symposium "Preserving the Astronomical Sky";	Ibid., annex III, Sect. II
						(c) IAU/ COSPAR/ United Nations Special Workshop on Education in Astronomy and Basic Space Science	Ibid., annex III, Sect. VIII
2000	Toulouse (France)	Europe	Centre National d'études Spatiales	80	34	Basic Space Science: satellites and networks of telescopes; tools for global participation in the study of the universe	A/AC.105/742

(*continued*)

United Nations/ESA workshops on Basic Space Science (1991–2004) and United Nations/ESA/NASA/JAXA workshops on the International Heliophysical Year 2007 and Basic Space Science (2005–2009)							
Year	City	Target region	Host institution	No. of participants	No. of participating countries	Workshop title	Report
2001	Reduit (Mauritius)	Africa	University of Mauritius	65	28	Basic Space Science: exploring the universe; sky surveys, space exploration and space technologies	A/AC.105/766
2002	Cordoba (Argentina)	Latin America and the Caribbean,	Centro Espacial Teófilo Tabanera of the Comisión Nacional de Actividades Espaciales (CONAE)	75	24	Basic Space Science: the World Space Observatory and the Virtual Observatories in the Era of 10-metre Telescopes	A/AC.105/784
2004	Beijing (China)	Asia and the Pacific	China National Space Administration (CNSA)	75	28	Basic Space Science and the International Heliophysical Year 2007	A/AC.105/829
2005	Abu Dhabi and Al-Ain (United Arab Emirates)	All regions	United Arab Emirates University	150	39	International Heliophysical Year 2007 and Basic Space Science	A/AC.105/856
2006	Bangalore (India)	All regions	Indian Institute of Astrophysics (IIA)	150	30	International Heliophysical Year 2007 and Basic Space Science	A/AC.105/882
2007	Tokyo (Japan)	All regions	National Astronomical Observatory of Japan (NAOJ)	75	28	International Heliophysical Year 2007 and Basic Space Science	A/AC.105/902
2008	Sozopol (Bulgaria)	All regions	Solar-Terrestrial Influences Laboratory of the Bulgarian Academy of Sciences	200	48	International Heliophysical Year 2007 and Basic Space Science "First Results of IHY 2007"	
2009 (planned)	Jeju (Republic of Korea)	All regions	Korea Astronomy and Space Science Institute (KASI)	–	–	Basic Space Science and the International Heliophysical Year 2007	–

Projects initiated or pursued through the UNBSSI (1991–2008): More detailed descriptions of the projects can also be found in Wamsteker et al. (2004)				
Year	*Workshop*	*Projects initiated*	*Achievements*	*Report or World Wide Web site*[1]
Since 1991	Starting with the UN/ESA Workshop in Bangalore, India	Telescope and planetarium equipment donation programme of the Government of Japan facilitated through UNBSSI	Telescopes donated to seven countries Planetarium equipment donated to twenty countries	See Annex III and IV of A/AC.105/902 for further details
1992	UN/ESA Workshop in Costa Rica and Colombia	Education and career development in basic space science Brochure "ISY92: Planetarium; A Challenge for Educators"	Establishment of an astronomical observatory for Central America in Honduras Donation of computer equipment by the European Space Agency (ESA): Cuba, Ghana, Honduras, Nigeria, Peru and Sri Lanka Establishment of a 5.5-m radio telescope in Colombia	Galactic emission map (Colombia): *http://aether.lbl.gov/www/projects/GEM* "ISY92: Planetarium; A Challenge for Educators": *http://www.unoosa.org/pdf/publications/planetariumE.pdf*
1993	UN/ESA Workshop in Nigeria	Southern African Large Telescope (SALT) (South Africa) Establishment of an inter-African astronomical observatory and science park on the Gamsberg in Namibia	The construction of SALT has been completed and is now in its commissioning phase	Southern African Large Telescope (South Africa): *http://www.salt.ac.za* Inter-African astronomical observatory and science park (Namibia): *http://www.mpia-hd.mpg.de/Public/PUBREL/booklet01.html*

(continued)

Projects initiated or pursued through the UNBSSI (1991–2008): More detailed descriptions of the projects can also be found in Wamsteker et al. (2004)				
Year	*Workshop*	*Projects initiated*	*Achievements*	*Report or World Wide Web site*[1]
1994	UN/ESA Workshop in Egypt	Kottamia Telescope (Egypt) Egyptian drill project for the Russian Mars 2001 mission	In 1995 an agreement was signed with Zeiss Company (Germany) to refit the Kottamia telescope with a new optical system. On-site installation was completed in May 1997.	Kottamia Telescope (Egypt): *http://www.nriag.sci.eg/*
1995	UN/ESA Workshop in Sri Lanka	World space observatory (WSO/UV)	Evaluation of the feasibility of a world space observatory Inauguration of a telescope facility (Sri Lanka)	ACCIMT telescope facility (Sri Lanka): *http://www.accimt.ac.lk/space/sfunctions/sfacilities/sfacilities.htm*
1996	UN/ESA Workshop in Germany	Discussions on a Network of Oriental Robotic Telescopes (NORT) Developing astronomy and space science worldwide Education and research using small astronomical telescopes	Foundation of the Working Group on Space Science in Africa 100-metre Effelsberg radio telescope Two air shower detectors, one situated in the northern hemisphere (United States of America) and one in the southern hemisphere (Argentina)	Working Group on Space Science in Africa: *http://www.saao.ac.za/~wgssa/* Network of Oriental Robotic Telescopes (NORT): *http://www.saao.ac.za/~wgssa/as2/nort.html* Pierre Auger cosmic ray project: *http://www.taridar.cnea.gov.ar/~auger*
1997	UN/ESA Workshop in Honduras	Observation of near-Earth objects	First issue of the newsletter *African Skies/Cieux Africains* published Inauguration of the Central American Astronomical Observatory in Honduras	Observatorio Centroamericano de Suyapa (Honduras): *http://www.unah.hn* Space Guard Foundation (Italy): *http://spaceguard.rm.iasf.cnr.it/*

(*continued*)

Projects initiated or pursued through the UNBSSI (1991–2008): More detailed descriptions of the projects can also be found in Wamsteker et al. (2004)				
Year	*Workshop*	*Projects initiated*	*Achievements*	*Report or World Wide Web site*[1]
1998	UN/ESA Workshop in Jordan	Operation of the Maragha Astronomical Observatory in Jordan Baquaa radio telescope Hands-on astrophysics Astrophysics for university physics courses	Development of the hands-on astrophysics materials Development of an astrophysics education module	Hands-on astrophysics: *http://hoa.aavso.org/* Astrophysics education module for university physics courses: *http://www.seas.columbia. edu/~ah297/un-esa/astrophysics/index.html* Maragha Astronomical Observatory (Jordan): *http://www.aabu.edu.jo*
1999	UNISPCE III: workshop dedicated to basic space science in Austria	International Astronomical Union (IAU)/Committee on Space Research (COSPAR)/United Nations Special Environmental Symposium "Preserving the Astronomical Sky" IAU/COSPAR/United Nations Special Workshop on Education in Astronomy and Basic Space Science	Initiation of the development of education curriculum on space and atmospheric science	Education curriculum on space and atmospheric science, *http://www.unoosa.org/pdf/sap/centres/scicurrE.pdf* (ST/SPACE/17)
2000	UN/ESA Workshop in France	Satellites and networks of telescopes Tools for global participation in the study of the universe	WSO/UV assessment study completed	World space observatory/UV: *http://www.seas.columbia.edu/~ah297/un-esa/wso.html* *http://wso.inasan.ru/index.html*

(*continued*)

Projects initiated or pursued through the UNBSSI (1991–2008): More detailed descriptions of the projects can also be found in Wamsteker et al. (2004)				
Year	*Workshop*	*Projects initiated*	*Achievements*	*Report or World Wide Web site*[1]
2004–2009	Basic Space Science and IHY Workshops in China, United Arab Emirates, India, Japan, Bulgaria and South Korea	Low-cost, ground-based, worldwide instrument arrays for achieving the goals of the International Heliophysical Year	Establishment and operation of 17 networks of low-cost, ground-based, worldwide instrument arrays	See Annex I of A/AC.105/902 for a detailed list of the instrument arrays and their points of contact
		Involvement of developing countries in the analysis of data obtained from space missions: data analysis concepts	Identification of five data analysis concepts	See Annex I of A/AC.105/902 for a detailed list of the instrument arrays and their points of contact

[1]The reports are available in all official languages of the United Nations from the Official Document System of the United Nations at http://documents.un.org/

Appendix III IHY Publications

Scientists and educators have produced a great deal of information on IHY activities and results. At the end of 2008, there were already hundreds of papers written on IHY and over 1000 presentations given at scientific meetings. A list of IHY meetings is given in Appendix IV, and this section details publications that were dedicated to IHY. A listing of all IHY papers and presentations is too lengthy for this publication; individual papers and reports that did not appear as part of an IHY-themed book or special issue are not included in this appendix. The following gives examples of publications that were exclusively developed for IHY science and public outreach.

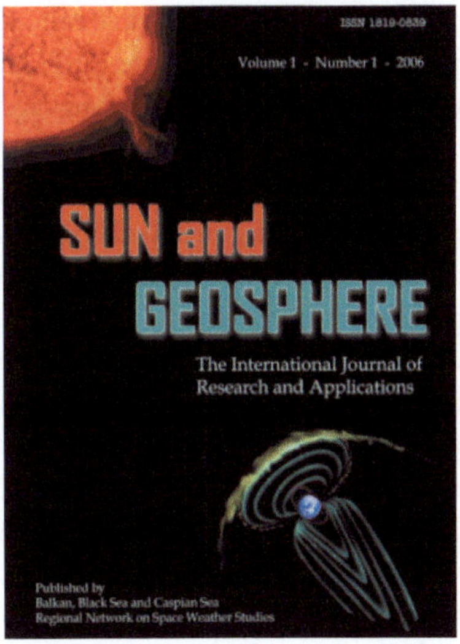

1. "Sun and Geosphere" Journal

"Sun and Geosphere" was founded in 2005 by the IHY Balkan, Black Sea and Caspian Sea Regional Network on Space Weather Studies (see Sect. 7), in response to the need for a scientific medium to communicate the results of the IHY.

It is a peer reviewed journal featuring all areas of heliophysical research. The Editor-in-Chief is Elchin Babayev of Shamakhy Observatory in Baku, National Coordinator of IHY-Azerbaijan. The journal is published online, with some versions in print. Its ISSN reference number is ISSN 1819-0839.

Past issues, submission, and prescription information is available from the *Sun and Geosphere* web page: *http://shao.az/SG/*.

2. U.N. Publications and Resources

The United Nations Office of Outer Space Affairs (OOSA) and the U.N. Basic Space Science program has produced a series of publications including the results

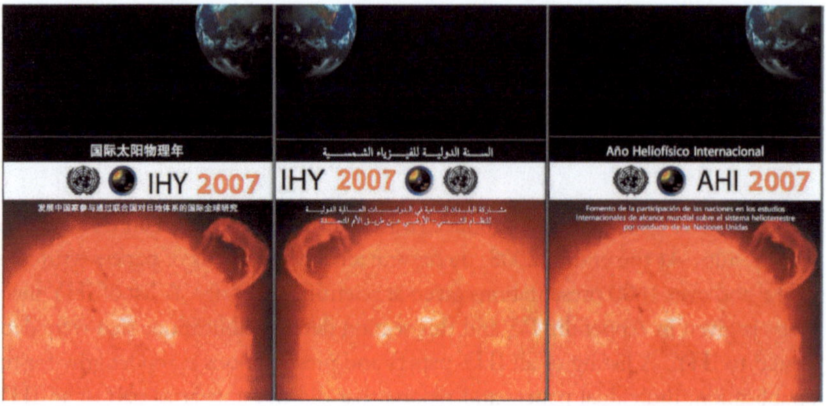

Fig. 1. *The IHY pamphlet, shown (left to right) in Chinese, Arabic and Spanish. It is also available in English, French and Russian.*

of previous workshops, curriculum books for education in astrophysics and space science, and a planetarium guidebook. These publications are a valuable resource because their topics focus on the development of space science research and education worldwide. They are translated into the six U.N. official languages: Arabic, Chinese, English, French, Russian and Spanish. Because many international scientific meetings are conducted in English, a language barrier may exist that affects the viability of space scientists in certain regions of the world. Because most of the UNBSS publications and curricula are available in multiple languages, their impact can easily extend beyond English-speaking circles.

Several additional IHY products are available in multiple languages. The most recent version of the IHY pamphlet, shown in Figure 1, was published in all six official languages of the United Nations. A number of outreach products (see Sect. 5) and web pages have also been developed in multiple languages, in keeping with the U.N. goal of improving the availability of these resources to the world.

Each IHY/UNBSS workshop results in a special report published by the United Nations. They are available in each of the six official U.N. languages, and can be accessed on the website of the United Nations Office of Outer Space Affairs. The reports available at the time of publication are as follows:

- "Report on the United Nations/European Space Agency/NASA Workshop on the International Heliophysical Year 2007" (Abu Dhabi and Al-Ain, United Arab Emirates, 20–23 November 2005) *A/AC.105/856*.
- "Report on the Second United Nations/National Aeronautics and Space Administration Workshop on the International Heliophysical Year 2007 and

Basic Space Science" (Bangalore, India, 27 November–1 December 2006) A/AC.105/882.

- "Report on the Third United Nations/European Space Agency/NASA Workshop on the International Heliophysical Year 2007 and Basic Space Science" (Tokyo, Japan, 18–22 June 2007) *A/AC.105/902*.
- "Report on the Fourth UN/ESA/NASA/JAXA Workshop on IHY 2007 and Basic Space Science" (Sozopol, Bulgaria, 2–6 June 2008) *A/AC.105/919*.

A complete list of the Basic Space Science Workshop publications can be found in Appendix II.

3. Special Issues

The October 2006 issue of *African Skies/Cieux Africains* was dedicated to IHY activities and opportunities in Africa (see Figure 2). From the introduction: "The IHY is a golden opportunity for space science institutes throughout Africa to initiate new collaborative programmes". This special issue featured campaign plans, national programs and observatory activities, with contributions from all over Africa.

Dr. Cherilynn Morrow coordinated a special issue of *Advances in Space Research* (Volume 42, Issue 11, December 2008) featuring IHY education opportunities and international outreach programs. The special issue was based on the IHY educational sessions at the 36th COSPAR Scientific Assembly in Beijing, China held July 16–23, 2006. Special session 2-0008-06 was called "2007: The International Heliophysical Year – A Great Opportunity for Space and Earth Science Outreach". A splinter meeting on "Globalizing Space Science Education and Outreach" was also hosted by Dr. Morrow, and participants voiced strong support for a special publication on the challenges and success stories of international and multicultural outreach efforts.

Another special issue of *Advances in Space Research* (Volume 41, Issue 8, 2008), called "Heliophysical Processes: Towards the International Heliophysical Year 2007" features papers on IHY international science programs, reviews of IHY science topics, theoretical modeling, and comparative studies with other planets.

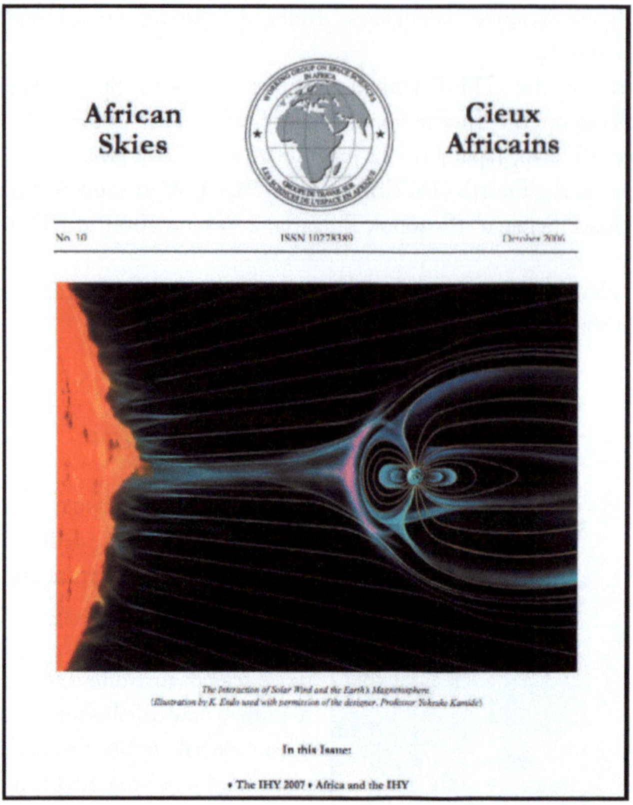

Fig. 2. *The October 2006 special IHY-themed issue of African Skies/Cieux Africains.*

IHY's partner organization ICESTAR also had a special issue in the Journal of Atmospheric and Solar-Terrestrial Physics (Volume 70, Issue 18, pages 2233–2396 (December 2008) titled "Transport processes in the coupled solar wind-geospace system seen from a high-latitude vantage point, Greenland Space Science Symposium 2007").

The proceedings form the 2nd European Assembly June 18–22, 2007 in Turin, Italy were published in special issues of the journal "Surveys In Geophysics Also, proceedings of the conference "The Sun, the Heliosphere, and the Earth" conference, held in Bad Honnef, Germany from 14 to 18 May 2007, will be published in online journal ASTRA.

The Belgian Federal Space Policy Office publishes a monthly magazine called "Science Connection", and a special issue on IHY is due to be released in early 2009. The Bulletin of the American International Union and the Bulletin of the Astronomical Society of India have regularly featured articles and reports of IHY's activities and results.

Fig. 3. *Left: The October 2007 issue of Astronomie magazine celebrated the fiftieth anniversary of space flight with a special issue featuring the Sun's power. Right: Sterne und Weltraum released a special issue in December 2007 for the IHY (see banner in lower right corner).*

Many online publications were produced, and there were even video news magazines that featured articles on IHY. The Czech Republic science and technology news magazine *PORT* had a broadcast on IHY on November 4, 2007 (see Sect. 7).

German space scientists have taken the opportunity to work with the popular astronomy magazine "Sterne und Weltraum" ("Stars and Space") to publish a

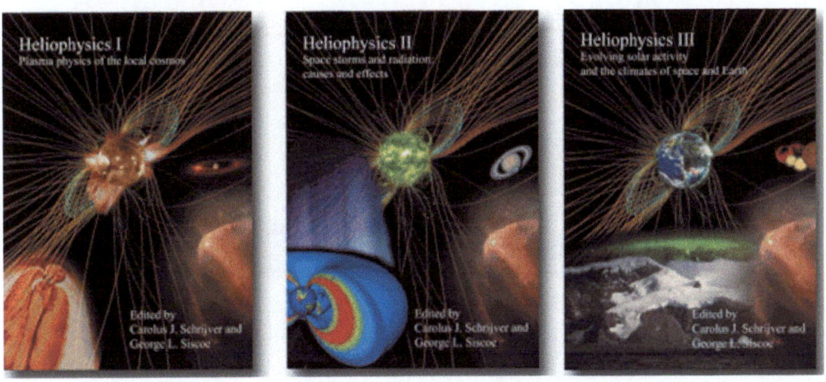

Fig. 4. *The first textbooks on the field of heliophysics are based on the lectures and resources of the LWS/IHY summer schools. Books I, II and III correspond to the schools in 2007, 2008 and 2009, respectively.*

special issue on the sun and heliosphere (Figure 3). This has been published in December 2007 under the title "Unsere Sonne–Motor des Weltraumwetters" ("Our Sun – the engine of space weather"). The magazine contains 23 articles on

Fig. 5. *IHY-Italy calendar for 2007–2008, featuring children's artwork from all over Europe.*

the sun, solar activity, planetary magnetospheres, the heliosphere, solar-terrestrial connection and space weather. These articles have been written by 28 "German scientists actually working in the field". Additionally, the German "Astronomie" magazine had a special issue titled "Kraftwerk Sonne", or "The Sun's Power" in October 2007, highlighting the fiftieth anniversary of the IGY and observations in space (Figure 3).

4. Heliophysics textbooks

As part of the Living With a Star Summer Schools on Heliophysics, a series of textbooks, the first on heliophysics, are being produced. Each book features the curriculum established for a school in a given year (2007, 2008 and 2009). The first book is titled "Heliophysics I: Plasma Physics of the Local Cosmos". The second book, Heliophysics II, is subtitled "Space Storms and Radiation: Causes and Effects", and Heliophysics III will be subtitled "Evolving Solar Activity and the Climates of Space and Earth". The books will be published by Cambridge University Press (Figure 4).

5. IHY/COST296 Calendar

IHY-Italy developed a calendar themed "Vivere Con Una Stella" or "Living With Our Star" that featured children's heliophysics-themed artwork from the IHY-Europe program. It is a two-year calendar (2007–2008) spanning all of IHY. The calendar was developed by the National Institute of Geophysics and Volcanology in partnership with COST 296 (Figure 5).

Appendix IV IHY Events, Conferences and Seminars

 = IHY Conference or Workshop  = IHY team meeting  = Event with a special session on IHY

Africa IHY Events		
	* Aug. 17–20, 2005	28th Annual Conference of the Nigerian Institute of Physics (NIP)/World Year of Physics Ile-Ife, Nigeria
	* Oct. 27, 2005	IHY-Nigeria National Planning Workshop Ife-Ife, Nigeria
	* Apr. 24–26, 2006	Hermanus Magnetic Observatory Stakeholders Workshop IHY Session Hermanus, South Africa
	* July 10–14, 2006	First IHY SCINDA Workshop Sal, Cape Verde
	* July 19–21, 2006	2nd National Workshop of IHY-Nigeria: "Space Weather: Impacts on Communications and National Economy" Ijebu Ode, Nigeria
	* Aug. 16–19, 2006	29th Annual Conference of the Nigerian Institute of Physics (NIP) Nsukka, Nigeria
	* Oct. 25–27, 2006	IHY & IPY Regional Workshop Cape Town, South Africa
	* May 1, 2007	IHY Open and Outreach Day at North-West University Potchefstroom, South Africa
	* July 19–21, 2007	Third IHY-Nigeria National Workshop Nsukka, Nigeria
	* Aug. 5–7, 2007	First National Conference of the Nigerian Astronomical Society (NAS) Nsukka, Nigeria
	* Aug. 15–18, 2007	Thirtieth Annual Conference of the Nigerian Institute of Physics (NIP) Lagos, Nigeria

(*continued*)

	Africa IHY Events	
	* Nov. 10, 2007	Geophysical Information for Teachers (GIFT) Workshop: "The IHY" Addis Ababa, Ethiopia
	* Nov. 11, 2007	Second IHY Scinda Workshop Addis Ababa, Ethiopia
	* Nov. 12–16, 2007	IHY-Africa Space Weather Science and Education Workshop Addis Ababa, Ethiopia
	* Nov. 13, 2007	Working Group on IHY Education Meeting Addis Ababa, Ethiopia
	* Dec. 3–5, 2007	African Regional Conference of the International Academy of Astronautics Abuja, Nigeria
	* June 17–18, 2008	2008 Africa Array Workshop Johannesburg, South Africa
	* Nov. 10–22, 2008	African Regional IHY School Nsukka, Nigeria
	* June 7–8, 2009	Third IHY Scinda Workshop Livingstone, Zambia
	* June 8–12, 2009	IHY-Africa 2009 Workshop Livingstone, Zambia
	Eastern Asia and the Pacific IHY Events	
	* July 10–13, 2004	IHY-Asia Pacific Planning Meeting Ooty, India
	* Sept. 11–13, 2004	IHY-China Planning Meeting Beijing, China
	* Oct. 26–29, 2004	IHY-Japan Core Group Meeting Kiyosato, Japan
	* June 20–24, 2005	Asia/Oceania Geosciences Society (AOGS) Meeting Singapore
	* June 23, 2005	Asia-Pacific IHY Regional Planning Meeting Singapore
(3)	* July 16–23, 2006	36th COSPAR Scientific Assembly Beijing, China
	* July 16–23, 2006	"Globalizing Space Science Education and Outreach" Splinter Meeting Beijing, China

(*continued*)

Eastern Asia and the Pacific IHY Events		
	* Oct. 9–12, 2006	2006 IHY Asia/Pacific Regional Planning Meeting & International Space Weather Meridian Circle Program Workshop Beijing, China
	* Nov. 27–Dec 1, 2006	The Second UN/NASA Workshop on the IHY and Basic Space Sciences Bangalore, India
	* Jan. 13, 2007	IHY-India Public Outreach Team Meeting Bangalore, India
	* Mar. 14–16, 2007	CAWSES/IHY Workshop Nagoya, Japan
	* Mar. 22–24, 2007	Thai National Astronomy Meeting and 1st SE Asian Astronomical Network Meeting Bangkok, Thailand
	* May 7–10, 2007	IHY Workshop and IGY Gold Meeting Nainital, India
	* May 19–24, 2007	Japan Geoscience Union Meeting 2007 Chiba City, Japan
	* May 26, 2008	IHY-Nepal Talk Program at the Balmeeki Campus Balmeeki, Nepal
	* June 18–22, 2007	Third UN/ESA/NASA Workshop on the IHY and Basic Space Sciences Tokyo, Japan
	* July 21, 2007	IHY-Nepal Talk Program at the Takshashila Academy Takshashila, Nepal
	* July 30–Aug. 3, 2007	4th Asia Oceania Geosciences Society (AOGS) Meeting Bangkok, Thailand
	* Aug. 8–19, 2007	IHY Pavilion at National Science & Technology Fair 2007 Bangkok, Thailand
	* Aug. 10, 2007	IHY-Nepal Talk Program at AIM Institute Lalitpur, Nepal
	* Aug. 14, 2007	IHY-Nepal Astronomy Colloquium Lalitpur, Nepal
	* Nov. 11–14, 2007	IUCAA Science Festival Pune, India
	* Dec. 10–22, 2007	First IHY Asia-Pacific School Kodaikanal, India

(*continued*)

Eastern Asia and the Pacific IHY Events		
	* Dec. 22, 2007	EDUSAT Observing Educational Program Ten locations in India
	* Apr. 24, 2008	IHY-Nepal Talk Program at the Nepal Academy of Science and Technology Lalitpur, Nepal
	* Oct. 20–31, 2008	2nd IHY Asia-Pacific School Beijing, China
	* Nov. 10–13, 2008	Fifty Years after IGY-Modern Technology and Earth and Solar Sciences Ibaraki, Japan
	* Dec. 15, 2008	Australian Institute of Physics Congress Adelaide, Australia
Eastern Europe & Central Asia & Balkan/Black Sea/Caspian Sea Region IHY Events		
	* June 6–8, 2005	Balkan, Black Sea and Caspian Sea Regional IHY Coordination Meeting Sozopol, Bulgaria
	* Mar. 30–Apr. 1, 2006	Annual Meeting of the Balkan, Black Sea and Caspian Sea Regional Network on Space Weather Studies Antalya, Turkey
(2)	* Aug. 14–25, 2006	International Astronomical Union (IAU) General Assembly Prague, Czech Republic
	* Aug. 18, 2006	Meeting of the IAU Working Group on IHY Prague, Czech Republic
	* Sept. 13–16, 2006	2nd International Symposium on Space Climate: Long-term Change in the Sun, and Its Effects in the Heliosphere and Planet Earth Sinaia, Romania
	* Sept. 24–29, 2006	8th Hvar Astrophysical Colloquium Hvar, Croatia
	* Sept. 26–30, 2006	International Conference "Fifty Years of Romanian Astrophysics" Bucharest, Romania
	* Mar. 21, 2007	"The Universe" IHY Lectures Stara Zagora, Bulgaria
	* May 2–5, 2007	International Workshop on "Flows, Boundaries and Interactions" Sinaia, Romania

(*continued*)

Eastern Europe & Central Asia & Balkan/Black Sea/Caspian Sea Region IHY Events		
	* May 3–5, 2007	Young Scientists International School on "Heliosphere and Galaxy" Bucharest, Romania
	* Mar. 30–Apr. 1, 2006	SWEETS Exhibition at the Middle East Technical University Ankara, Turkey
	* Sept. 16–19, 2007	International Conference: 50th Anniversary of International Geophysical Year (IGY) and Electronic Geophysical Year (eGY) Suzdal, Russian Federation
	* Nov. 5–11, 2007	International IHY Symposium and Sputnik 50th Anniversary Celebration Zvenigorod, Russian Federation
	* June 2–6, 2008	Fourth UN/ESA/NASA/JAXA Workshop on the IHY and Basic Space Science: "First Results from the IHY 2007" Sozopol, Bulgaria
	* Sept. 22–26, 2008	9th Hvar Astrophysical Colloquium Hvar, Croatia
	* Sept. 29–Oct. 3, 2008	Forecasting of Radiation and Geomagnetic Storms by Networks of Particle Detectors" (FORGES-2008) ITZZ Nor Amberd, Armenia
Western Europe IHY Events		
	* Apr. 7–11, 2003	Spring 2003 EGU-AGU-EUG Joint Assembly Nice, France
	* Dec. 8, 2003	IHY UK Planning Meeting London, United Kingdom
	* Mar. 2, 2005	United Nations Committee on the Peaceful ises of Outer Space Adopts IHY Work Plan Vienna, Austria
	* Apr. 25–29, 2005	European Geosciences Union (EGU) 2005 General Assembly Vienna, Austria
(2)	* July18–29, 2005	International Association for Geophysics and Aeronomy (IAGA) Scientific Assembly Toulouse, France
	* July 22–23, 2005	First International IHY Planning Meeting Toulouse, France
	* Jan. 10–13, 2006	First IHY European General Assembly Paris, France

(*continued*)

Western Europe IHY Events		
	* Sep 9–14, 2006	ISSS Advanced School in Space Environment L'Aquila, Italy
	* Jan. 5, 2007	"In the Fire of the Sun" IHY Exhibition Inauguration Zurich, Switzerland
	* Feb. 5–9, 2007	Heliosphere Impact on Geospace: Kick-off Meeting of ICESTAR/IHY/IPY Helsinki, Finland
	* Feb. 19–20, 2007	IHY Opening Ceremony and Exhibition at the United Nations STSC Meeting Vienna, Austria
	* Feb. 19, 2007	IHY EPO Subcommittee Meeting Vienna, Austria
	* Feb. 20, 2007	IHY Science Meeting Prague, Czech Republic
	* Mar. 8–9, 2007	IHY Cross-Border School Conference "Discover the Stars at Armagh" Armagh, Ireland
	* Apr. 15, 2007	National Astronomy Meeting ssx2 Lancashire, United Kingdom
(2)	* Apr. 15–20, 2007	European Geosciences Union (EGU) 2007 General Assembly Vienna, Austria
	* Apr. 25, 2007	"Comparing Earth and Venus" Workshop Brussels, Belgium
	* Apr. 25, 2007	IHY-Germany Opening Ceremony Regensburg, Germany
	* May 4–9, 2007	Greenland IPY Space Science Symposium Kangerlussauq, Greenland
	* May 14–18, 2007	IHY Science Conference: "The Sun, The Heliosphere and the Earth" Bad Honnef, Germany
	* May 19, 2007	Kickoff Meeting for Heliospheric CIPs Bad Honnef, Germany
	* May 30, 2007	General Scientific Meeting of the Belgian Physical Society (BPS) Antwerp, Belgium
	* June 12, 2007	United Nations Workshop on IHY Vienna, Austria

(*continued*)

Western Europe IHY Events		
	* June 12, 2007	"The Sun Day" Exhibition for IHY-Finland Several sites in Finland
	* June 18–22, 2007	2nd IHY European General Assembly Torino, Italy
(3)	* July 2–13, 2007	International Union for Geophysics and Geodesy (IUGG) XXIV General Assembly Perugia, Italy
	* Sept. 12–13, 2007	IHY Workshops at the British Association of the Advancement Festival of Science York, United Kingdom
	* Sept. 24–27, 2007	International Symposium: "Solar Extreme Events of 2007" Athens, Greece
	* Sept. 28, 2007	IHY Tent at "Open Door Weekend" at Space Pole Brussels, Belgium
	* Oct.1–5, 2007	IHY Exhibit at Space50 Jodrell Bank, United Kingdom
	* Nov. 5–Dec16, 2007	IHY Exhibition at the Observatoire de Paris Paris, France
	* Nov. 19, 2007	IHY-Czech Republic Team Meeting Prague, Czech Republic
	* Apr. 13–18, 2008	European Geosciences Union: General Assembly 2008 Vienna, Austria
	* Sept. 15–19, 2008	IAU Symposium 257: Universal Heliophysical Processes Ioannina, Greece
	* Nov. 10–13, 2008	The Second IHY Conference on "The Sun, The Heliosphere and the Earth" Bern, Switzerland
(2)	* Aug. 23–30, 2009	2009 IAGA Scientific Assembly Sopron, Hungary
Latin America and Caribbean IHY Events		
	* Dec. 8–9, 2005	First Latin American Regional Planning Meeting for IHY 2007 Sao Paulo, Brazil
	* Mar. 8, 2006	Second Latin American Regional Planning Meeting for IHY 2007 Rio De Janiero, Brazil

(*continued*)

Latin America and Caribbean IHY Events		
	* Mar. 15, 2006	Second IHY-Latin America Regional Meeting Rio de Janiero, Brazil
	* Apr. 17–22, 2006	Latin American School & Workshop Lima, Peru
	* Sept. 24, 2006	Third IHY-Latin America Regional Planning Meeting Buenos Aires, Argentina
	* Sept. 25–29, 2006	IHY Latin American Summer School 2006 Buenos Aires, Argentina
	* Nov. 15, 2006	Third IHY-Latin America Regional Meeting Puerta Vallarta, Mexico
	* May 21–25, 2007	2007 AGU Joint Assembly Acapulco, Mexico
	* May 23, 2007	Geophysical Information for Teachers (GIFT) Workshop: "The IHY" Acapulco, Mexico
	* July 12–17, 2007	VIII Latin American Conference on Geophysics Mérida, Mexico
	* July 15, 2007	Fourth IHY-Latin America Regional Meeting Mérida, Mexico
	* Nov. 24–Dec 2, 2007	21st Feria Internacional del Libro Guaralajara, Mexico
	* Feb. 14–20, 2008	IHY Latin American Summer School 2008 Sao Paulo, Brazil
	* Aug. 12–14, 2009	IAU General Assembly Rio de Janeiro, Brazil
North America IHY Events		
	* Oct. 10–19, 2002	2nd World Space Congress (COSPAR) Houston, TX, USA
	* Dec. 9, 2002	Fall 2002 AGU IHY Town Meeting San Francisco, USA
	* Apr. 20–22, 2004	IHY US Planning Workshop Sacramento Peak, NM, USA
	* Sept. 2, 2004	IHY/eGY Discussion Session Boulder, CO, USA
	* Oct. 19–21, 2004	IHY UNBSS Science Organizing Committee Planning Meeting Greenbelt, MD, USA

(*continued*)

North America IHY Events		
(4)	* Dec. 13–17, 2004	Fall 2004 American Geophysical Union (AGU) Meeting San Francisco, CA, USA
	* Feb. 16–18, 2005	IHY North American Community Science Planning Workshop Boulder, CO, USA
(3)	* May 23–27, 2005	Spring 2005 American Geophysical Union (AGU) Meeting (joint with AAS/SPD) New Orleans, LA, USA
	* June 26–July 1, 2005	Joint CEDAR-GEM Joint Workshop Santa Fe, NM, USA
	* July 31–Aug. 4, 2005	Optics & Photonics/SPIE 50th Annual Meeting San Diego, CA, USA
(5)	* Dec. 5–9, 2005	Fall 2005 American Geophysical Union (AGU) Meeting San Francisco, CA, USA
	* May 23–26, 2006	Spring 2006 American Geophysical Union (AGU) Meeting Baltimore, MD, USA
(5)	* Dec. 11–15, 2006	Fall 2006 American Geophysical Union (AGU) Meeting San Francisco, CA, USA
	* July 23–Aug. 2, 2007	Polar Aeronomy and Radio Science Summer School Fairbanks, AK, USA
	* July 30–Aug. 7, 2007	IHY/ILWS Heliophysics Summer School Boulder, CO, USA
	* Nov. 15, 2007	Sputnik Golden Anniversary Celebration with the U.S. Russian Embassy Washington, DC, USA
	* Jan. 23–29, 2008	Polar Gateways Conference: An Interdisciplinary Science and Global Outreach Workshop Arctic Circle Sunrise 2008 Barrow, AK, USA
	* July 13–20, 2008	37th Committee on Space Research (COSPAR) Scientific Assembly Montreal, Canada
	* June 16–20, 2008	CEDAR Workshop Zermatt, UT, USA
	* July 20–31, 2008	Polar Aeronomy and Radio Science Summer School Fairbanks, AK, USA

(*continued*)

North America IHY Events		
	* Aug. 5–7, 2008	Canadian Space Agency Educator Workshop Quebec, Canada
	* Aug. 25–29, 2008	Whole Heliosphere Interval Data & Model Assessment Workshop Boulder, CO, USA
	* Nov. 10–14, 2008	AGU Chapman Conference on Universal Heliophysical Processes Savannah, GA, USA
	* Dec. 15–19, 2008	AGU 2008 Fall Assembly San Francisco, CA, USA
Western Asia IHY Events		
	* Nov. 20–23, 2005	IHY United Nations/ESA Workshop on Basic Space Science Al-Ain, United Arab Emirates
	* May 14, 2006	IHY-Yemen Kickoff and National Committee Meeting Sana'a, Yemen
	* June 18, 2006	IHY-Yemen 2nd National Planning Committee Meeting Sana'a, Yemen
	* Aug. 14, 2006	IHY-West Asia Regional Meeting Amman, Jordan
	* Mar. 15, 2008	IHY-West Asia Regional Meeting Sharjah, United Arab Emirates
	* Apr. 29–May 9, 2007	IHY-Yemen Miniature Physics Mosaic Exhibition Sana'a, Yemen
Worldwide/Multinational IHY Events		
	* Mar. 29, 2006	Total Solar Eclipse Starting in Brazil, traversing Northern and Saharan Africa and ending in Mongolia: 12 IHY Viewing Stations along the path of the eclipse
	* Mar. 29, 2006	Africa Science Education Expedition to the African Total Solar Eclipse
	* Apr. 12, 2006	"Yuris Night" World Space Party: Celebrating the 45th Anniversary of Human Space Flight
	* Apr. 12, 2007	"Yuris Night" World Space Party: Celebrating the 46th Anniversary of Human Space Flight
	* Mar. 21, 2007	Sun-Earth Day 2007 * June 10, 2007 "Open Doors" Day at IHY Observatories and Museums

(continued)

Worldwide/Multinational IHY Events		
	* Mar. 20, 2008	Sun-Earth Day 2008 * Mar. 20, 2008 Whole Heliosphere Interval Campaign Begins
	* Apr. 12, 2008	"Yuri's Night" World Space Party: Celebrating the 47th Anniversary of Human Space Flight
	* Apr. 16, 2008	Whole Heliosphere Interval Campaign Ends
	* Apr. 12, 2009	"Yuri's Night" World Space Party: Celebrating the 48th Anniversary of Human Space Flight
Online IHY Events		
	* Nov. 13–17, 2006	Joint CAWSES/IHY Virtual Conference: The State of the Sun-Earth System During Extreme Space Weather, "Return to the Auroral Oval for the Anniversary of the IGY"
	* Oct. 4–10, 2007	World Space Week Berkeley Online Education Conference for IHY

Appendix V IHY and the International Science Years

1. Brief history of the IGY

The International Heliophysical Year proposal follows a tradition of international cooperation in scientific research that began in the 19th century, which consisted of the First and Second International Polar Years, followed by the International Geophysical Year in 1957.

The First International Polar Year was the idea of an Austro-Hungarian Naval Lieutenant Karl Weyprecht (c.f. V. Heathcote, Neils H., Annals of the International Geophysical Year, 1, 1959). Weyprecht (Figure 6) had just returned from a polar expedition where he commanded one of the research vessels. In January 1875 at the Academy of Sciences in Vienna, Weyprecht expressed his ideas to establish an international collaboration to obtain a set of simultaneous observations, extending over a considerable time period, at various locations around the Arctic. The concept was presented again in September 1875 at the 4th Meeting of the Association of German Naturalists and Physicists at Graz. In 1877, a detailed program was prepared and submitted to the International Meteorological Congress. In 1879 the International Meteorological Congress met in Rome and recognized the importance of the proposal. On October 1–5, 1879 the 1st International Polar Conference (IPC) met at Hamburg. It was determined that a minimum of eight Arctic stations was needed, to obtain observations of a duration of at least 1 year. The IPC was established, with representatives from Austria, Hungary, Denmark, France, Germany, The Netherlands, Norway, Russian Federation, and Sweden. Dr. G. Neumayer of Hamburg was the first Commission president. In July 1880 the 2nd IPC met at Bern, Switzerland. There an Italian representative joined the existing representatives, and Prof. H. Wild became the second president. On August 1, 1881 the 3rd IPC met at St Petersburg. The United States joined the group, and a program of observations was adopted. The First International Polar Year began August 1, 1882 and continued for 13 months to September 1, 1883. Scientific results and observational data were published in the Bulletin of the International Polar Commission. In 1884 and 1891, the 4th and 5th Polar Conferences were convened. Weyprecht did not live to see the culmination of his grand concept. He died on March 29, 1881.

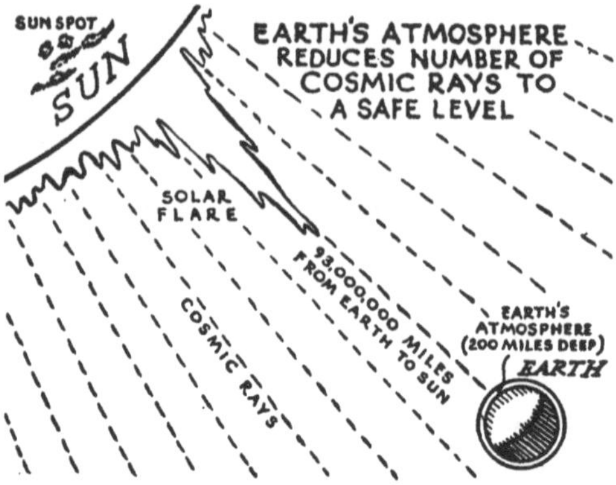

Fig. 6. *Left: Lieutenant Karl Weyprecht first proposed the International Polar Year in 1875 after returning from an Arctic expedition (from Chapman, 1959). Right: during the IGY it was known that events on the Sun influence the Earth, but the exact mechanism was unknown. The discovery of the solar wind was a big step (from Hyde, 1957).*

In 1927 Dr. J. Georgi at Deutsche Seewarte in Hamburg suggested that a *Second International Polar Year* be conducted on the fiftieth anniversary of the first (Laursen, V., Annals of the International Geophysical Year, 1, 1959, 211). A proposal was submitted to the International Meteorological Committee, and then forwarded to Reseau Mondial and Polar Meteorology. In June 1928 an informal organizational meeting was held in London to discuss plans for the event. Finally, in 1929 the Meteorological Conference of Directors in Copenhagen endorsed the plan for the cooperative study of magnetic, auroral and meteorological phenomena. Also in 1929 the International Cloud Commission passed a resolution for an international year for clouds coinciding with the Polar Year. The Commission for the Polar Year 1932–1933 was appointed to prepare detailed plans for the observations to be made and the methods for making them. A collaboration was established between the Commission for the Polar Year and the International Union of Geodesy and Geophysics. In August 1930 the first meeting of the Commission for the Polar Year took place in Leningrad, to further refine proposals for the Polar Year. In December 1930 at a meeting in London the Commission prepared a detailed report containing proposals, for research programs in meteorology, terrestrial magnetism, atmospheric electricity, aurora, and aerology. At a subsequent meeting in September 1931 the Commission for the Polar Year, despite being urged to delay due to poor economic conditions worldwide, decided to go ahead with the Polar Year program. On 1 August 1932 the Second International Polar Year began. It continued until 1 September 1933.

The Commission introduced the concept of "International Days". The scientific objective was to study phenomena on the largest possible scale with simultaneous observations, the same as the last previous polar year. The most significant new development that affected how the program was conducted was the advent of radio communication.

In 1950, a proposal for the *International Geophysical Year,* 25 years after the Second Polar Year, was brought before the Mixed Commission on the Ionosphere, which endorsed it. The Mixed Commission on the Ionosphere was formed by the International Council of Scientific Unions (ICSU) under the sponsorship of the International Union for Scientific Radio (URSI) with the cooperation of the International Astronomical Union (IAU) and the International Union for Geodesy and Geophysics (IUGG). The IUGG drew up a tentative program, and adopted a resolution to transmit it to the International Council of Scientific Unions (ICSU), which sponsored the event. All bodies endorsed the proposal by 1951.

During times when the Sun was especially active, on a day not designated as a World Day (Figure 7), alerts were issued. These could be followed by the

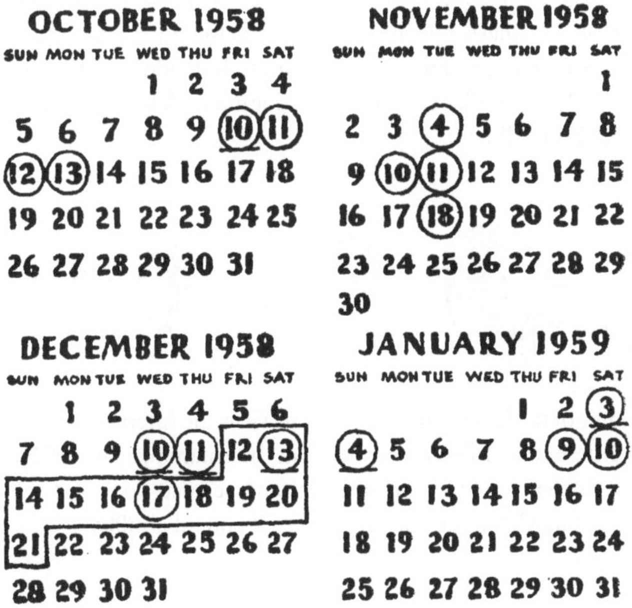

Fig. 7. *IGY calendars were marked to indicate World Days (circles) and World Meteorological Intervals (box). (from Hyde, 1957).*

declaration of Special World Intervals that followed alerts. These could be called with an 8 h notice. Rocket and balloon launches might take place, and other programs of study might be intensified. World Meteorological Intervals consisted of 10 consecutive days, four times a year, usually near the beginning of seasons, for intensive study, rocket campaigns, etc. Data were collected at three centers (US, Europe, and Soviet Union) and made available to all nations.

The IGY was a tremendous success. The newly developed space-flight capability was used to discover and explore Earth's radiation belts, to study the magnetosphere, and to provide the first observations of the emission from the Sun's corona. Public interest in the scientific results of the IGY was high. The IGY provided a forum and a backdrop for discussing the importance of geospace influences on Earth. Space physics, and many of its current institutions were born during the IGY.

2. The Celimontana Declaration: IHY and the other International Science Years

The following declaration was authored by the leaders of the four International Years occurring on the 50th anniversary of the IGY, who gathered in September 2005 to establish formal cooperation between the four international years as well as the many associated programs and activities. Individual statements from the International Year of Planet Earth, International Polar Year, and Electronic Geophysical Year follow.

The leaders of the Electronic Geophysical Year, the International Heliophysical Year, the International Polar Year, and the International Year of Planet Earth, meeting with representatives of international science bodies at the Home of Geography, Villa Celimontana, Rome, on 7 September 2005,

note that we

- share a fundamental motivation to understand, in a descriptive and predictive sense, many facets of our planet;

- share a common recognition that understanding our planet requires not only an understanding of linked systems and processes that interact on many time and space scales, but also an understanding of this planet as home to inhabitants of wondrous variety and complexity;
- share the goal of extending to all citizens a sense of substantial human influence and dependence on many systems of planet earth and a view of shared knowledge of the planet as a fundamental right and responsibility;
- recognise the 50th anniversary of the International Geophysical Year (2007–2008) as a unique opportunity to raise public awareness and increase scientific capability;
- recognise a mutual interest and benefit in sustaining cooperation among our programs; and

declare that we will

- maintain vigorous and open communication to realise such benefit;
- define and implement joint activities in areas of common scientific interest, as well as in education, outreach, capacity building, and development of data and information systems; and
- communicate our collective efforts, results and products to the scientific community, governments, and the public.

Signed:

(Eduardo F. J. de Mulder) for the International Year of Planet Earth
(David J. Carlson) for the International Polar Year
(Joseph M. Davila) for the International Heliophysical Year
(Daniel N. Baker) for the Electronic Geophysical Year

planetearth®
Earth Sciences for Society

The International Year of Planet Earth (2007–2009) aims to contribute to the improvement of everyday life, especially in the developing countries, by uncovering the underused societal potential of the world's Earth scientists, as expressed in the Year's subtitle Earth sciences for Society. Ambitious outreach and science programmes constitute the backbone of the International Year, politically endorsed by all 191 member states of the United Nations Organisation when it proclaimed 2008, the central year of the triennium, as a UN Year of Planet Earth.

Its desired outcomes are raising public awareness and enhancing research. The project was jointly initiated in 2001 by the International Union of Geological Sciences (IUGS) and the United Nations Educational, Scientific and Cultural Organisation's (UNESCO) Earth Science Division. Since initiation, the project has attracted 12 Founding Partners that actively support the initiative either in kind or in cash, 23 Associate Partners and collected support by the geoscientific communities in most nations.

The Outreach Programme lies close to the heart of the International Year because of its prime aim to generate interest and greater awareness among the general public, decision makers and politicians about the effective application, for the betterment of human society, of the widely available wealth of information in the hands of the Earth science community. The publication of the Year's first brochure: Planet Earth in our hands was a step in that direction, quickly followed by the brochure on Outreach, bringing Earth sciences to everyone.

The scientific themes selected for the International Year of Planet Earth were all determined on the basis of their relevance to Society. The selected themes are: *Groundwater*, *Hazards*, *Earth and Health*, *Climate change*, *Resources*, *Megacities*, *Deep Earth*, *Ocean*, *Soil*, and *Earth and Life*. Brochures addressing each of these themes are available as hard copy or can be downloaded from the web site (see below).

Success or failure of the International Year's ambitions will depend to some considerable extent on how these are realised at national and local levels. For that reason, the Year's national committees are perhaps the most important structural components of the Year of Planet Earth because their activities will be most clearly visible to the public. Every country is encouraged to create such a committee.

2.1. International Polar Year (IPY)

The International Polar Year 2007–2008 (IPY), jointly and officially sponsored by the International Council of Science (ICSU) and the World Meteorological Organisation (WMO), has generated enormous scientific interest and enthusiasm. More than 1000 teams of investigators from more than 50 countries submitted preliminary plans for IPY research. The initial expressions of interest have evolved into coordinated international proposals, each proposal gathering the interests and efforts of hundreds of researchers to focus on a crucial area of polar science. The ICSU–WMO IPY Joint Committee, charged with oversight and guidance of the IPY, recently endorsed 139 of these coordinated efforts covering research, data and information services, and education and outreach as the basis for a full rich IPY programme. The IPY development process remains open to additional expressions of interest and coordinated proposals, up to a final submission date of 31 January 2006.

An IPY planning chart (available at *http://www.ipy.org*) based on coordination proposals endorsed so far shows the breadth and variety of proposed IPY research by geographic region and science topic. IPY research will cover all aspects of the past, current and future state of the cryosphere and of linkages between the cryosphere and global processes. IPY research will build on IGY history, and on its obvious partnership with IHY, by exploring space from polar regions and by using space-based tools to monitor polar regions. IPY will build connections among geophysical, biological and social sciences as it focuses on biogeochemical systems and processes and as it embraces human perspectives to address all facets of social dynamics and cultural resiliency.

Many national and international planning and funding processes have responded to IPY activities. Many nations have announced substantial new IPY funds, have allocated substantial resources such as ships, and have solicited new proposals based on IPY themes and plans. It already seems certain that IPY will represent the largest coordinated international research programme focused on this planet in the last 50 years. The vast IPY research effort will present a new set of challenges and opportunities in data exchange and data services, requiring close collaboration with eGY, and will represent a timely opportunity to engage public attention through formal and informal education and a wide range of outreach activities. A recent IPY leaflet published in several languages, available on the IPY website, represents the first of many IPY outreach products.

2.2. Electronic Geophysical Year (eGY)

The following eGY declaration continues to be endorsed by a wide range of scientific organizations and institutions. The elements in the declaration form the basis for the 2007–2008 eGY activities. Please refer to the eGY website for a complete description of eGY programs, plans and opportunities: *http://www.egy.org*.

2.2.1. Declaration for a Geoscience Information Commons

"Knowledge is the common wealth of humanity" – *Adama Samassekou, Convener of the UN World Summit on the Information Society*

Preamble: The Electronic Geophysical Year (eGY) joins with the International Council for Science, the World Summit on the Information Society, and many other bodies in recognizing that knowledge is the common wealth of humanity. We have a shared responsibility to create and implement strategies to realize the full potential of digital information for present and future generations. In the 21st century and beyond, access to digital information and new technologies for information integration and knowledge discovery will influence the free and productive development of societies around the world. Providing ready and open access to the vast and growing collections of cross-disciplinary digital information is the key to understanding and responding to complex Earth system phenomena that influence human survival. In the geosciences, as elsewhere, the issues of concern are as follows.

- Article 1: Data access
  Earth system data and information should be made available electronically with interoperable approaches that facilitate open access.
- Article 2: Data release
  Owners, custodians, and creators of Earth system data should work together to share their digital information with the world community, though in a manner that respects propriety requirements and security constraints.
- Article 3: Data description
  Providers and users of Earth system data and information should share descriptions of structure, content, and contexts to facilitate interoperability and the discovery of relationships within and between information resources.
- Article 4: Data persistence
  Data and information about the Earth system should be preserved and sustained in forms that are both software and hardware independent so as to be openly accessible today and in the future.
- Article 5: Data rescue
  Effort should be made to identify and rescue critical Earth system data and ensure persistent access to them.
- Article 6: Common standards and cooperation
  Standards for interoperability should be identified, created, and implemented through international collaboration.

- Article 7: Capability building
  Communities with advanced information technology and communications capabilities should contribute to developing such capabilities elsewhere to reduce the digital divide.
- Article 8: Education and public outreach
  Students, scientists, decision-makers, and the public should be informed about and be enabled to contribute to our understanding and management of Earth system phenomena that impact human survival.

Acknowledgments

We thank the many IHY team members whose efforts made this publication possible. The completion of this report occurred during a period of exceptionally high activity for IHY. During the week of the finalization of this report, there were three major IHY events happening on three different continents: the African IHY Regional School in Nigeria, the Fifty Years After IGY Symposium in Japan, and the Chapman Conference on Universal Processes in the USA. Despite busy schedules, our team members found time to contribute to the production of this report. We kindly acknowledge the following authors for their role in this publication:

Mr. R. Wickramatunga who developed the colorful UN IHY brochures that have been translated into the six official languages of the UN.
Mrs. Ayoni Oyeneyin who did the mailing of ten thousands of IHY brochures, flyers, reports etc. in the period of time from 2005 to 2009.

G. Aburjania, J. Acharya, R. Albrecht, H. M. K. Al-Naimiy, A. Andic, S. Avakyan, E. Babayev, K. S. Balasubramaniam, J.-P. Barriot, C. Barton, H. Basurah, A. O. Benz, S. Berdyugina, A. Beshu, M. Bianda, S. Bijoor, O. Bock, J. A. Bonet, D. Boteler, V. Bothmer, J.-L. Bougeret, J. Bradford, A. R. Breen, C. Briand, B. Bromage, R. Calfas, D. Carlson, A. Chilingarian, E. Christian, M. Cohen, M. Connors, E. W. Cliver, N. U. Crooker, N. B. Crosby, J. M. Davila, C. Davis, C.E. DeForest, E. DeMulder, E. Donovan, E. Drobnes, C. Dumitrache, G. Eichhorn, E. O. Famutimi, Q. Fan, F. Farnik, M. Faulknor, N.J. Fox, B. Fraser, M. Fujimoto, T. Fuller-Rowell, M. Galand, P. T. Gallagher, T. Garner, L. Gentile, K. Georgieva, H. Ghalila, S. Gibson, M. Gigolashvili, N. Gopalswamy, L. M. Green, K. Groves, E. Guemene Dountio, W. Guevara Day, S. Gurubaran, G. Haines-Stiles, R. A. Harrison, S. Hasan, D. M. Hassler, H. Haubold, B. Heber, L. Hidalgo, R. Hippler, U. Inan, Y. Kamide, J. Kasper, K. Kauristie, G. Kayo, A. Kebede, K. Kecskemety, D. Knudsen, M. Kojima, J. Kozyra, B. Kramer, T. W. Kraupe, D. Lamb, E. Lewis, V. Lewnes, W. Liu, W. Macek, G. Maeda, A. Mahrous, J. Makela, C. Mandrini, I. Mann, P. K. Manoharan, J. Manuel, G. Maris, P. Martinez, C. Mazaudier-Amory, I. McCrea, L.-A. McKinnell, K. McWilliams, D. Milling, M. Moldwin, C. Monstein, C. Morrow, K. Munakata, G. Munyeme, S. Nait Amor, C. Ng, I. S. Nurgaliev, K. O. Obrou, M. Odondi k'Orowe, N. Ostgaard, A. Ozguc,

A. Paranjpye, W. Peterson, M. Petitdidier, S. Petty, S. Poedts, M. Potgieter, E. Pryse, M. C. Rabello-Soares, A. B. Rabiu, A. Rager, R. Ramelli, R. Rankin, J.-P. Raulin, J. Rodriguez-Pacheco, D. Rosa, H. O. Rucker, Jennifer Rumburg, T. Sakurai, D. Scherrer, B. Schmieder, W. Schmutz, R. Senseney, G. Shepherd, K. Shibata, K. M. Sigsbee, Pitan Singhasaneh, G. Siscoe, R. Smith, R. Stamper, J. O. Stenflo, A. V. Stepanov, A. Stoev, P. Stoeva, L. Strachan, A. H. Sultan, Q. Sultan, D. Summers, B. J. Thompson, L. Trichtchenko, Y. Tulunay, R. Vainio, A. Van Ballegooijen, I. S. Veselovsky, B. Vrsnak, C. Wang, J. Wang, A. Warmuth, S. Watari, D. F. Webb, R. Wesenberg, A. Yau, E. Yizengaw, K. Yumoto, F. Zuccarello.

IHY Outreach Coordinators

IHY Regional Coordinators

IHY Schools Program

IHY Secretariat

IHY's Goals and Objectives

IHY's Heritage

ITNE Instrument Program

Legacy plan

Low Cost Ionosonde for IHY/UNBSS

Low-Frequency Radio Antenna Arrays

MAGDAS Instrument Program

Multilingual Resources and Translations

Muon Detection Network Instrument Program

Open Doors Day

Opening Ceremony

Passport to Knowledge (P2 K)

RENOIR Instrument Program

AUGO: Athabasca University Geophysical Observatory

AUTUMN: Athabasca University THEMIS UCLA Magnetometer Network

AWESOME: Atmospheric Weather Educational System for Observation and Modeling of Effects

BAFTA: British Academy of Film and Television Arts

BITEC: Bangkok International Trade & Exhibition Centre

BSS: Basic Space Science

CALLISTO: Compound Astronomical Low-cost Low-frequency Instrument for Spectroscopy and Transportable Observatory

CANOPUS: Canadian Auroral Network for the OPEN Program Unified Study

CARISMA: Canadian Array for Realtime Investigations of Magnetic Activity

CAS: Chinese Academy of Sciences

CASLEO: Complejo Astronómico El Leoncito

CAWSES: Climate and Weather of the Sun–Earth-System

CCD: Charge Coupled Device

CCSTI: Centre de Culture Scientifique Technique et Industrielle

CeBIT: Centrum der Büro- und Informationstechnik

CEDAR: Coupling, Energetics and Dynamics of Atmospheric Regions

CGSM: Canadian GeoSpace Monitoring

Basu, Sunanda

Basurah, Hassan

Benna, Medhi

Benz, Arnold O.

Berdyugina, Svetlana

Bewsher, Danielle

Bijoor, Sheila

Bonet, José Antonio

Bradford, John

Breen, Andrew R.

Briand, Carine

Chilingarian, Ashot

Chitre, S. M.

Choudhuri, A. R.

Cobabe-Amman, Emily

Crosby, Norma B.

Davila, Joseph M.

EA: Equatorial Anomaly	Hajjar, Rojer
EAP: Equatorial Anomaly Peak	Haq Sultan, Abdul
EDUSAT: Educational Satellite	Harrison, Richard A.
EGU: European Geophysical Union	Hasan, Siraj
EISCAT: European Incoherent Scatter Scientific Association	Hassler, Donald M.
ELF: Extremely Low Frequency	Heber, Bernd
EMIC: Electromagnetic Ion Cyclotron	Heinzel, Petr
ENIT: Ecole National d'Ingénieurs de Tunis	Howard, Art
EOARD: European Office of Aerospace Research & Development	Inan, Umran
EPO: Education and Public Outreach	Kasper, Justin
ePOP: Enhanced Polar Outflow Probe	Kathiravan, C.
ERNE: Energetic and Relativistic Nuclei and Electrons	Kauristie, Kirsti
ESA: European Space Agency	Kebede, Abebe
ETH: Eidgenössische Technische Hochschule	Kecskemety, Karoly
EUV: Extreme Ultraviolet	Kojima, M.
FAA: Federal Aviation Association	Kramer, Busaba
FLR: Field Line Resonance	Kudela, Karel
FLTSAT: Fleet Satellite	Macek, Wieslaw
FMI: Finnish Meteorological Institute	MacDowall, Robert
FMT: Flare Monitoring Telescope	Maeda, George
FPI: Fabry-Perot Interferometers	Mahrous, Ayman
GAIA: Global Auroral Imaging Access	Makela, Jonathan
GIFT: Geophysical Information For Teachers	Mammar, Fodil
GISTDA: Geo-Informatics and Space Technology Development Agency	Mandrini, Cristina
GNSS: Global Navigation Satellite System	Mann, Ian
GOES: Geostationary Operational Environmental Satellite	Mann, Ian
GPS: Global Positioning System	Manoharan, P. K.
HI: Heliospheric Imager	Maricic, D.

HMO: Hermanus Magnetic Observatory

IAFE: Instituto de Astronomía y Física del Espacio

IAGA: International Association of Geomagnetism and Aeronomy

IAU: International Astronomical Union

IBEX: Interstellar Boundary Explorer

ICESTAR: Interhemispheric Conjugacy Effects in Solar Terrestrial Aeronomy Research

ICME: Interplanetary Coronal Mass Ejection

ICSU: International Council of Scientific Unions

IGf: Instituto de Geofisica

IGY: International Geophysical Year

IHY: International Heliophysical Year

IIA: Indian Institute of Astrophysics

IMAGE: Imager for Magnetopause to Aurora Global Exploration

IMPACT: In-situ Measurements of Particles and CME Transients

INAF: Italian National Institute for Astrophysics

INAO: Iraqi National Astromical Observatory

INPE: Instituto Nacional de Pesquisas Espaciais

INTAS: International Association for Cooperation with Scientists from the former Soviet Union

IPC: International Polar Conference

IPRSS: Ionosphere Prediction, Radio and Space Services

IPS: Interplanetary Scintillation

IPY: International Polar Year

IR: Infrared

IRSOL: Institute of Solar Researches

Maris, Geogeta

Markova, Eva

Maymoni, Jammal

McKinnell, Lee-Anne

Milling, David

Moldwin, Mark

Monstein, Christian

Morrow, Cherilynn

Munakata, Kazuoki

Munyeme, Geoffrey

Mweene, Habatwa

Nair, G. Madhavan

Nait Amor, Samir

Najid, Nour-Eddine

Neumayer, G.

Obridko, V.

Odondi K'Orowe

Okeke, P. N.

Ostgaard, Nikolai

Ozguc, Atila

Paranjpye, Arvind

Parnell, Claire

Peticolas, Laura

Petitdidier, Monique

ISAS: Institute of Space and Astronautical Sciences

ISC: International Schools Committee

ISSI: International Space Science Institute

ISTP: Institute of Solar–Terrestrial Physics

ISWI: International Space Weather Initiative

ISWMCP: International Space Weather Meridian Circle Program

ITM: Ionosphere/Thermosphere/Mesosphere

ITNE: Ionospheric Tomography Network of Egypt

IUCAA: Inter-University Centre for Astronomy and Astrophysics

IUGG: International Union of Geophysics and Geodesy

IYPE: International Year of Planet Earth

IZMIRAN: Institute of Terrestrial Magnetism, Ionosphere and Radiowave Propagation

JAXA: Japanese Aerospace Exploration Agency

JHU: Johns Hopkins University

JKUAT: Jomo Kenyatta University of Agriculture and Technology

JOULE: Joule Sounding Rocket

K12: Kindergarten through 12th grade

KFKI: Central Research Institute for Physics

LEO: Low-Earth Orbits

LEP: Lightning-Induced Precipitations

LIDAR: Light Detection and Ranging

LIPS: Laser-Induced Plasma Spectroscopy

LSAMA: Laboratoire de Spectroscopie Atomique, Moléculaire et Applications

LWS: Living With a Star

MAGDAS: Magnetic Data Acquisition

Poedts, Stefaan

Pontin, David

Popescu, Miruna

Popescu, N. A.

Potgieter, Marius S.

Rabello-Soares, Maria Cristina

Rabiu, A. Babatunde

Raulin, Jean-Pierre

Rodriguez-Pacheco, Javier

Rosa, D.

Rucker, Helmut

Ruffolo, David

Rusin, Vojtech

Sabat, Hanna

Sabbah, Ismail

Scherrer, Deborah

Schmieder, Brigitte

Shah, Rishi

Shaltot, Musalam

Shibata, K.

Smith, Roger

Soonthornthum, Boonrucksar

Stamper, Richard

Stenflo, Jan O.

Stepanov, Alexander

McMAC: Mid-continent Magnetoseismic Chain

METU: Middle East Technical University

MHD: Magnetohydrodynamic

MIDEX: Medium-Class Explorer

MIT: Massachusetts Institute of Technology

NAOJ: National Astronomical Observatory of Japan

NARIT: National Astronomical Research Institute of Thailand

NASA: National Aeronautics and Space Administration

NASO: Nepal Astronomical Society

NASRDA: National Space Research and Development Agency

NAST: Nepal Academy of Science and Technology

NECOP: Nigerian Environmental and Climatic Observing Program

NICT: National Institute of Information and Communications Technology

NIMS: Navy Ionospheric Monitoring System

NORSTAR: Northern Solar Terrestrial Array

NRCan: Natural Resources Canada

NRF: National Research Foundation

NSF: National Science Foundation

OOSA: Office of Outer Space Affairs

P2 K: Passport to Knowledge

PCB: Printed Circuit Board

PCI: Peripheral Component Interconnect

PECS: Program for European Cooperating States

PICAM: Planetary Ion CAMera

PICASSO: Portable Ionospheric Camera and Small-Scale Observatory

Stoev, Alexey

Stoeva, Penka

Sumners, Carolyn

Thompson, Barbara

Tulunay, Yurdanur

Vainio, Rami

Van Allen, James

Vila, Paul

Vršnak, B.

Walker, Dave

Warmuth, Alexander

Watari, S.

Webb, David

Weyprecht, Karel

Yizengaw, Endawoke

Yumoto, Kiyohumi

Zherebtsov, G.

Zuccarello, Francesca

PMOD/WRC: Physikalisch-Meteorologische Observatorium Davos and World Radiation Center

PO: Public Outreach

PolarDARN: Polar Dual Auroral Radar Network

PPARC: Particle Physics and Astronomy Research Council

RAL: Rutherford Appleton Laboratory

RENOIR: Remote Equatorial Nighttime Observatory for Ionospheric Regions

RF: Radio Frequency

RISR: Resolute Bay Incoherent Scatter Radar

RMKI: Research Institute for Particle and Nuclear Physics

ROT: Rate Of TEC

SAA: South Atlantic Anomaly

SAMBA: South American Magnetometer B-Field Array

SAMNET: Sub-Auroral Magnetometer Network

SAS: Slovak Academy of Sciences

SATCOM: Satellite Communications

SAVNET: South Atlantic VLF Network

SCAR: Scientific Committee on Antarctic Research

SCINDA: Scintillation Network Decision Aid

SCOSTEP: Scientific Committee on Solar–Terrestrial Physics

SEGMA: South European GeoMagnetic Array

SERC: Solar Energy Research Center

SERSES: Sun–Earth System and Space Weather

SEVAN: Space Environment Viewing and Analysis Network

SGO: Sodankylä Geophysical Observatory

SID: Sudden Ionospheric Disturbance

SOHO: Solar and Heliospheric Observatory

SPD: Solar Physics Division of the American Astronomical Society

SSA: Space Science and Astronomy

SST: Swedish Solar Telescope

STE Lab: Solar–Terrestrial Environment Laboratory

STEL: Solar–Terrestrial Environment Laboratory

STEP: Solar Terrestrial Energy Program

STEREO: Solar Terrestrial Relations Observatory

STFC: Science and Technology Facilities Council

STIL: Solar–Terrestrial Influences Laboratory

STPP: Solar Terrestrial Physics Program

STSC: Science and Technical Subcommittee

SuperDARN: Super Dual Auroral Radar Network

SWEETS: Space Weather and Europe – an Educational Tool with the Sun

TCD: Trinity College Dublin

TEC: Total Electron Content

TGF: Terrestrial Gamma Ray Flash

THEMIS: Time History of Events and Macroscale Interactions during Substorms

TID: Transient Equatorial Disturbance

TIGER: Tasman International Geospace Environment Radar

TLE: Transient Luminous Events

TNAM: Thai National Astronomy Meeting

UCLA: University of California Los Angeles

UH: University of Helsinki

UNAM: Universidad Autónma de México

UNBSS: United Nations Basic Space Science

UT: University of Texas

UT: University of Turku

VES: Venus Express

VGMO: Virtual Global Magnetic Observatory

VLF: Very Low Frequency

VTT: Vacuum Tower Telescope

VxO: Virtual Observatory

WAAS: Wide Area Augmentation System

WDC: World Data Centre

WHI: Whole Heliosphere Interval

WMO: World Meteorological Organisation

WSW: World Space Week

XUV: Extreme Ultraviolet

List of figures

Figures

1 IHY Overview

2 Leadership, Coordination & International Programs

3 Science: Universal Heliophysical Processes and Coordinated Investigation Programmes (CIPs)

4 IHY/UNBSS Observatory Development Program

5 Education and Public Outreach

6 "IGY Gold" History initiative

7 IHY National Committee Reports

Appendix

MIX
Papier aus verantwortungsvollen Quellen
Paper from responsible sources
FSC® C105338

FSC
www.fsc.org

If you have any concerns about our products,
you can contact us on
ProductSafety@springernature.com

In case Publisher is established outside the EU,
the EU authorized representative is:
Springer Nature Customer Service Center GmbH
Europaplatz 3, 69115 Heidelberg, Germany

Printed by Libri Plureos GmbH
in Hamburg, Germany